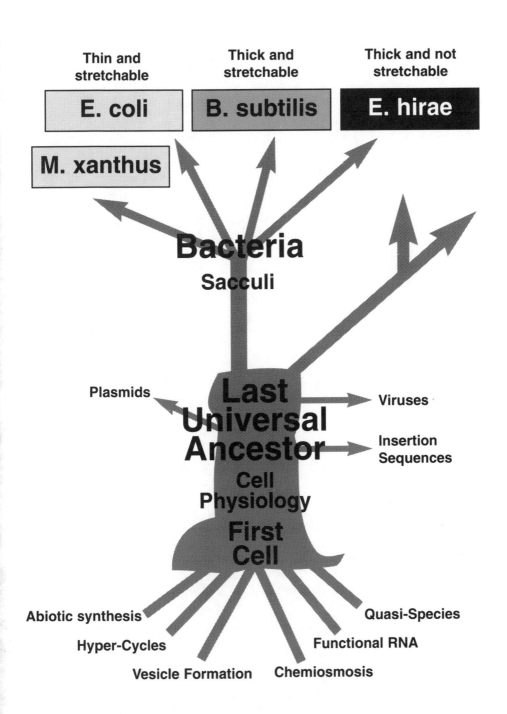

Thin and stretchable — E. coli

Thick and stretchable — B. subtilis

Thick and not stretchable — E. hirae

M. xanthus

Bacteria
Sacculi

Last Universal Ancestor

Plasmids

Viruses

Insertion Sequences

Cell Physiology

First Cell

Abiotic synthesis

Hyper-Cycles

Vesicle Formation

Chemiosmosis

Functional RNA

Quasi-Species

PRE-CELLULAR 'LIFE'

BACTERIAL
GROWTH
and FORM

ARTHUR L. KOCH
INDIANA UNIVERSITY
BLOOMINGTON INDIANA

CHAPMAN & HALL

New York • Albany • Bonn • Boston • Cincinnati • Detroit • London • Madrid • Melbourne
Mexico City • Pacific Grove • Paris • San Francisco • Singapore • Tokyo • Toronto • Washington

Art direction: Andrea Meyer, emDASH inc.
Cover design: Saeed Sayrafiezadeh, emDASH inc.

Copyright © 1995
Chapman & Hall

Printed in the United States of America

For more information, contact:

Chapman & Hall
115 Fifth Avenue
New York, NY 10003

Thomas Nelson Australia
102 Dodds Street
South Melbourne, 3205
Victoria, Australia

Nelson Canada
1120 Birchmount Road
Scarborough, Ontario
Canada, M1K 5G4

International Thomson Editores
Campos Eliseos 385, Piso 7
Col. Polanco
11560 Mexico D.F. Mexico

Chapman & Hall
2-6 Boundary Row
London SE1 8HN
England

Chapman & Hall GmbH
Postfach 100 263
D-69442 Weinheim
Germany

International Thomson Publishing Asia
221 Henderson Road #05-10
Henderson Building
Singapore 0315

International Thomson Publishing-Japan
Hirakawacho-cho Kyowa Building, 3F
1-2-1 Hirakawacho-cho
Chiyoda-ku, 102 Tokyo
Japan

1 2 3 4 5 6 7 8 9 10 XXX 01 00 99 97 96 95

Library of Congress Cataloging-in-Publication Data

Koch, Arthur L.,
    Bacterial Growth and Form / Arthur L. Koch.
        p.   cm.
    Includes bibliographical references and index.
    ISBN 0-412-02871-9
    1. Bacterial growth.   I. Title.
    QR84.5.K63   1995
    589.9'031—dc20   94-38640
                                           CIP

British Library Cataloguing in Publication Data available

Please send your order for this or any Chapman & Hall book to **Chapman & Hall, 29 West 35th Street, New York, NY 10001, Attn: Customer Service Department.** You may also call our Order Department at 1-212-244-3336 or fax your purchase order to 1-800-248-4724.

For a complete listing of Chapman & Hall's titles, send your requests to **Chapman & Hall, Dept. BC, 115 Fifth Avenue, New York, NY 10003.**

Contents

A Profile of Arthur L. Koch
Professor of Biology, Indiana University

Arthur Koch calls himself an ecologist. Those who know him recognize Arthur as one of the true renaissance scientists of the several past decades. His accomplishments span evolution, cell biology, prokaryotic genetics, bacterial growth, bacterial transport mechanisms, mutagenesis, and more recently the development of the concepts of surface stress on microbial morphology. His book is concerned with all of the above. It seems that all of Arthur's past research has led to the genesis of this book.

Arthur was born in Saint Paul, Minnesota, and served in the Pacific Theatre in World War II as a member of the U.S. Navy. After military service he gained a B.S. in chemistry from Caltech and a Ph.D. in biochemistry from the University of Chicago. After four years in the Division of Biology and Medicine at the Argonne National Laboratory, Arthur joined the faculty at the University of Florida School of Medicine. In 1967 Arthur became a professor of Biology at Indiana University, where he remains.

Arthur has won numerous awards, including a Guggenheim Fellowship. He has been a generous donor of ideas to many of us. His enthusiasm for studying the growth of bacteria is readily apparent in casual conversations. Even though Arthur has a superior intellect, he has always encouraged and brought out the best in others.

Arthur Koch's appetite for prokaryotic cell biology emerges in all chapters of this book. The theme of surface stress, like a theme in a great symphonic work, reoccurs time after time. The surface stress theory constitutes a significant achievement in bacterial physiology. In this book, the role of surface stress is defined with respect to cell morphologies, to the regulation of autolysins, to the method that some bacteria use to precisely bisect their septa, to the method some bacteria use to maintain constant wall thicknesses, to prospective sites for new antibiotics, to the method that a prokaryote uses to segregate its growing chromosome(s), to the dynamics of cell wall turnover, and to flagellar movement and gliding motility.

The surface stress theory states that microbial morphologies are predictable when just a few facts are known about their division patterns. The principles of engineering applied to the construction of dams or boilers are also applicable to the construction of a bacillus or a coccus. The concept "make before break" in reference to assembly and turnover of cell walls of bacteria is analogous to the concept of constructing cofferdams before permanent dams are built. The equations developed for surface tension are the same for a bacterium as for a soap bubble or a chemical reactor unit. In his work, Arthur has relied on equations developed by D'Arcy Wentworth Thompson. His thesis is that bacterial growth (and of course that of all walled microorganisms) depends on how the bacterial surface responds to turgor pressure. When a bacterium takes in nutrients, it must either expand its surface area or develop a higher turgor pressure. The way in which the bacterium responds to turgor ultimately dictates its shape. This is, of course, an oversimplification, but it permits the many variations of shapes found in the microbial world. The fundamental equation

$$P\Delta V = T\Delta A,$$

where P is pressure (or force per unit of surface), ΔV is volume change, $P\Delta V$ is the work of expansion, T is surface tension (a unit of force per distance), and ΔA is the change in surface area, can be modified to accommodate spheres, rods, or capped rods under pressure. The text reveals how a simple equation can undergo permutations, become complicated, and ultimately be applied to all kinds of microbial shapes. It is challenging to follow the author's arguments about turgor and shapes. The reward is a refreshing understanding of the relationships between turgor pressures, surface expansion, and the forces that determine microbial morphologies.

As surface stress theory has emerged these past few years, several corollaries have developed. Surface stress states that a new wall must be added before an old wall is turned over by autolysins (make before break). Surface stress requires that new wall units be inserted into the pre-existing wall, to become functional as wall only when they are crosslinked and able to bear turgor. Surface stress demands that surface expansion (elongation) of rod-shaped bacteria occur by the diffuse intercalation of new wall into old wall in cell cylinders. When this new wall is crosslinked and able to bear pressure it is thought to stretch. The addition and subsequent stretching of wall at many sites ensures elongation of the cell. Surface stress rules out the elongation of cell cylinders of rod-shaped bacteria by equatorial growth zones. Surface stress can now account for the growth patterns of *Escherichia coli, Bacillus subtilis,* the blunt-ended *B. anthracis,* and streptococci. Koch has provided the only serious work to explain the morphologies of bacteria first described by van Leeuwenhoek over three hundred years ago.

Another theme intercalated into this book is one that has been popularized by Koch; namely, that bacteria undergo periods of feasting, only to be confronted with periods of fasting. The feast-famine lifestyle has resulted in evolutionary adaptations and to a large extent makes certain patterns of bacterial metabolism and behavior predictable. Starvation results in small cells, causing the ratio of the surface area to volume to increase, optimizing the uptake of nutrients. The surface stress theory allows for an understanding of modulation of surface-to-volume ratios at various growth rates and defines the need to couple surface expansion with the environment of the bacterium.

We now invite you to partake of the richness of this book. We both have been blessed to know Arthur personally. Those of you who will never be able to meet Arthur will get to know him well by studying this book. At the same time you will gain a new appreciation of microbial physiology.

R. J. Doyle, Professor University of Louisville
Lolita Daneo-Moore, Professor Temple University

Preface

In this book, the pronoun "we" is used frequently. It is neither the royal nor the editorial "we"; rather, it means you and I. We are going to consider bacteriological problems together. As much as possible, I will try not to lecture; we will be thinking our way through the important biological problems. I assume that you already know some microbiology. My most important goal in writing the book is to make accessible the relevant thinking from fields of science other than microbiology that are important to microbiology. The book is written for people who already have a fascination with bacteria.

This book consists of topics that are largely omitted from microbiology textbooks. It contains a good deal of my own work, both experimental and theoretical, together with a lot of speculation. It is, however, only one-third of a text on microbial physiology. The book by Neidhardt, Ingraham, and Schaechter (1990) covers another third. The final third is covered in the recent textbook by White (1995).

A list of all those who contributed to the ideas developed in this book would number more than one hundred. The short list follows: Ronald Archibald, Dick D'Ari, Terry Beveridge, Ian Burdett, Angelika Busch, Marta Carparrós, Steve Cooper, Lolita Daneo-Moore, Paul Demchick, Ron Doyle, Don Gilbert, Frank Harold, Jean van Heijenoort, Michael Higgins, Joachim-Volker Höltje, Kathryn Koch, Herb Kubitschek, Harold Labischinski, Stine Levy, Neil Mendelson, Nanne Nanninga, David Nickens, Tom Olijhoek, Greg Payne, Miguel de Pedro, Suzanne Pinette, Harold Pooley, Tina Romeis, Elio Schaechter, Heinz Schwarz, Uli Schwarz, Gerry Schockman, Elaine Sonnenfeld, Marcus Templin, John Thwaites, Frank Trueba, David White, Steve Woeste, and Conrad Woldringh. This list does not include three orthopedic surgeons who kept me quiet enough to start to write this book.

I would never have written this book if it had not been for Greg Payne. He encouraged me, worked over my English, and did not worry about deadlines.

Many people, both known and unknown, helped me by correcting my grammar, correcting my ideas, correcting my organization, and in the production of the book. These include David White, Greg Payne, Lisa LaMagna, Torrey Adams, George Hegeman, William Baldwin, and Frank Harold. A young undergraduate wrote to tell me that I wrote very clearly and explained things very well. That is a lie, but I answered by saying that she was correct, but only because there had been twenty-seven drafts that had been corrected, criticized, complained about, and rejected by friends and enemies. This book has been no different, it needed all the help it could get.

PROLOGUE

Thinking about Bacteria

KEY IDEAS

Prokaryotes are simple, but sophisticated.
Prokaryotes occupy important niches.
"Minimalist" organisms do nearly everything needed to grow.
Many problems are solved by being small.
Prokaryotes have exquisite regulatory systems, different from those of eukaryotes.
Crucial to the biology of all life forms are osmotic pressure and turgor pressure.

The purpose of this book is to outline and defend an approach to thinking about what bacteria were, what bacteria are, and how they do what they do. Particular emphasis is directed to their ability to establish their shapes as they grow and divide. In this prologue I will try to convince you that the study of bacteria and their life strategies is important even during what is being called the Golden Age of the Developmental and Cell Biology of Eukaryotic Organisms.

The major point is that prokaryotes [both eubacteria and archaebacteria (archaea)] do many of the same things that eukaryotes do, but with simpler equipment that is utilized in extremely sophisticated ways. This idea is interwoven with

xiii

others, such as that the simpler equipment reflects a more primitive strategy that, as illustrated in the first chapter, can tell us a good deal about the origin of life. A concept that is raised at many points in the book is that prokaryotes place reliance for their morphogenesis on the deft use of biophysical principles, whereas eukaryotes depend on the special properties of their mechano-proteins. A stress-resistant wall and a cytoskeleton are both ways to combat osmotic pressure, and these two ways may have led to the evolutionary separation and divergence of prokaryotes and eukaryotes.

What Prokaryotes Can and Cannot Do

In strict ecological usage "habitat" means the collection of physical and biological components of the environment of the organism that permit it to prosper (see Ricklefs 1990). The habitat of the eastern blue bird, *Sialia sialis,* is fields and not forests, for example. The "niche" is the job situation that an organism has been adapted to fill. The downy woodpecker, *Picoides pubescens,* to take another example, drills the bark of trees looking for grubs and stores seed in cavities it has dug. The *competitive exclusion principle* of ecology (Gause's principle) asserts that two organisms cannot occupy the same habitat and niche; i.e., they cannot coexist utilizing exactly the same set of resources. Although sometimes eukaryotes appear to be coexisting in the same niche and using a very similar habitat, when carefully examined it is found that almost always they are occupying different niches (see Ricklefs 1990); on the other hand, multiple occupancy seems to be common among prokaryotes (Milkman 1973). Probably the latter finding means that we do not know enough about microbial ecology. Very rarely are niches shared among eubacteria, archaebacteria, or eukaryotes. Most of these apparent exceptions are only superficial because a little further knowledge about the organisms is enough to indicate that they really have different strategies for exploiting different parts of the environment. Thus, all organisms have become specialized to a very high degree, but sometimes the strengths and strategies of one kingdom outweigh the evolutionary fine-tuning adaptations in another kingdom. For example, soil fungi and actinomycetes coexist in the soil—they differ dramatically in much of their lifestyle, but find regions and circumstances within the same gram of soil where each can excel. The combined actions of organisms of all kingdoms are symbiotic when viewed on a large scale. Just imagine the consequences of the sudden destruction of all organisms of only one of these kingdoms. Microbiologists frequently point out how important bacteria are for higher organisms. But the loss of the eukaryotes would be catastrophic for the prokaryotes as well. The synergisms and mutualisms between kingdoms are expressed often even on the microscopic scale.

The strengths of the eukaryotes, of course, are that they have large cells, engage in phagocytosis, form multicellular structures with the development of tissues, reproduce biparentally as diploids, move long distances rapidly (animals), grow above their environment (higher plants), and grow in an extreme range of osmotic environments (particularly fungi). Of course not all eukaryotes do all these things. One might construct a similar list for prokaryotes. They are small enough that internal plumbing is not required, they are metabolically very versatile so that one carbon source can be sufficient for all needs, they can utilize unusual energy sources (and they may require only CO_2 as a carbon source), they can grow in unusual (extreme) environments (high/low temperatures, acidic/basic media, high-salt environments, in the presence of toxic metals, etc.), and they can sometimes overcome the severe environment of a functioning mammalian immune system. This list overlaps with a corresponding one for plants.

Bacteria are the Simplest Organisms That Grow and Reproduce in a Self-Contained Way

All organisms exploit their environment; many organisms exploit other living creatures in their environment. Viruses exploit living cells for almost all their needs, supplying only some genetic information for the production of new viruses. We humans as a "high-tech" society exploit everything, inanimate or animate. We always did, and we are getting better at it. Here we focus on the simplest classes of organisms that have members that do "it" (i.e., the pursuit of life) almost all by themselves—these are the "minimalist" eubacteria and archaebacteria. Such organisms require the minimum number of elements from their environment. They can grow on a few inorganic substances and a source of exploitable energy. By the study of these minimalist prokaryotes we come closer to understanding early forms of life, those that existed between the First Cell and the Last Universal Ancestor (sometimes called the "progenote"). The First Cell was neither a minimalist nor a chemoautotroph, but probably today's minimalists exhibit many of the features of the Last Universal Ancestor. The Last Universal Ancestor is by definition the organism that gave rise to descendants leading on one hand to the eubacteria and on the other hand to the parent of the eukaryotes and archaebacteria. We take as given that the First Cell was dependent on abiotic resources for virtually all its needs (substrates and energy sources). It had very few functions but was capable of evolving new ones. Evolution progressed through a succession of successful saprophytic anaerobes and in so doing produced a versatile intermediary metabolism. As the organisms progressively depleted their environment of resources that they could utilize, they evolved ever better metabolic enzymes and pathways and ever better membrane transport systems to utilize substances that were available. Many of the metabolic characters present in today's bacteria must have evolved before the splitting into three kingdoms (domains). Such evolution would have depended on the availability of resources. Conversely, other special metabolic characters of bacteria had to evolve later when certain resources began to be generated biologically.

In the case of modern bacteria that need only a single carbon source, the metabolic capabilities to do the necessary organic chemistry to make the full range of building blocks require several hundred enzymes. Yet, other modern microorganisms are largely dependent on their environment for up to 100 different small molecular weight compounds. Both strategies represent specializations and elaborations from the primitive early state. There is a trade-off between maintaining a pathway of biosynthesis and maintaining an active transport mechanism. The balance point depends on the history of the cell line, but actually it is a minor gain to be able to do without a number of enzymes if one then has to manufacture the extra transport capabilities to import the end product. Both the cell that is metabolically versatile and the cell that commandeers preformed environmental resources present different aspects of our general theme—and in both cases they require lots of cytoplasmic membrane for adequate absorptive area. There is a need for a high surface-to-volume ratio. This ratio depends on cell geometry and on the cell's being small. These features bring us back to the consideration of how the bacteria can be so small and efficient and still be versatile and capable of rapid growth.

The Importance of Being Small

Prokaryotic cells are small compared to eukaryotic cells. Although many grow as separate independent cells, there are important interactions among some bacteria. These interactions are usually few and not essential for the biology of most prokaryotes. Again, this is a major simplification compared to eukaryotes, which effectively are large, either because they have large cells (some protozoa) or because they are multicellular. Bacteria do not depend on a well-controlled multicellular structure as do fungi, plants, and animals.

One reason to study bacteria is that they are small enough not to need the paraphernalia that large eukaryotic cells do—vacuoles, endoplasmic reticulum, and other indoor plumbing. (For a detailed treatment of the diffusion problem in bacteria see Koch 1971, 1985, 1990; Koch and Wang 1982.) The paradigm for bacterial growth and division includes the quintessential aspects of living things without the complications of the special adaptations needed for the big business of the larger "higher forms." We will see that bacteria, however, have plenty of special adaptations of their own.

I must reemphasize that although bacteria have had as long to evolve as have higher organisms, and that although they have perfected and diversified their basic strategies at many levels to occupy and retain many niches that are unavailable to eukaryotes, still the thread of simplicity of many aspects runs through them all.

The Various Kinds of Cell Regulation

Every cell can coordinate most of its processes in concert with its other cellular processes and with external circumstances. The key breakthrough in our understanding was the discovery in the late 1950s of feedback inhibition of critical enzymes and repression of messenger RNA synthesis. Since that time we have learned about catabolite repression, attenuation, translation, modulation, termination, regulatory proteolysis, repressors that bind to two DNA domains, mechanisms that delete special regions of the chromosome to yield new functions, inversion of genes to give new functions, regulation by phosphorylation, autophosphorylation, and separate sigma factors controlling different sets of genes. We have learned about the effects of gene order, transposons, insertion sequences, transformation, transduction, and restriction-methylase systems. Global regulatory systems responding to various challenges are present in bacteria; these are the SOS, the heat-shock, the oxygen, and alarmone regulon responses. Many of these things are unique to eubacteria, some are not. Those that are unique may represent new evolution since the evolutionary split, but others may have been superseded by other mechanisms in the other kingdoms. As we (remember, you and I) consider the bacterial approach to morphology, we must be certain to keep these possibilities in mind.

Now is a Good Time to Overview Growth

An encyclopedic amount is now known about prokaryotes: their biochemistry, physiology, and molecular biology. The DNA and protein sequences, regulatory circuits for transcription control, controls at the level of protein synthesis, and controls at the level of protein function are coming to be well understood (see reviews in Neidhardt *et al.* 1987, 1995). So this is the time to apply these results to the problem of prokaryote morphology. A good deal is known about morphological processes in eukaryotes (see Alberts *et al.* 1989; Darnell, Lodish, and Baltimore 1990). This includes knowledge about cytoskeletal structures consisting of microtubules, actin fibers, and intermediate filaments. On the other hand, there are no creditable claims for their existence in prokaryotes. I believe that this is simply because counterparts do not exist in bacteria. However, some workers assume that such mechano-proteins are essential for life and will be found when properly searched for. A purpose of this book is to present and defend the argument that primitive, simpler mechanisms utilizing no mechano-proteins could to grow and divide and that this is the current situation in modern prokaryotes. It is wise to keep the eukaryotic paradigms constantly and clearly in mind in order to see the similarities and differences, although the mechanisms in eukaryotes are often fundamentally different from those of prokaryotes.

Philosophy of Presentation to be Used in this Book

Because evolution goes from the simple to the complex, I shall also follow this order. This means that Occam's razor will be used many times, and I will presume the simplest explanation until it is absolutely necessary to adopt a more complex scenario. Please note that this order is the reverse of the usual direction in which knowledge is gained. When the genetics and physiology of a process have been studied, it is the complex regulatory part that is most evident at first. As an example, consider what happens when one tries to study microbial enzyme evolution in the laboratory. In the usual situation, one forces an organism, by strong selection, to utilize an unusual substrate. Almost always when the first sparce, slow growth occurs, it is because a regulatory gene of a previously unknown system has mutated to become constitutive. Often times one has to "peal off" yet additional regulatory apparatus before the selective pressure becomes manifest directly on the structural enzyme that one wished to study in the first place.

Another aspect of this problem is that the cell very frequently has multiple ways (parallel pathways) of accomplishing the same thing. For example, there is evidence for backup mechanisms in the cell division process. Yet another complication is that the cell has various safety features, like the SOS system to protect itself after DNA damage. It also has a heat shock regulon, a response to oxidizing conditions, and other features. All too often workers have stumbled on such features that then hide from view the underlying basic process.

For these reasons, I will speculate about what were the original, the basic, the essential processes for life to evolve, but then I will try to build into that framework the supernumerary regulatory structures in the final edifice.

This Book's Goal: To be the Interface Between Morphology and Chemistry

The study of growth brings together biochemical, biophysical, and physiological studies to supplement ultrastructural studies. Other disciplines like thermodynamics, diffusion theory, and kinetics provide key concepts. The interrelations of facts, disciplines, and theories are crucial in this endeavor. The theme of this book is: How do bacteria grow and how did they evolve to achieve their form and function?

BACTERIAL
GROWTH
and FORM

1

From the First Cell to the Last Universal Ancestor

KEY IDEAS

*The First Cell and its precursors needed the exogenous substrates supplied by abiotic
   generation and utilizable energy resources supplied by a primitive chemiosmosis.*
*A primitive cell arose that had a few informational molecules; these favored the cell's
   replication and Darwinian evolution.*
Early evolution proceeded simultaneously on many fronts.
*A primitive gene generated a phospholipid-like material inside the cell that fostered
   cell enlargement and cell division.*

Many side branches were lost before the Last Universal Ancestor evolved.
Most metabolic pathways developed during this "monophyletic" epoch.
Many regulatory mechanisms were developed.
New energy-transducing systems evolved, permitting the world biomass to increase:
  Methanogenesis
  Anoxygenic photosynthesis
  Oxygenic photosynthesis
  Fermentation/respiration.
Evolution of a stress-bearing wall led the eubacteria to branch off.
Evolution of the cytoskeleton led the eukaryotes to branch off.
Archaebacteria may have split off because of their development of methanogenesis.

This chapter considers what must have happened to create the First Cell and what must have developed between the time when the First Cell arose and the time when the Last Universal Ancestor gave rise to multiple lines of descent. The First Cell was the first entity that could improve itself and its lot by Darwinian evolution. The Last Universal Ancestor is defined operationally as the latest organism that had various descendants that evolved into eubacteria, archaebacteria (archaea), eukaryotes, and organelles of eukaryotes. The parent of subcellular genetic elements, such as viruses, insertion sequences, and plasmids, has been little considered, but arguments can be made that its development was fairly late. The Last Universal Ancestor probably had many of the characteristics that we now find retained in bacteria and, of course, many of the features of most living organisms. During the intervening "monophyletic" epoch between the First Cell and the Last Universal Ancestor, the processes shown in the central part of Fig. 1.1 were being perfected, but had not reached the state in which we now find them in various organisms today. In this diagram, no phylogeny is implied except that within the epoch many mutations and changes occurred, but each improved organism totally displaced its predecessors and the side shoots, and that multiple branching occurred after the Last Universal Ancestor to produce a diversity of types of biological descendants.

The First Cell

If civilization were destroyed, it is doubtful that high technology could ever be reestablished. This is because the raw resources like tin and copper needed by mankind to move into the bronze age have been mined from the easy places and the technology to recover them from more difficult places would no longer exist. This notion, due to Garrett Hardin, applies as well to the origin of life. Life could not start again as readily if this planet were autoclaved, because life forms have destroyed suitable habitats, polluted the world, and consumed exploitable resources. Whether or not organisms started on this planet, it is a good bet, as originally proposed by Oparin (1936) and Haldane (1929), that everything the

Early Life

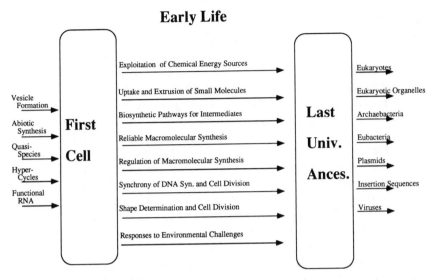

Figure 1.1. Major categories of processes that must have developed during the monophyletic period from the time of the First Cell until the Last Universal Ancestor. Before the First Cell there was a buildup of chemical diversity, but evolution in the adaptive sense of Darwin was not occurring. The main part of the figure lists the various kinds of processes evolved before the Last Universal Ancestor. The Last Universal Ancestor is defined as that individual from which all current life has been derived, or at least, the three urkingdoms; these have been renamed: Eucaryae, Archaea, and Bacteria. Fully effective mechanisms for all eight types of processes were present in the Last Universal Ancestor, although they were further modified after the major splits.

organisms initially needed could be gratuitously acquired from their environment (see Miller and Orgel 1973). With further evolution, better processing was possible and fewer non-negotiable demands need be made upon the environment.

Oparin and Haldane's insights about the origin of life were ideas concerning the presumption that what was needed was readily available in the environment. Three-quarters of a century ago, they focused on the biochemical processes that were known in their day and on this basis imagined what must have been. Their emphasis on "the art of the possible" was also the central idea of Horowitz (1945) and of all later workers. Horowitz gave a rational explanation of how life, once going, could develop metabolic pathways. The main idea in all the scientific speculations has always been the same: in getting started, what was available—for free—was all that could be used to form life.

Thus, as a matter of dogma, we will assume that the first cell lived in a "broth" containing many abiotically produced substances, both simple and polymeric. Perhaps it was not as rich as chicken soup, but it contained the array of substances needed to make informational molecules. We must further assume that the First Cell had access to a simple source of energy, and was largely concerned with the replication of a few, perhaps only one, informational molecule (see Koch 1985a). This informational molecule had to have a function on which "survival of the fittest" could act. Further evolution could then lead to a more effective function and duplication plus modification could lead to multiple functions. Could Darwinian evolution have been implemented before cells arose? Could individual units of acellular life exist in a restrictive environment; e.g., in neighborhoods such as surfaces of clays, etc? After all, the action of the "survival of the fittest" paradigm is necessary to "bootstrap" life. Therefore, I doubt that noncellular life is capable of evolving adaptively because it is difficult to imagine that the gene function and the resultant product could remain localized together to increase their survival under particular conditions while entire copies of such systems (genes and gene products) were being replicated and needed to be dispersed to new habitats. Consequently, a process that is the equivalent of cell division or propagule formation cannot easily be attributed to a life form commencing on clays as postulated by Cairns-Smith (see below).

Subsequent to the First Cell, mutation, duplication, and diversification would lead to better utilization of the environment and would in turn depend on the evolution of effective transport systems, on metabolic pathways to produce needed substances from available substrates, on accurate and appropriate syntheses of macromolecules, on accurate cell division mechanisms, and so forth. (See Fig. 1.1.) Through many refinements at all these levels, the Last Universal Ancestor arose. This organism was the branch point and evolutionary lines diverged from it. Before this time every stage advanced by displacing previous life forms as a simple consequence of the competitive exclusion principle. But from the Last Universal Ancestor came diversity, and all the variety of present day life forms resulted.

The First Cell's Properties

It is not germane to our goal to catalogue the production of abiotic compounds or their abiotic linking together to make biomolecules (see Miller and Orgel 1973), but it is germane to follow the thread of cellular morphology. Likewise, it is not appropriate to consider pre-life forms on clays (Cairns-Smith 1986), in atmospheres or at extraordinary temperatures (Woese 1979). This book emphasizes that life as it started was a very simple, but cellular, phenomenon. The argument against describing pre-cellular entities as living is quite simple: only cells (i.e., something surrounded by a phospholipid-like bilayer and containing genetic material) could retain their functional products in order to be able to evolve in the sense of improving existing functions or developing new ones to do new things and exploit new niches and habitats in the Darwinian sense (Koch 1985a). Without the cell to contain the superior product, selection would not favor the cell, and hence the gene, that made the superior product. This is a view different from that of other students of evolution, but its assumption allows the bypassing of the features on the left-hand side of Fig. 1.1, such as Eigen's quasi-species and hypercycles (Eigen and Schuster 1979). These would function in the pre-cellular time, however, because evolution via these processes only leads to better competitors among similar molecules and not to the achievement of a qualitatively different function or an access to exploiting a new strategy.

To believe that living cells came into being before adaptive evolution was able to function requires imagining that spontaneous generation of membrane-bound vesicles in an abiotic world occurred. Such an assumption is easily credible. The Miller-type experiments where gas mixtures are energized in an apparatus that traps nonvolatile water-soluble compounds (see Schopf 1983) produces a high yield of water-insoluble material. Given a variety of organic compounds generated in non-specific ways, some will be amphiphiles. Mechanical agitation of a suspension of such chemicals yields vesicles spontaneously. In fact, a mixture of long-chain fatty acids and long-chain alcohol molecules (Hargreaves and Deamer 1978) has bilayer-forming properties equivalent to those of phospholipids.

So it is reasonable that vesicles existed, but the successful vesicle that was to become the First Cell needed a way to divide and partition genes and gene functions to daughter cells. Cell division is quite complex process, but it is easy to imagine a paradigm for a primitive cell division process that is essentially a default circumstance and trivially simple (see Koch 1985a). Its only necessary requirement is that one of the first genes had the function to assemble an amphiphilic compound *inside* the cell (see Fig. 1.2). Such molecules, equivalent to phospholipids, would enter the inner leaflet of the bilayer, causing invagination and

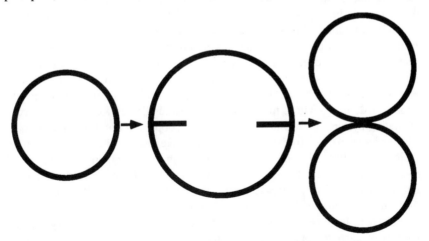

Figure 1.2. Primitive method of cell division. The First Cell had to carry out all essential features, but had very few of the mechanisms functioning in modern cells. It is argued that a single metabolic process that caused the formation of an amphiphile inside the cell would lead to cell division by introduction of new polar lipids into the inner leaflet of the bilayer. The figure is redrawn from Koch (1985a).

cell division based on physical forces within the bilayer. As the content of the inner leaflet increased, invagination of the inner leaflet would have occurred. Moreover, as a primitive cell grew in molecular content, the increased turgor pressure would increase the rate of flipping of the inner leaflet components to the outer leaflet, producing spontaneous division. Of course, this division process would have been imprecise and very irregular, but it could have worked before any cell division machinery was perfected.

From such a beginning, cellular biology could evolve. One of the processes that would evolve would be a precise and better way to enlarge the cell surface and to divide. This innovation involves the active accomplishment of what originally was spontaneously accomplished abiotically and gratuitously.

Other things had to be available to jump-start life. The one that comes to the mind of a molecular biologist is a linear molecule to serve as a template and a source of molecular information. Nucleic acids will not be discussed nor will the role of catalytic RNA be developed; I am not the person to deal with these issues. The subject is a currently active field. I will develop, however, another key issue: the primitive generation of energy.

The First Cell's Energy Supply

The supposition of the availability of a "free-lunch" for the First Cell and its precursors voiced above also must apply for cellular energy. There must have been free energy in the original sense of Gibbs free energy available to the cell to use in an arbitrary way. A cell cannot be given energy, because energy is something that the cell must transform via an exergonic process into a chemical form that it can use. However, the set of substrates to carry out the exergonic process must have been easily and commonly available and the transformation process must have been very simple. Even though fermentation, respiration, and photosynthesis have been "touted" in the literature for this role, they are too complicated, in my view, to be considered as candidates for the first energy-generating system. Fermentation requires the cooperation of a dozen enzymes. If the pathway evolved from the back to front as suggested in the general paradigm of Horowitz (1945), then the cells along the evolutionary path must have evolved glycolytic enzymes and transport systems for a number of phosphorylated intermediates. This is unlikely because these substances, at least in modern cells, do not readily go across cell membranes. A respiration system, as in modern cells, is unlikely to have been an early source because many metabolic enzymes and a series of carriers coupled to each other would have been needed and these would have needed to be coupled to the energy generation of ATP and to function in a precise arrangement in the membrane.

Likewise, photosynthesis, even bacterial (anoxygenic) photosynthesis, would be beyond the means of the First Cell: photosynthesis employs a photopigment and proteins of clever structure to separate positive and negative charges necessary for the energy transformation part of the process. Photosynthesis is very sophisticated and has special environmental requirements, such as having light in the right quantity and quality. Only light that is not too energetic (short wavelength) to destroy the photopigments could be tolerated.

So the argument has gone full circle, the original energy transformation had to be something very much simpler that essentially produced energy spontaneously. Tom Schmidt and I have suggested that it was something akin to a unit process that is part of respiration, methanogenesis, and photosynthesis; i.e., chemiosmosis (Koch and Schmidt 1991). Chemiosmosis is the process whereby the cell has a mechanism that couples an electron flow down an electrochemical gradient to create a proton flow up its electrochemical potential to drive protons without an accompanying anion outside of the cell. This electrogenically produced protonmotive force is then used for transport events and for the synthesis of ATP in modern organisms.

Spontaneous chemiosmosis would occur if four conditions were simultaneously met: (i) there was a reductant outside the cell, (ii) there was an oxidant inside the cell, (iii) electrons could come from the reductant, pass readily through the membrane that encloses the cell, and reduce the oxidant, and (iv) protons passed only slowly through the membrane. Simple as these conditions are, it would seem that these are stringent requirements. A potential candidate system for the power supply of the most primitive cell is shown in Fig. 1.3 (Koch 1985a; Koch and Schmidt 1991). It involves the reaction of H_2S and Fe^{++} on the outside of the cell to produce insoluble iron pyrites and a pair of protons and electrons. This much had been suggested earlier by Wächtershäuser (1988a,b, 1993). If these compounds reacted to form molecular hydrogen, then energy trapping would not occur; likewise, if some reductant molecule consumed the electrons on the outside of the membrane, no energy could be trapped. However, if the First Cell and its nonliving precursors had a lipid bilayer with metal impurities that could conduct electrons across the vesicle membrane, it could prevent energy wastage due to the formation of molecular hydrogen on the outside of the primitive cell. If the electrons on reaching the inner face reduced a substance that readily had penetrated the membrane, such as CO_2 or CO, protons would be consumed, and consequently a protonmotive force would be generated. This in turn could favor essential transport events and possibly the production of high-energy phosphate bond compounds.

Thus, the availability of the reactants and some trace metal-organic compounds may have been sufficient to lead naturally to the development of a protonmotive force in a closed bilayer vesicle. For this simple extracytoplasmic chemiosmotic process, the most primitive cell and its immediate nonliving precursor would contribute nothing towards the process. Of course, all benefits result from the spontaneous exploitation of an environment that is not at chemical equilibrium. For a cell that had only a cytoplasmic membrane and the glimmer of a genome, this source of free energy could have been practical.

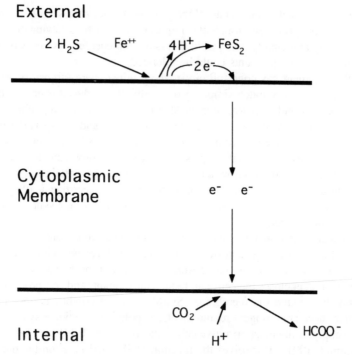

Figure 1.3. A possible early chemiosmotic energy source. A spontaneous source of utilizable free energy could be produced by the mechanism depicted. The mechanism depends on: (i) having a totally surrounding bilayer, (ii) having a reductive process on the outside of a phospholipid-like membrane; i.e., the formation of insoluble FeS_2, (iii) metal impurities in the membrane that can conduct electrons to the inside, and (iv) a reduction process consuming the electrons with an available source of oxidant; e.g., CO_2. Such a process results in a protonmotive force permitting work to be done as in modern chemiosmosis.

Between the First Cell and the Last Universal Ancestor

Once an organism had arisen, it grew and its descendants evolved. As it happened, much of cellular biology and biochemistry was created before the biota of this planet became diverse. This section deals with global aspects of the evolution during this period.

Saltations in the World Biomass

When the First Cell arose the world biomass was infinitesimal. Gradually, the global biomass would have increased, but even at the time of the Last Universal Ancestor it may have been limited by the available useful abiotically produced materials. For a long period the biomass would increase slowly with the rate of geological and photochemical production of abiotic organic compounds. With the evolution of ways to transform available chemical species into more immediately useful ones, the biomass would increase. Still, the biomass would not increase very fast because there would be biotic and abiotic degradation of some compounds into nonutilizable forms. Moreover, some cells would die and their resources not be returned to the available pool because there were as yet no hydrolytic enzymes nor any effective way to secrete degradative proteins the way saprophytes do today. Probably the biomass only very gradually could increase as new metabolic enzymes and new transport systems were perfected to make more efficient use of the limited sources of organic compounds, i.e., as metabolic methods to deal with them were devised.

The development of all but the first process listed in Fig. 1.1 would have little effect on the world's biomass, though great effect on the qualitative nature of that life. If a cell took up molecules and rearranged them, made macromolecules, regulated the processes, properly manufactured double-stranded nucleic acid, and divided *and* did that all poorly, the biomass would be substantially the same as if it did it well. [In ecological terms (see Ricklefs 1990), increasing the carrying capacity of the system increases the biomass, increasing the rate of growth does not.] The rate of genome replication would also not change very much. Only if the cells devised ways to exploit some new resource (say a previously recalcitrant carbon compound present in low concentration or a different energy source) would the living protoplasm of the planet increase. An increased number of organisms would raise the chances for new mutations to speed the improvement of every aspect of cellular life. Working against this and slowing evolution is genomic repair, which slows the effective mutation rate.

Naturally, when anoxygenic (non-oxygen-producing) photosynthesis and methanogenesis arose, the biomass of the planet would have risen greatly and explosively. Representatives of the earlier forms of anoxygenic photosynthesis are not evident in the record of isotope ratios (Hayes 1983), but early methanogens have left their mark, presumably because they became numerous and produced abundant fossil biomass. With the development of oxygenic photosynthesis, there would have been another abrupt and very extensive increase in biomass. Importantly for the speed of evolution, again there was an increase of replication of genomes in the planet per year. This would have drastically increased the rate of evolution. Moreover, the number of populations in which evolution could take place would increase because the geographical locations where independent evolution was taking place would increase. These world changes, due to access to new energy sources, would have increased the organic matter very much, but after a while there would have been a decrease in living biomass, because dead organisms would abound, storing not only the organic chemicals, but important metals, nitrogen, and phosphorus. This accumulation of both living and dead biomass would favor the evolution of the saprophytes, herbivores, and carnivores. Finally with the evolution of an oxygen-containing atmosphere, a world ecosystem with abundant total biomass and a rapid turnover arose. Its present amount may or may not persist, considering the current works of mankind.

The Speed of Evolution

Between the First Cell and the Last Universal Ancestor evolutionary pressure led to the development of the processes categorized in Fig. 1.1. Each of these involves many genes and subprocesses. Evolution must have occurred simultaneously for the multiple genes coding for the multiple proteins involved in many pathways.

NO SEX

A key point is that the evolution was not going on in multiple parallel pathways independently because there was no way to achieve lateral transfer of genetic material from one substrain that had one improvement into another that had made a different improvement (Koch 1994b, 1995b). Of course lateral gene movements via plasmids and viruses occur quite commonly in modern organisms. One might think that DNA transformation would have served as a default. But, it too, requires very special conditions and prior evolution to make a cell competent to be transformed. Genetic transfer during this phase could not have occurred because all the mechanisms for moving genes from organism to organism exhibited by today's organisms are highly evolved and very special processes. Although one might have no problem conceiving that code-bearing macromolecules would be present in the environment because of the destruction of their originating organisms, there appears to be no mechanism to transfer nucleic acids into a "living" vesicle. Deamer and Barchfeld (1982) have shown that one can dehydrate and rehydrate vesicles and transfer macromolecules to the inside of vesicles, but presumably drying would be lethal if the vesicle were a primitive recipient cell.

So evolution during the monophyletic epoch would have been very slow; the job of the next section is to count out the factors that would have sped and slowed the process. But to start thinking about the problem of evolution, think of the primitive biosphere as a continuous culture apparatus seeded with one strain and run indefinitely. Mutations would arise and, if successful, replace the original strain. The kinetics of this process must be investigated before we consider the variations due to secondary causes.

SIMPLE SELECTION AND REPLACEMENT

Consider a mutation that arises at a rate varying from very fast down to 10^{-10} per cell generation (10^{-10} per cell generation is the rate that streptomycin resistance develops in *Escherichia coli,* a process that is much too slow to be clinically relevant). Truly radical and revolutionary changes occur at much smaller rates and will be considered below. We can treat the ratio X of the number of mutants relative to the wild type in the population as a continuous variable (see Koch 1972). We will need the analytical solution to the kinetic problem. Assume that the actual number of parental ("normal," wild-type cells) is N and the number of mutant cells is M. M and N have net growth rate constants (birth rates minus death rates) of μ_m and μ_n and mutation rates of m_m and m_n (in all cases the subscript indicates the species produced). If time is designated by t, then we can write the increase of N due to net growth by:

$$dN/dt = \mu_n N,$$

the increase by mutation from the M cells by:

$$dN/dt = m_n M,$$

and the loss by mutation to M by:

$$dN/dt = -m_m N.$$

Of course, all three effects operate simultaneously, so we must add them up to get the total.

$$dN/dt = \mu_n N + m_n M - m_m N. \qquad \textit{Rate for wild type}$$

(Do not panic when the symbol dN/dt is used; it just means the change in the number of normal cells occurring during a small interval of time. We will use the methods of calculus with this notation here, but if a reader is lacking in familiarity with this tool, he or she can become refamiliarized by reading the first sections of Chapter 3. The goal throughout the book is to set up any mathematics logically and simply and then, if necessary, jump to the solution.)

In just the same way we can set up a rate equation for the mutant form:

$$dM/dt = \mu_m M + m_m N - m_m M. \qquad \textit{Rate for mutant type}$$

Now we have two equations, four constants, and three variables, N, M, and t. We can combine these two equations and reformulate in terms of the ratio of the two types. After all, that is what we are really interested in. By defining $X = M/N$, remembering that:

$$dX/dt = d(M/N)dt = [N(dM/dt) - M(dN/dt)]/N^2,$$

and doing a little bit of algebra, we obtain one equation:

$$dX/dt = -m_n [X^2 - (\mu_m - m_n - \mu_n + m_m)X/m_n - m_m/m_n].$$

This is the differential form of the solution to the problem. To get an answer as to how evolution would progress in a given circumstance, we would need the values of the four constants, μ_m, μ_n, m_n, and m_n, and the initial value of X, which we may call X_0. Now our minicomputer starts with the value X_0, adds to that dX/dt calculated from the equation above, obtains a new value of X, and uses it again and again to plot out the entire time course. With it we could see how important growth rates, mutation rates, and initial proportions are in any given hypothetical situation. However, to get an analytical solution is not hard (see Koch 1972), and with that we can speed up calculations on our minicomputer. If r_1 and r_2 are the roots of the quadratic equation:

$$X^2 - [(\mu_m - m_n - \mu_n + m_m)/m_n]X - m_m/m_n = 0,$$

then the solution is:

$$(X - r_2)/(X - r_1) = (X_0 - r_2)/(X_0 - r_1)e^{-\mu_n(r_2 - r_1)t}. \qquad \textit{Evolution equation}$$

This equation was evaluated and used to simulate the time course of evolution from the four constants and the initial proportions (see Fig. 1.4). [Take note, if you set this up on your computer, that some special forms of the solution are needed in particular cases, as for example when r_1 and r_2 are equal (Koch 1972).]

This solution has been useful in understanding the role of silent mutations in speeding evolution, in understanding periodic selection, and unequal crossing-over of tandemly duplicate genes (Koch 1974, 1980a).

PROCESSES AFFECTING THE RATE OF EVOLUTION

Evolution by single steps can lead to improvements as long as each replacing form is more "fit" than its predecessor. But at some point this kind of gene evolution would cease because only multiple intragenic changes would yield a superior type. There is an alternative, however, that will occur when the environment fluctuates (Koch 1972) (see Fig. 1.5). The mechanism arises naturally when the gene's function at some time has a vital, limiting role, and first double copies and then multiple copies increase in the population, while at other times copies are expendable; during the latter periods an unneeded copy of the gene might become silent. During the periods where it is not serving a functional role, it could accumulate multiple random mutations that usually would not, but rarely would, represent radical improvements. Of course, these improvements would be only relevant when the selective pressure for gene function was reapplied and the gene became reactivated. Many cycles of this kind were no doubt needed to perfect the processes we see in every organism. Obviously, this phenomenon greatly speeds and extends evolution of single genes.

Other phenomena would have greatly slowed evolution. One is periodic selection; this is the phenomenon in which mutations in one gene, even if favorable, may not replace the original forms because of an independent takeover of the population by a favorable mutation in some other gene at some other loci.

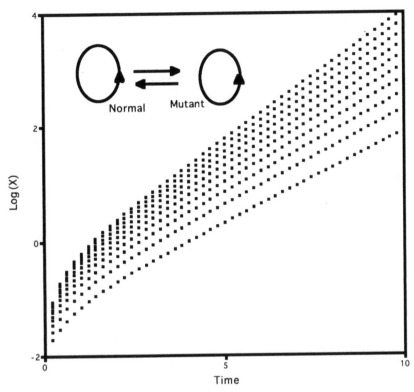

Figure 1.4. An example of the time course for the change of two alternate alleles in a prokaryotic population. The curves were calculated by the *Evolution equation* developed in the text for the case where mutation in both directions can occur and differential growth rates for the two forms can be different. The computer program uses alternate code in the cases in which the *Evolution equation* becomes indeterminant. The figure is illustrative, but the important use of the program is as a module in a more complex program to understand factors that affect the rate of haploid evolution. The parameters for this example are that the initial ratio of species was 1; $\mu_m = 2$; $\mu_n = 1$; $m_n = 0.1$; and $m_m = 0$.

Usually, the takeover by the favorable mutation in the second gene carries with it the parental or prototype form at other loci (see Koch 1974, for review). During the monophyletic epoch under consideration, where evolution of thousands of genes is going on simultaneously, the interference on one gene caused by all evolutionary advances in others would have been very large.

A second phenomenon results because during extremely severe selection a gene may become tandemly duplicated by a rare nonhomologous crossing-over event. Then the formation of a gene family by unequal crossing-over would lead to the expansion of the numbers of repeats to satisfy the cell's needs. Later, when selection is relaxed, a contraction in the number of repeats would occur. Such cycles also defeat progress because they are independent of improvement

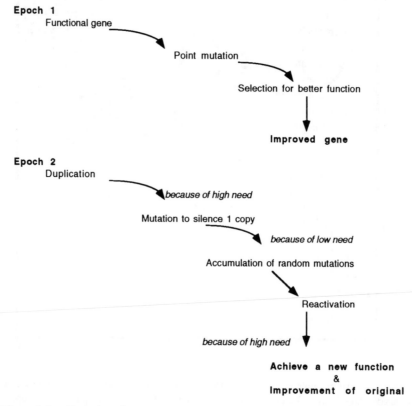

Figure 1.5. Evolution of a prokaryotic gene. Two basic processes naturally occur once the machinery for inheritance, mutation, and selection are in place after the First Cell has arisen. These are both cyclical repetitious processes, but only one cycle is shown. During Epoch 1 only the Darwinian mode functioned and evolution was by replacement of one form by another that grew faster. This process of single step changes finds the nearest local growth optimum, but will not evolve towards nonlocal higher maxima. With the development of the ability to regulate a gene's action and form tandem duplications, life entered Epoch 2 in which cycles of duplication, silencing, multiple mutation, and reactivation occurred and could act in a fluctuating environment to create more radical changes.

of the basic structure (see Koch 1980a). Regulatory mutations can also act so that evolution of the structural gene is similarly impeded.

SALTATORY MUTATIONS

Truly rare, but extremely favorable, "saltatory" mutations cannot be treated by the above mathematics. They follow probablistic mathematical laws and are limited by different factors. These factors include:

(i) The size of the population of cells. As long as the biomass is small, evolutionary progress by big jumps would be slow because of lack of opportunity for the mutations to occur.

(ii) The mutation rate. Before the evolution of precise protein synthesis and DNA repair processes had been achieved, chromosomal replication would be error-prone. The rate of trying out new mutations would then have been larger than today. On the other hand, it is possible that the total crop of defective cells could be so large that the growth of the population may have been reduced and thereby curtailed.

(iii) The availability of cells in the total population that are of an appropriate genotype so that they can be helped by the rare saltation mutation. This is a key point. If the rare saltation mutation occurs in a genetic background that cannot be aided by it, the mutation will be lost. The need for genetic preparation is likely to be the common case; throughout evolution we can expect that mutation and selection at a locus or loci are needed to produce a kind of organism for which the radical mutation could be useful (see Koch 1993a). This is an area where further application of the evolution equation will be fruitful.

THE BOTTLEBRUSH NATURE OF THE EARLY PHYLOGENETIC TREE

During the epoch from the First Cell to the Last Universal Ancestor, mutations and improvements on many fronts were occurring. The Last Universal Ancestor had incorporated an extremely large set of refinements and extensions compared to the metabolic equipment of the First Cell. The evolutionary tree during this period must have consisted of a series of stages of universal ancestors and their many cousins and looked like a bottlebrush. All the cousin lines evolved, however, into dead ends. Many of these might have been functionally superior to the main line in some respects, but because of some fatal flaw, the results of competition, or chance left no descendants. What distinguishes any Universal Ancestor was that its ancestors were the winning competitors at all earlier stages. Although the Last Universal Ancestor certainly was not endowed with the modern form of all the evolutionary features common to the eukaryotic, prokaryotic, and archaebacterial kingdoms, the "unity of biochemistry" suggests that it probably had developed most of them during the monophyletic period.

Development of Processes During the Monophyletic Epoch

In the next several subsections, some of the processes listed in the central part of Fig. 1.1 will be briefly presented. Others are beyond the scope of this book.

THE ORIGIN OF INFORMATIONAL MOLECULES

If the first requirement for life is an energy supply, then tied for first place are the facility to remember what worked before and the ability to do it again.

In today's organisms (with very special exceptions) this is accomplished with the double-stranded deoxyribonucleic acid (DNA) holding in its complementary sequence of bases the information for its precise replication and the facility to transcribe and then translate the information into functional protein. If a protein is useful, then the cognate DNA prospers. If the DNA mutates then most times the protein is less useful and the DNA, the cell, and the organism flounder. Very rarely, the protein is better and the new version of DNA supplants the old through competition. The above is a modern translation of what Darwin had to say. It is the basis of evolution that can be innovative.

Of course there are good *a priori* and *a posteriori* reasons to insert RNA into this scheme between DNA and functioning protein. The *a priori* reason is that RNA, it now appears, might have been the first informational macromolecule and also the first that could carry out directed functions. The range of functions attributable to RNAs is clearly limited and, consequently, the switch to the present system must have been a tremendous improvement. The reality of the Central Dogma of Biology in which DNA information is transcribed into RNA which in turn is translated into protein is clear; how it was established on earth is the largest puzzle in the field of the origin of life. But it must have happened during the monophyletic epoch. An *a posteriori* reason for a cell's utilizing an RNA intermediate is that by controlling the rate of RNA production the cell can readily control the production of different proteins.

In terms of the Watson-Crick structures, we can readily understand why the code pairs a short planar pyrimidine with a long planar purine. We can also understand that it can work to uniquely pair bases only when they are presented in a standard orientation and spacing provided by the double strand of sugar phosphates. In addition, one of them must proffer a hydrogen atom to another base that has an electronegative atom that can accept it to form a hydrogen bond. The major bonding of the proton, of course, remains with the nitrogen atom to which it was attached. A rule of length and a rule about donor-acceptor pairing is sufficient to correspond to the rules of adenine-pairs-with-thymine and guanine-pairs-with-cytosine. These rules are unique only if we restrict interest to these four nucleic acid bases. But what, for example, originally kept hypoxanthine from substituting for guanine?

Entirely different systems ought to have been able to work as well and may function in life elsewhere in the universe. Rigid positioning and stable double-stranded structures would seem to be absolutely required wherever chemical life exists. But why not a code based only on pyrimidines or only on purines, as double stranded helices of chains of these can be made in the laboratory? Why not a code with both purines and pyrimidines, but one incompatible with our system because they pair, for example, diaminopyrimidine and diaminopurine, respectively, with xanthine and uracil?

Why is only one system of coding working in today's life forms? Even the exceptional cases only emphasize the universal nature of the code. Possibly you have not worried about the absence of alternatives, but let us categorize the possible answers: (i) Life could have originated but once, and once a successful system functioned (the one we have), it prevented any other from developing. (ii) The needed chemicals for some of the other systems were not created abundantly by abiotic processes. (iii) Some of the alternative structures just cannot form or are not sufficiently stable. (iv) Life did start with one of these alternative structures, but could not sequester the resources; these were lost and then other styles of life could start after an earlier, but poorer one, was extinct. Elimination, for example, could have rested on an inability to form a sufficiently complex genome that was also sufficiently stable in the face of mutational pressure.

With evolving complexity a new problem was created. As the organisms came to have more gene functions, the need arose to segregate at least one of each kind of genetic factor into all daughter cells. The prokaryotic solution (and that of a stage in the lifecycle of many viruses) is to have all the genes lined up in a circular chromosome that is replicated as a single entity; the eukaryotic solution is to have multiple linear chromosomes and to have available a highly evolved mechanism for mitosis (and meiosis). The solution of some protista (amicronucleate protozoa) is to have many copies of each gene in the macronucleus and to replicate from one end of the macronucleus to the other all genes indiscriminately, but systematically. Evidently, a system is needed, so we could ask about the life form on this planet with the poorest system. A candidate group is certain viruses, such as vesicular stomatis virus; its genome consists of several disconnected parts. The replication works because generally many viruses invade a cell. It does not matter if late in the process many viruses are formed that do not have a full genetic complement because different defective particles complement each other when multiple infection is probable. But this strategy is unworkable for a cell, because essentially every cell must be independently capable of growing and not depend on the cooperation of its neighbors for supplying genetic functions.

Without divine intervention, something like random replication of genes or parts of genes took place in the most primitive organisms. Quite frequently this led to an inviable organism and consequently led to a selective pressure to evolve an accurate segregation mechanism. There are two aspects of the problem. If there are too few copies of a gene, by chance the organism will lose an essential function and die, whereas if there are too many, Darwinian evolution will not work (for more about this, see Koch 1984a) and progress will be stifled. Consequently, there would have been a strong selective pressure to string all the kinds of genes in a linear array on what could be called the first chromosome. Now, at least the problem has been simplified in that it is necessary only to segregate at least one replicon copy into each new cell. Before segregation mechanisms were developed, the goal of reducing the number of chromosomal copies to a manageable level (say five or six chromosomal copies) would suffice. This is large enough to ensure that at least one complete chromosome is present in a high proportion of the newborn cells, while small enough to retain the ability to carry out Darwinian evolution at a useful rate. Finally, there evolved a simple mechanism for reliable segregation and the reduction of the number of copies of the entire single chromosome to one (when measured before chromosomal replication). Understanding how this could have happened may come from a more complete understanding of how the copy number of various plasmids is maintained (see Koch 1995b). Only later would there have been need of the other regulatory mechanisms for precise gene replication and segregation. The important one for prokaryotes would be to form a single linkage group of all genes.

EVOLUTION OF BASIC BIOCHEMISTRY

Basic intermediary metabolism must have almost totally evolved before the separation into the different kingdoms, and this, no doubt, is the reason for the "unity of biochemistry." Only fifty years ago we had very little insight into metabolism, although perceptive studies had been going on for fifty years before that. Rapid progress waited for the end of World War II and the return of war-weary veterans. It was American funding inspired by the success of the Manhattan Project, the commercial availability of the Warburg apparatus, spectrophotometer, and pH meters, and the beginnings of microbial genetics that launched the new era.

What did we learn? That nature used a successful trick over and over again. She repeatedly used dehydrogenation, hydrolyses, hydration and dehydration, transfers of groups, internal rearrangements, and the use of cofactors and coenzymes as handles and temporary repositories. We learned that ATP was the energy currency the cell used to force reactions in the needed direction. Although the vital ingredient of cellular finance was neither "high-energy," nor "phosphate," nor a "bond," the Gibbs free energy of hydrolysis of ATP had an ideal amount of energy in a packet to take care of the cell's anabolic processes. This energy could be manipulated to drive syntheses and to power transport processes.

The concept arose that enzymes could exist that could catalyze almost any organic reaction. Of course, an enzyme catalyzes a reaction only in the direction of equilibrium. But the cell was "smart" enough to know that by carefully picking which enzymes to form, a metabolic pathway could be constructed and pushed in the desired direction. This mainly meant choosing which compounds were to be phosphorylated by ATP and which were to be hydrolyzed by water, etc. In this way the cell could drive syntheses of needed small molecules, syntheses of macromolecules, and transport events. Oxidation states are mainly adjusted by the cell, then and now, via hydrogenation and dehydrogenation with a panoply of hydrogen carriers. That is all there is; the rest of the metabolic map consists of minutiae. The slight changes that different organisms used to achieve profound changes in their strategies of life and alteration to adapt to different ecological niches is interesting, but the bulk of details of metabolism solidly indicate the "unity of biochemistry," reflective of the metabolic machinery of cells developed during the epoch between the First Cell and the Last Universal Ancestor.

So what compounds were needed? In large amounts a score of amino acids found in normal proteins, nucleic acid bases, certain carbohydrates, and phospholipids. Each amino acid could be used to extend a peptide chain and each one had side groups that could function differently in formation of the three-dimensional proteins and in their function. These groups had to be able to make more or less unique attachments to other amino acids and could serve as catalysts and structural components. Of course, other compounds that are developed for the formation or in the degradation of protein amino acids are present in cells.

Information transfer requires nucleic acid bases that have aromatic rings with enough structural planarity and rigidity to make a rigid platform for amino and oxygen functions to make meaningful specific hydrogen bindings to serve as informational molecules. Moreover, the nucleic acid bases serve essential roles in many cofactors and coenzymes because their rigid structure can be recognized precisely by enzymes. The carbohydrates, ribose and deoxyribose, are needed for the backbone; i.e., for the noninformational parts of the informational molecules.

The ancestral cell needed structural elements for membranes and walls. Amphiphiles such as lecithin, which forms the phospholipid cytoplasmic membranes, were certainly needed. Whether at the end of the monophyletic epoch a way was devised to secrete carbohydrates that then could be polymerized or secrete proteins that would fold into functional forms is not clear. In addition, the minimalist organism needed to be able to make and breakdown various amino acids, nucleic acid bases, carbohydrates, and a few other compounds to be storage forms for future needs for energy and carbon (and nitrogen), and some complex prosthetic factors and cofactors. And that is substantially it.

UPTAKE AND EXTRUSION SYSTEMS

The uptake and extrusion processes of modern cells depend on special intrinsic membrane proteins whose structures traverse the bilayer and serve to transport molecules either as carriers or as channels. Channels require helical structures with multiple looping of protein chains through the membrane. Perhaps the most primitive transport device would have been akin to valinomycin or a proton conductor (Fig. 1.6). This type of antibiotic has hydrophobic properties and the ability to enfold an electric charge in the form of a potassium ion or a proton. Such molecules can enter the membrane and diffuse in either direction, both with and without a K^+ or H^+ ion. Such carriers or channels serve to take potassium or protons down an electrochemical gradient. Either molecule is deleterious to modern cells and is the reason that valinomycin is a poison and pentachlorophenol an herbicide. But something like the macrolide antibiotics could have been useful as a first transport process. Such kinds of carriers would have been slow, but at least they would not require the complex structure of modern carriers or channels, and in the first instance they might have been produced abiotically. Starting from

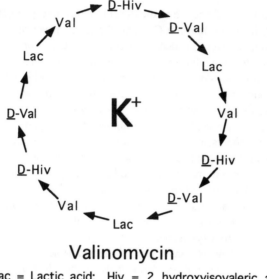

Valinomycin

Lac = Lactic acid; Hiv = 2 hydroxyisovaleric acid

Figure 1.6. The ways in which valinomycin is able to move charge through the lipid bilayer. Valinomycin, free and ligated to K^+, is lipid soluble, but is capable of enclosing the K^+ and thus is able to go across the lipid bilayer with or without the K^+. When it takes the cation across the membrane it affects the membrane potential.

the earliest system for active transport, elaborations of the protonmotive force-driven carriers discussed above would occur. Diversification into the synthesis of special carriers and channels could then proceed leading to special membrane proteins for efficient transport.

By the time that the Last Universal Ancestor was evolved, nearly all these synthetic processes must have been developed. Then the big push would have been for the tapping of energy in sizeable amounts from the environment and resisting the osmotic challenge that resulted from the success of organisms in intermediary metabolism.

EVOLUTION OF THE CELL'S SHAPE

The First Cell's shape, when suspended in an unstirred environment and not adhering to a surface, would have been spherical as long as the osmotic pressure inside the cell was a little greater than that in the environment. It would have been quite irregular while dividing. These cells would have been of diverse, and not fixed, shapes, particularly if the cells became quite large and were suspended in a more or less turbulent fluid.

The Transition from a Monophyletic World to One With High Diversity

After the time of First Cell, there would have been strong selection for generating energy effectively, concentrating chemical species from the environment, extruding others into the environment, forming adequate amounts of various precursors for the formation of informational macromolecules, synthesizing protein accurately, and replicating/segregating the genome. Although development of these processes must have been long and laborious, a consequence of such success had to be that the internal osmotic pressure became quite a bit greater than the surrounding environment. This created a new problem. This challenge called for either some continuing "boat-bailing" process or some reinforcing structure to resist the influx of water.

The most basic solution to the turgor pressure problem would be to form a strong elastic network around the cell. The higher the pressure, of course, the stronger the surrounding fabric must be. In today's eubacteria this fabric is the peptidoglycan (also called murein) layer. It is the chemical structure ideally suited for its function (see Chapter 4). It has a multiplicity of functional groups on each of the units that make it up, so that it can form a resilient cell covering that surrounds the cell and is connected entirely by covalent bonds. The closest analogy to a man-made product is the synthetic fabric Spandex©, which stretches easily but then stretches no more and is very strong.

Before we consider the first strategy for resisting turgor pressure, let us consider the strategy of higher plants and animals. For plants, one-dimensional cellulose chains are held together by many weak hydrogen bonds; covalent bonds in one direction and hydrogen bonds in the other directions. This may have been the prototype in the middle part of the monophyletic period, after the ability to manufacture and then remotely assemble a structure outside the cell proper was devised and when resistance against only weak osmotic forces was needed. This is not an effective strategy for a small free-living modern bacterial cell—this type of wall is not strong enough. Today's plants succeed in this strategy by having the sheer mass of cellulose arranged as fibrils that follow helical paths around the cells, swaddling and reinforcing them, and by embedding the fibers in a lignin matrix. Additionally, the wall between two cells partially supports the turgor pressure of both. Yeast and some fungi have a wall structure that is the logical equivalent of fiber glass with the same superior physical properties. In the yeast case, the composite is formed of β-1,3 and β-1,6 glucan fibers embedded in a mastic of chitin.

Some animals, whether they be single cell protozoa with a contractile vacuole or multicellular "higher organisms" with a kidney, have devised pumps to secrete water. Mammalian cells excrete sodium ions and some compensating anion to lower the internal osmotic pressure.

It is evident from this overview that making a strong elastic fabric covering may be a much simpler way of solving the problem. The exoskeleton solution is still complex because of the difficulty of secreting the pieces and assembling them outside of the cell membrane. The complexity is enhanced because of the necessities of enlarging a structure while it is stress-bearing and dividing into daughter cells. This problem is even more difficult because its solution had to be based solely on elementary biophysical and biochemical processes. Without mechano-proteins, like the actin and tubulin of eukaryotic cells, there is little alternative to the solution of secretion of linear molecules bearing side branches containing functional groups that permit external crosslinking, as in the strategy of most eubacteria today.

Can any of the strategies of modern organisms provide a suggestion for the early evolution of the ability to resist osmotic pressure at a time before either peptidoglycan-like material or cytoskeletal elements were available? Here is a suggestion modeled on the way modern Gram-positive organisms divide. They simply make a bisecting wall and then split it. All the modern organism needs to do is to be able to produce and incorporate wall material in a localized region of the cell where the septum is to form. Originally, it could be carbohydrate fibers like chitin or cellulose, something akin to peptidoglycan, or perhaps some insoluble inorganic material like the iron pyrites mentioned above. In order to divide the cell into two, new wall must arise in a particular location, say in a belt around the cell's middle. The ability to steer proteins and biochemicals to regions of the membrane must have been developed later for more accurate cell division, but to begin with, a crude method causing zonal aggregation of wall material could work well enough. Then, spontaneously, a septum will start to form. If the turgor pressure was sufficient inside, the external part of the intact septum will be the most stressed part and will gradually split, stretch, and distort due to the stress distribution. Daughter cells will form to create a new region of external cell wall. If such a covering had useful mechanical properties it would aid in resisting the turgor pressure as well as be a key part of the primitive cell division process. Although a simple mechanism, it does require the cellular ability to specify spatially to some degree the band where the deposition reaction is to take place. There is an incidental corollary: this method would work only with coccus-shaped cells. Consequently, the Last Universal Ancestor may have had to be a coccus. With that as an evolutionary start, development of the exoskeleton of the cell's structure would have evolved and been part of the eubacterial descendants of the Last Universal Ancestor, although later in many lines of descent this "split and stretch" mechanism would have been discarded and replaced by a variety of alternatives.

The three extant kingdoms had to result from two decisive bifurcations as two events quite close together in time. As distinct from the earlier improvement leading to the Last Universal Ancestor, these splits had to create organisms that did not compete with each other. That require one thing: development of abilities to exploit different (and novel) resources. Only then would the lines of descent no longer be in competition with each other. If the ability to capitalize on new resources was the key, what were the innovative tricks? Any answer is speculative; however, almost from first principles we could argue that there was one new factor that lay behind each split that formed new kingdoms. Although these breakthroughs had to open new niches and habitats, once the split was established, further evolution could modify or even eliminate the feature that caused the split in the first place.

What were the innovative tricks?

(i) For eubacteria. If you read (or reread) this section after going through the rest of the book, then you may have a predilection to think in terms of the turgor pressure relations within cells. Did the scion of the Last Universal Ancestor that developed a stress-resistant external wall to support a high turgor pressure lead to eubacterial descent? Is a strong, completely surrounding wall the major breakthrough? It would permit cell constituents to be concentrated to facilitate rapid growth. In addition, it would be important in invading fresh water (oligotrophic and low salt) environments. It happens that a stress-resistant wall can be made simply, and the cell that carries out this strategy can afford to be small and simple. It can remain haploid and depend on a wide dispersal of cells as result of rapid and efficient generation of propagules. Efficiency demanded loss (or lack) of introns, rapid step time during protein synthesis, etc. The general plan of bacteria was and is to occupy a habitat and exploit it to maximize the number of cells for dispersal by physical and, later, biological circumstances. Of these features only the turgor-resistant wall was new and unique, whereas the other features were probably extensions of characters already present.

(ii) For eukaryotes. Again the reader of the rest of this book will suggest that the key development was the cytoskeletal components that allowed the cell to solve the osmotic problem. This was a big breakthrough: It solved the same problem as did a stress-resistant outer wall, but with a quite different array of further evolutionary possibilities. Although a system of struts could support only a small turgor pressure, the flexibility in shape and facility to render larger cells feasible were very critical. The eukaryotes could afford to be more permeable to ions and to have a flexible membrane so that they could engage in pinocytosis and phagocytosis. In later developments many branches chose to maintain life histories that did not depend as much on rapid or efficient growth, but on being able to occupy different habitats.

(iii) And, finally for archaea. Were they merely left over and belatedly developed a different kind of stress-resistant wall or ways to survive in very salty environments? Or do they represent a positive evolutionary step? Because we know less about the archaebacteria, these questions are imponderable. I will make a tentative suggestion that the breakthrough was the elaboration of a metabolic system that permitted methane production from carbon dioxide and some reductant at the level of molecular hydrogen. Probably the Last Universal Ancestor had developed parts of the overall process, but only with the achievement of metabolic steps for the complete process could a whole new set of habitats be occupied with a correspondingly large increase in biomass. Although the overall process of carbon dioxide reduction with a reductant at the level of molecular hydrogen has a negative free energy, some of the steps have a positive free energy. This means that the chemoautotrophic pathways of the Last Universal Ancestor may have evolved further chemiosmotic mechanisms similar to the one shown in Fig. 1.3 and then inverted the membrane components orientation for some of them so that the combination completed modern pathways and effected an important extension in terms of providing energy to increase the world's biomass. Later adaptations of methanogens led to several lines specializing in different ways. It is now thought that the archaea and eukaryotes are more closely linked to each other than to the eubacteria. If so, then the evolutionary breakthroughs were first the exoskeleton, splitting of the eubacteria, then the methanogenesis (and growth at high temperature) splitting off the archaea, and finally a cytoskeleton allowing the development of eukaryotes.

After these fundamental bifurcations, great diversification of the entire world ecosystem took place because, first, numerous highly sophisticated biological mechanisms of control, regulation, and function had been evolved that could be adapted to allow organisms to do many things, and second, new organisms created new niches for other organisms. Although the processes listed in the central part of Fig. 1.1 must have been largely developed before the Last Universal Ancestor, further improvement in these matters continued, giving distinctive character to the processes as carried out by organisms of the various kingdoms.

Abundant, Exploitable Free Energy was Only Available After the Last Universal Ancestor

Where and when was free energy available in large consumable amounts? Was the Last Universal Ancestor photosynthetic, methanogenic? Presumably the answer is no, or these characters would be more widely expressed. That the Last Universal Ancestor obtained energy photosynthetically is denied by the limited distribution of the ability to perform the purple bacteria type of photosynthesis. Similarly, methanogenesis is not distributed broadly even among the archaebacteria.

Was the Last Universal Ancestor capable of a range of anaerobic respirations? If the chemiosmotic process indicated in Fig. 1.3 was active in the primordial cell, then the evolution of membrane proteins with ferredoxins and other metal-containing proteins would give a selective advantage at every stage. The evolution of membrane-linked processes very likely also occurred to bring the Last Universal Ancestor within "shooting distance" of both photosynthesis and methanogenesis. Basically, anaerobic respiration requires only a reductant, such as H_2S or H_2, and a range of oxidants, such as sulfate, nitrate, or nitrite. The key requirement for an ecologically useful process is that although the reductant and oxidant thermodynamically should react spontaneously, they must be available in the environment to react and yet have not reacted. The failure to have reacted to equilibrium may be because the nonbiological reactions are slow or because one chemical species is continuously generated by geological (and today by biological) processes. In the modern microbial world there are a large number of chemoautotrophs that exploit a large range of reductants and oxidants wherever the two coexist in order to generate energy in a biologically utilizable form. Thus, they follow the tradition suggested for the First Cell and have made improvements in terms of the range of substrates and efficiency in their use. Many lithotrophs are also autotrophs and use only, or mainly, CO_2 as the source, while others are more eclectic. The current total biomass of these forms, however, is determined by the geological situation and the oxidants and reductants that the rest of the organisms of the world afford them.

The concept that seems to emerge from these points is that the big important splits to form kingdoms did not take place on the basis of development of new ways to exploit vast untapped energy resources (although methanogenesis may be an exception). The history on this planet seems to be that life evolved monophyletically to yield organisms that were very efficient anaerobic respirers but had not completed and elaborated the membrane-linked processes in the Last Universal Ancestor to the point where complete methanogenesis or photosynthetic energy generation was possible when some other kind of evolutionary advance took place.

When anoxygenic photosynthesis and methanogenesis developed, the biomass on the planet greatly increased. Once oxygenic photosynthesis had evolved with its two linked photosystems that combined to derive enough energy from visible light to allow the splitting of water and the production of metabolic energy in very large quantities, abundant growth could take place in a greater range of habitats. Almost wherever in the world there was light, CO_2, and water, the floodgates were opened to allow the biomass of this planet to increase astronomically. That in turn provided the organic compounds for saprophytes, herbivores, and carnivores to arise and make a living in the form of generating ATP by combustion of organic matter with oxygen. Then the processes of respiration combined with oxidative phosphorylation achieved a flexible, versatile, compact, efficient energy-generating system needed for motile organisms and for plants at night. The result is that there is an oxidizing atmosphere, and almost the entire ecosystem, down to almost the smallest niche and habitat, is driven today by the separation of oxidizing potential from reducing potential by the oxygen-linked photosystems I and II.

2

The Contrast of the Cellular Abilities of Eukaryotes and Prokaryotes

KEY IDEAS

Eukaryotes have mechano-proteins:
    actin, tubulin, and the proteins of the intermediate filaments.
In eukaryotes, these have roles in chromosome separation, constriction and cell
    division, motility, and apical growth.
Prokaryotes do not have mechano-proteins.
Bacteria use turgor pressure as a source of power.
Prokaryotes have a sacculus that is equivalent to an exoskeleton.
Eukaryotic cells have a cytoskeleton serving some of the roles of the sacculus.
Eukaryotic cells can afford to be much larger than prokaryotic cells.
Eukaryotes use their flexible membrane for pinocytosis, phagocytosis, and movement.

The Cell Biology of Eukaryotes

The techniques of physiology, molecular biology, and ultrastructural cytology
have led to an understanding of the functions and many mechanisms of action

of eukaryotic cells and their parts. Although typical prokaryotes contain no membrane-bound organelles, the typical eukaryotic cell contains two types of organelles. One class of organelles, those derived from endosymbiont prokaryotes, includes mitochondria, chloroplasts, and (possibly) flagella. The other class is composed of extensions of the cell membrane: the nuclear membrane, the endoplasmic reticulum, the Golgi apparatus, lysosomes, peroxisomes, vacuoles, and other inclusion bodies. Actually, some prokaryotes do have invagination of their cytoplasmic membrane, called mesosomes. Some mesosomes are artifacts, but some are regions with enhanced numbers of membrane proteins needed for photosynthesis and chemoautotrophy. These are the analogues of the second class of eukaryotic organelles.

Organelles in this second category serve a number of functions. Their nature suggests an evolutionary pathway that might have started with the infolding of membrane, in order to increase the membrane surface. This increase would have been needed by chemoautotrophs existing before and after the Last Universal Ancestor and by phototrophs. Not only have the infoldings probably led to development of the endoplasmic reticulum, they led to the other laminar, tubular, and vesicular structures, including the double nuclear membrane. These, through further evolution, have differentiated into the various subtypes and developed novel functions.

One can presume, with Stanier *et al.* (1970), that the primitive eukaryotic cell had a flexible membrane that was needed for pinocytosis and phagocytosis and therefore had the ability to take up molecules, fluids, and particles. The mechanism of this process is becoming known. The ability to consume particles may have helped lead to the evolution of bigger cells. The ability to take up particles as big as bacterial cells not only required bigness, but led to the requirement for digestive vacuoles. But bigness led to difficulty in the efficient exchange of nutrients and waste products and mandated a plumbing system, now provided by the endoplasmic reticulum. Later, in metazoans, the endoplasmic reticulum was used for efficient secretion of proteins into the environment for digestive functions.

The Eukaryote's Mechano-Proteins

The cytoskeleton present in all eukaryotic cells is formed of three classes of mechano-protein elements: actin in actin filaments, tubulin in microtubules, and five subclasses of proteins in intermediate filaments. These filamentous cell structures are composed of protein(s) that are capable of self-assembling, and in at least the first two cases, are able to do mechanical work. For the first two, a mechano-enzyme (either myosin, dynein, or kinesin) hydrolyzes ATP to supply the necessary power. Accessory proteins are needed, sometimes for assembly and sometimes for interaction with other parts of the cell.

Actin constitutes as much as 5% of the protein of many nonmuscle cells; it is clearly vital to the eukaryote way of life. Myosin in muscle cells is present in the thick filaments, however in nonmuscle cells smaller assemblies are reversibly formed as needed. Aggregation is regulated by the state of phosphorylation of the mechano-protein. Controlled in this way, for example, a contractile actin-myosin ring forms and powers the squeezing of the mitotic cell to cause the cytoplasmic division in the animal cell as part of mitosis or meiosis. Stress fibers of actin and myosin may connect different parts of the cell membrane or sometimes may connect to the nuclear membrane; these fibers define the cell's shape and movements. The abundant actin filaments in most cells form a network, called the cell cortex, just inside the cell surface. Some filaments are crosslinked to others by proteins such as filamin. Cytoplasmic streaming in plant cells is due to actin and myosin function.

Tubulins are the second class of fibrous mechano-proteins. They are present within the cell in microtubules, the filaments with the largest diameter. Tubulins are less abundant than the actins. They are the motive agents of ciliary movement. The microtubules are paired hollow tubes formed of tubulin dimers. These proteins assemble in the right environment (GTP is required) into paired tubules with dynein arms that move one pair of tubules relative to another pair. Dynein is a mechano-enzyme; it is a complex of nearly a dozen proteins, and catalyzes active movement, again powered by ATP hydrolysis. Most microtubules grow from a pair of centrosomes, which become located on opposite sides of a dividing nucleus. Kinesin and cytoplasmic dynein are analogues of myosin in that they hydrolyze ATP to move vesicles unidirectionally along a microtubule. Sometimes dynein causes movement in one direction and kinesin in the other.

The third class of mechano-proteins of the cytoskeleton are the intermediate filaments; these consist of four or five types. It may be that they can support tension but cannot create it. At least it is not clear how they could perform the work of contraction or extension. They may be the equivalent of cables and guy wires.

From this brief review, it appears that all the known mechano-proteins belong to one of three categories, and that they require a mechano-enzyme (myosin, dynein, or kinesin) to do work paid for by the hydrolysis of high-energy phosphate bonds of ATP.

Absence of Mechano-Proteins and Mechano-Enzymes in Prokaryotes

Although a variety of structural mechano-proteins/enzymes are used by eukaryotes to transduce biochemical energy into mechanical work, such mechano-proteins do not appear to be present in bacteria. There were some earlier claims that an actin-like protein is present in bacteria. The protein, however, was later identified as the elongation factor T_u (or EF-Tu). This factor serves an essential role in prokaryote protein synthesis and has several additional roles. Although EF-Tu possesses some sequence homology to actin, its aggregates do not contract in an energy-dependent way as do actin-myosin complexes *in vivo*. EF-Tu, like actin, is present in many copies per cell. A similarity has been noted between EF-Tu and the G-proteins (Woolley and Clark 1989); the G-proteins are involved in signal transduction in both prokaryotes and eukaryotes, but have not been associated with contractile properties.

There are structural similarities of prokaryote proteins with eukaryote mechano-proteins and enzymes (Koch 1991a), but not functional relationships These do suggest that evolution that generated the latter was on-going before the Last Universal Ancestor. Here are some of the similarities. For example, a protein called Pep M5 of Group A streptococci is similar to fragments of myosin and tropomyosin molecules. The flagellar filament of *Salmonella* has a similarity to actin, and the α-6 gene of *Acinetobacter baumannii* to myosin. There are epitope similarities between surface antigens of *Streptococcus mutans* and *S. pyogenes* with mammalian myosin. It has been found that certain osmophilic archaebacteria are inhibited by drugs interfering with tubulin and actinomyosin. There are tubulins in *Spirochaeta bajacaliforniensis* and an ameboid cytoskeleton-like structure regulated by Ca^{2+}. In addition, in studies too numerous to mention, it has been found that many genes for mechano-proteins and mechano-enzymes can be cloned in *Escherichia coli* and form the eukaryotic protein in the bacterium without causing appreciable difficulties for bacterial growth. Although by itself any one of the successful clonings of eukaryote gene is not significant, collectively, this is evidence against the function of such mechano-proteins/-enzymes within the bacteria.

Although prokaryotes do mechanical work during cell movements, they do not use mechano-proteins but utilize an alternative source, the protonmotive force. Chemotaxis is powered by chemiosmotic protonmotive force and not high-energy phosphate bonds, and gliding motility also appears to utilize the chemiosmotic force and again is not coupled to contractile proteins.

Conceivable Roles for Mechano-Proteins/Enzymes in Prokaryotes

Because we know about the importance of mechano-proteins/enzymes in eukaryotes, it is tempting to think that they might function as contractile elements in prokaryotes. Let us examine some of the places where mechanical work is needed in all cells and the known sources of energy.

(1) Mechanical force is needed to ratchet the messenger RNA on the ribosome in three nucleotide steps for each amino acid incorporated during protein synthesis. Ratcheting does take place, powered by the consumption of guanosine triphosphate (GTP). In this process EF-Tu is involved, and although there is similarity of the details of prokaryotic and eukaryotic protein synthesis, there is little similarity of protein synthesis in either kingdom to the eukaryotic process of

muscle contraction. Actually, EF-G, not EF-TU, is involved in the translocation step, but there has been no proposed role for actin or tubulin type action in the translation process.

(2) Many enzymes, such as polymerases and rho factors, move along nucleic acid during their normal function. Rho factor is a protein involved in terminating transcription of certain prokaryotic operons. These movements involve ATP hydrolysis events and allosteric sites on the proteins. Proteins executing such movements are found in both eukaryotes and prokaryotes, and no doubt are present in archaebacteria as well. They appear to be devices that move along a linear DNA molecule, but do not do mechanical work on the DNA.

(3) Chromosome separation in eukaryotes requires mechano-proteins (both tubulins and actins), but they apparently are not required in prokaryotes. Perhaps the strongest suggestion of a mechanical action in bacteria, although extremely weak, is the observation of rapid movements of nucleoids as they separate (Sargent 1974). It is not, of course, evidence for mechano-proteins, but evidence for active movement. In the most popular model of chromosome segregation, the replicon model of Jacob *et al.* (1963), separation is powered by wall growth occurring in a zonal fashion in the central part of the cell. Zonal growth, however, is not consistent with the observed data on wall growth (see later). To take this into account the replicon model has been revised (Koch *et al.* 1981a), but still the motive power for chromosome separation comes from wall growth. At present, there are two research teams searching for a role of mechano-proteins in bacterial chromosomal segregation. Holland *et al.* (1990) have found some interesting relationships with calcium ion that mimic some of the features of muscle contraction. Inhibitors of eukaryotic mitosis have no effect on bacterial cell division. The second group is that of Hiraga (Niki *et al.* 1991, 1992). They believe that MukB is a protein that will ultimately prove to be a mechano-enzyme that functions in prokaryotes.

(4) Constriction and cell division is different in animals, plants, and fungi, but all eukaryotes use actin one way or another in the process. Although one purpose of this book is to support an alternative suggestion for bacteria, potential roles for mechano-proteins in constriction in Gram-negative cells can be imagined. (There is no needed role in the Gram-positive cells that form a stress-free septum and then split it.) In eukaryotes there is a band of fibrous proteins underlying the middle of the cell just inside the cytoplasmic membrane in the regions where constrictions will occur; this is not observed in prokaryotes. There is, however, a band of the protein FtsZ that forms where the constriction is to take place (Bi and Lutkenhaus 1991; Ward and Lutkenhaus 1985).

(5) Motility is powered by ATP hydrolysis in eukaryotes; the evidence, however, is quite convincing that both flagellar motility and gliding motility of prokaryotes are powered by a protonmotive force. Conversely, there do not appear to be any movements, such as occur in bacteria, of eukaryotic cells or their organelles that utilize the energy of the proton gradient.

(6) Growth in a hyphal, apical habit is characteristic of fungi and some bacteria such as *Streptomyces*. It would be beneficial for this type of growth to pump cell components towards the tips of the hyphae and to move resources in the reverse direction. Such movement of proteins, vesicles, and wall components is observed in the case of the eukaryotic fungi; in fact, it allows them to grow in length about an order of magnitude faster than their prokaryotic counterparts (see Chapter 11). This transport mechanism is known to involve cytoskeletal elements, but there is no evidence that they are used in eubacteria.

In summary, we have contrasted the energy forms used by eukaryotes and eubacteria for the same physiological purpose. The evidence is compelling that the energy needs are met quite differently in the two kingdoms. Only one of these six functions (nuclear segregation/cell constriction/division in Gram-negative organisms) could conceivably use unknown mechano-proteins/enzymes in order to function, but effort will be devoted in later chapters to go deeply into the alternative use of turgor pressure to achieve mechanical movements in pro-karyotes.

Cellular Turgor Pressure

Osmotic pressure leads to a nonintuitive phenomenon. The subject is introduced here, although it will be dealt with more extensively (see Chapter 5), to allow the comparison of ways in which different organisms deal with the osmotic pressure of their environment. Osmotic pressure depends on the number of particles and is independent of their molecular weight or the mass concentration. Osmotic pressure in an ideal system is given by:

$$\pi = RTc \qquad\qquad Osmotic\ pressure$$

where R is the gas constant, T is the absolute temperature (in Kelvins), and c is the particle concentration measured approximately in moles per liter, but more accurately in moles per kg of solvent (or osmolal). Frequently we will talk about the osmotic pressure differential, in which case c will be replaced by $c_1 - c_2$, where the subscripts refer to different sides of the membrane.

Presumably, therefore, a ribosome subunit contributes as much to the osmotic pressure as does a molecule of glucose. This statement is accurate when it is applied to more or less rigid particles because rigid particles, no matter how big or how small, contribute to the colligative properties of the solution to the same degree. A random coil, for example, a linear chain of glucose residues as found in amylose, would contribute to the osmolarity a value between 1 and the number of residues in the chain. The variation occurs because the real criterion has to do with the number of *independent* particles, and the remote parts of a portion of a flexible chain are essentially independent of each other. On the other hand, a polysome containing many 70s ribosomes behaves as if it were one enormous particle because it is essentially inflexible. In fact, there are not very many long flexible molecules within cells (not even mRNA is an exception). The low molecular weight solutes control the colligative properties of the cell and this determines the osmotic pressure. The small solutes of the cell include the metabolites of intermediary metabolism, largely organic acids and phosphate derivatives, and the magnesium and potassium ions needed as cofactors and for balancing the charge of nucleic acids. They also include the counterions needed to match the charges on the macromolecules.

Cells in a higher animal are bathed in an optimized fluid, the *internal milieu*. Therefore, most animal cells can afford to equilibrate many substances with their local environment. Other specialized bodily tissues, including the impermeable skin and the functioning kidney, create and maintain the internal milieu. The result is that animal tissues are nearly isoosmotic with the blood and lymph. Note, however, that there are facets of mammalian physiology that depend on minor deviations from this situation (e.g., capillary exchange). Also, each mammalian cell uses the energy of ATP to extrude three Na^+ ions for every two K^+ ions accumulated. To maintain electroneutrality some anion must also leave the cell. It is claimed that this process continuously tends to lower the total salt in cells. It only lowers the osmotic pressure slightly relative to the plasma, because salt continuously reenters the cell. This can be demonstrated because under some conditions an inhibitor of the Na/K pump causes animal cells to swell and burst. This electrogenic extrusion of sodium for potassium provides the individual cells in the animal with a ready store of energy in the in-to-out potassium flux coupled with the out-to-in sodium flux across the cell membrane. It is used, for example, by nerve cells for energy to power axon transmission. Plants are more like bacteria in their dealing with the environment rather than like animal cells dealing with their *internal milieu*. Root hairs take up ions and water, against high concentration gradients. Of course, this requires considerable energy expenditure.

Different bacteria can live and grow in different ranges of external osmotic pressures. Some organisms can live in essentially distilled water. Many other bacteria can grow in severely dehydrating environments. Halophiles require a high salt concentration, sometimes limited by the solubility of various salts. Salt tolerance is an important ecological and taxonomic trait of bacterial species.

A typical organism such as *E. coli* can grow in an external osmotic pressure range spanning from almost zero to a moderately high value of 1 osmolar. When enteric cells are grown in a salty or sugary medium they respond by accumulating osmoprotectants that raise the internal osmotic pressure. This serves as a technique to expand their range of habitat. Without this ability, an organism could grow only under a very restrictive span of osmotic strengths. Adaptation can result from a number of different mechanisms even in the same organism (see Csonka 1989; Csonka and Hanson 1991). Their role is to raise the osmotic strength of the osmotically challenged cell by accumulating or synthesizing a variety of substances so that they can serve to maintain the differential between the inside and outside osmotic pressure. All the molecules so accumulated are small, do not penetrate the lipid bilayer, and are compatible with native protein structures. Thus, the main contributors to the osmotic strength are soluble, small molecular weight components. Below we shall see that the mechanisms are the coarse controls and that the wall growth mechanism is the fine control. A critical question to be addressed by microbial physiologists is: What is the fraction of the osmotic pressure that arises from necessary cell constituents; i.e., if growth could take place when there is no external osmotic pressure what would be the internal osmotic pressure? Is this equal to the turgor pressure that cells achieve during exponential growth within their range of growth? Answers to these questions are currently being studied (Baldwin and Koch 1995). The lowest osmotic strength we can grow *E. coli* is 15 mosm, with is very much less than almost all minimal media but growth requires the presence of calcium ions.

How Can An Osmotic Pressure Differential be Created?

The development of a higher osmotic pressure inside a cell depends on its having a higher concentration of particles (ions and molecules) inside than those outside. Of course, if the molecules can penetrate through the cell membrane they will diffuse until there is no concentration gradient (more precisely, until the electro-chemical potential is the same on both sides of the membrane); in this state, there will be no osmotic pressure differential. Bacterial cells have a positive osmotic pressure generated by the continuing active accumulation of substances from the environment and aided by the impermeability of the cytoplasmic membrane. The active transport processes for small molecules in modern organisms depend on energy-generating systems to produce either phosphorylating reagents or to produce proton gradients across the cytoplasmic membrane; these in turn are coupled to transport events *per se*. When the proton gradient is employed, the active carriers are called symporters when the solute accumulation is linked with the flow of protons back into the cell. They are called antiporters if the carrier traverses the membrane with the proton and without the solute and returns without the proton but with the solute.

Consider an extreme case chosen from a first course in biochemistry. Glucose is actively pumped into a respiring cell while oxygen diffuses into the cell from the environment down its concentration gradient. Together they are converted into carbon dioxide and water, both of which can penetrate the membrane and flow rapidly outward down their respective concentration gradients until there is equal chemical potential on both sides. This chemical conversion requires the enzymes of the glycolytic pathway, the tricarboxylic acid cycle, and membrane-bound electron transport processes to cause oxidative phosphorylation processes on the inner membrane of the mitochondrion or the cell membrane of the prokaryote. After the uptake and consumption of a glucose molecule, the cell's osmotic pressure has gone down because of the consumption of the glucose. Numerically more important is that coupled to the combustion of glucose in the best case is the combination of 38 molecules of inorganic phosphate with 38 ADPs to produce 38 ATPs, a net reduction of 38 particles. Although this is going the wrong way, it is only temporary. Actually, when the subsequent hydrolysis of the ATP is coupled to cellular processes, not only will 38 more particles be recreated, but also some active transport events may be coupled, and if these are entry processes then the internal osmotic pressure of the cell increases.

Let us view this process more closely, but focus on a sugar that is not transported and phosphorylated at the same time by the phosphotransferase system (PTS). During chemiosmosis of lactose, the overall process is that water molecules have been split into negatively charged hydroxyl groups that remain in the cell and protons that are extruded. The hydroxyl ion raises the internal oxmotic pressure (because water itself will pass quickly through the membrane, and therefore be replaced from the environment). Consequently, the osmotic pressure both in and out is increased because of the additional negative ions now inside the cell and the protons outside. With the action of the membrane's symporters and antiporters, the proton potential is exchanged for gradients of other substances, but without changing the number of particles. Thus, in essence, proton extrusion is the prime event that increases the intracellular osmotic pressure, but the qualitative nature of the internal constituents is subsequently changed by the action of various porters.

With a higher osmotic pressure inside the cell than outside, water will tend to enter to achieve the same chemical activity on both sides. This causes the cell to expand until the elastic network of the wall is stretched to make a counter tension sufficient to balance the osmotic pressure differential. This means that turgor pressure is created inside.

Contrast this strategy with that of the multicellular animal body, in which food and water are taken in and processed, and what is excreted is basically arranged by the kidneys, lungs, and intestines with energy-consuming processes to make the internal environment and the contents of cells about 300 mosmolar. Most of the cells are slightly permeable to some salts, so essentially no osmotic pressure differential is maintained relative to the internal milieu.

The Differences Between Transport in Bacteria and Animal Cells

Bacteria use both chemiosmotic-linked symports and antiports for many solutes. They also use phosphate bond energy for transport events. So do eukaryotes, but the prokaryotes have the unique use of the phosphotransferase system (PTS), by which sugars are phosphorylated and transported at the same time. Eukaryotes, on the other hand, have a large range of possibilities afforded by specially regulated ion channels for large volume flow and by pinocytosis and phagocytosis for taking up water, solutes, and particles. Typically, then, prokaryotes are more specialized in accumulating small molecules and ions from a dilute solution, whereas cells in a higher animal are bathed in the internal milieu and in many cases are specialized in uptake and secretion of macromolecules. Therefore, most animal cells can equilibrate almost *in toto* with their local environment's osmotic pressure.

Synthesis of Protoplasm: The Growing Cell's Heavy Industry

Eukaryotes and prokaryotes differ very much in their nuclear organization, gene transcription, messenger translation regulation, and cellular differentiation. These topics are not germane to this book, but all these and yet other factors determine how fast protoplasm can multiply. At best, the most rapidly dividing eukaryotic animal cell in a rich culture medium can double in biomass every six h. Yeasts and other lower eukaryotes may divide once every 2 h. In contrast, some bacteria under favorable conditions can double every 15 min. These numbers present clear evidence of the superiority of the prokaryote's efficiency in carrying out macromolecular synthesis of cell constituents. This means that in an environment in which all small molecules were supplied, prokaryotes would overgrow the eukaryotes. Of course, it is the sophistication of the eukaryotes that gives them the advantage under many ecological situations.

This claimed speed of prokaryote growth could be countered also by pointing out that some eukaryotes can engage in replication of their genomes even more rapidly than bacteria can reproduce. In a fertilized drosophila egg the genome can double every 8 min. Having such speed is necessary for the fly who has a limited time to grow from egg to mature organism. For example, during the summer season the fruit fly, *Drosophila melanogaster,* must go from egg to adult and back to egg in only two weeks. This rapid replication can occur because the embryos's genomes are subdivided into many short individual replicons that replicate simultaneously. An additional requirement for this speed is that the egg cytoplasm has been previously prepared to have on hand the needed raw materials for chromosome replication. Growth is not limited from the outside by resources such as metabolic substrates or nucleic acid precursors (in fact no nutrients enter during this phase). Moreover, the egg has been supplied with adequate supplies of polymerases and other necessary enzymes.

A related phenomenon is that certain eukaryotic single-celled organisms have a lifestyle that requires them to synchronize their growth with the diurnal cycle and arrange their cell division events to occur in a circadian (i.e., about once per day) fashion. In these cases the cell must gauge its internal resources before it attempts cell division; when they are not adequate it must wait for another day. Another case is that for animal cells, protein synthesis and protein degradation are coupled and controlled by the organisms to regulate net growth and to control intracellular turnover and cell renewal. Thus, the entire organism may be in a steady state where the adult organism maintains a constant size.

Returning to prokaryotes, under a range of conditions it is a good first approximation to assume that growth is limited by the rate at which the elements needed for protein synthesis are formed and by the rate at which ribosomes can make protein (see Chapter 3). All the complex additional apparatus is almost nonlimiting compared with the constraints of the construction of ribosomes and the minimum step time for the addition of an amino acid to a growing protein chain. Thus, speed is the other side of the coin of efficiency. In both of these respects the prokaryote is much the superior of the eukaryote.

When does a prokaryote need to grow fast? The strategy of some bacteria requires that they quickly go from a starvation economy to rapid growth (see Koch 1971) when conditions become favorable. Although the control in these responses to the environment is important, at this point we should focus on the fact that the prokaryote's major goal in an adequate medium is single-mindedly to create the balanced constituents of protoplasm; this is not the case with higher or most lower eukaryotes. Many soil and aquatic prokaryotes however, when recently abstracted from their natural environment, do not grow or cannot grow rapidly in a nutrient sufficient medium, although such "oligotrophs" may later adapt to speedier growth in richer environments. These species specialize in occupying sparse habitats and using the rare resources at a time when there is no competition from other organisms.

The Importance of the Bacterial Cell Wall

One feature usually used to distinguish eukaryotes from prokaryotes is the lack of membrane-bound organelles in the latter. I take exception to this because I consider the envelope of gram-negative bacteria to be an organ, just like a liver or heart, that has many different parts that work together to carry out needed functions. So I use the term organelle for the Gram negative wall to mean the membrane bounded region stretching from the inner leaflet of the cytoplasmic membrane to the outer reaches to the lipopolysaccharide. It contains a volume accessible by water that is the periplasmic space and the periplasm, in analogy to the cytoplasm, is the fluid compartment that contains soluble diffusible proteins and membrane derived oligosaccharides and is essentially in thermodynamic equilibrium with the salts and small molecules of the surrounding medium.

3

Bacterial Growth

HOPE FOR FUTURE STUDIES

KEY IDEAS

"Balanced" growth implies that in a constant environment, after a long time, the cells in one generation will be just like those in succeeding generations.

A first-order process is one in which the rate depends only on the first power of the concentration of a single reactant.

No matter what proportions of cell components are present to start with, eventually they will all increase exponentially with the same specific growth rate μ as does the cell number.

The square root law is $\mu = \sqrt{k_x k_y}$, where k_x and k_y are the rate constants for the formation of two cell components, each controlled by the other.

The metabolic control cycle is limited by the formation of stable RNA from global protein and the formation of all proteins from RNA. This leads to the square root law.

During balanced growth the subpopulations of cells in any phase of growth are a constant proportion of the total.

The generation time or doubling time (time from division to division) of individual cells is quite variable. The age-at-division has a coefficient of variation (CV) in the neighborhood of 20%, larger than the CV for the cell size-at-division, which is usually 10% or less.

The rates of macromolecular syntheses have little statistical variability because of the large number of events involved.

The cell cycle has at least two discontinuous unique events, chromosome replication and cell division.

If growth in a constant, suitable environment could occur for a very long time, evolutionary changes would occur that would greatly simplify the organism, no matter how the environment was maintained constant.

In the chemostat, the specific growth rate calculated from the usual formula is $\mu = \mu_{max} S/(K_m + S)$ the value of μ that would be observed with a low-density batch culture that had the same steady-state medium concentration of the limiting nutrient as the chemostat, but had only very few bacteria so that consumption could be neglected.

The "Growth law" at the cellular level a priori should be exponential, except for minor corrections for chromosome and cell envelope synthesis. Although there are a number of valid experimental approaches, unequivocal experimental evidence is still lacking.

This chapter has a number of purposes. One is to introduce or remind the reader about the elements of calculus, statistics, and chemical kinetics. Another is to analyze the growth of cells and the population of cells both for the simplest possible case and for more realistic cases. Another purpose is to itemize ways to extract information about the cell cycle from the analysis of the steady-state distribution of the sizes of cells in growing populations. Incidentally and insidiously a large number of facts from microbial physiology will be presented.

Development of Growth Equations Without Calculus

The law of bacterial exponential growth is the cornerstone of much of microbiology; it is given in every beginning textbook. But some individuals find it difficult to grasp. Because of its importance, the derivation will be made here in a simple way to emphasize the assumptions involved. Let us start with the assumption that we have **a culture in "balanced" growth** (Assumption 1). This phrase means experimentally that the culture has been growing for many generations in a sufficiently dilute suspension so that the environment has not been altered during previous cell generations either by consumption of nutrients or by the production of toxic substrates. The supposition is that the organisms have come into balance with their environment. This implies that the cells in one generation will be just like those in the succeeding generation.

Imagine that we have N_0 cells now. How many will we have after a brief interval of time Δt? Let us assume that **the probability of production of new cells at any time is exactly proportional to the number we have at that time** (Assumption 2). This can also be formulated to state that the more cells that are present, the faster new cells will arise. A third, equivalent way to formulate this assumption is to say that a cell divides when it is ready to, independently of the behavior of any other cell in the culture.

Let us assume that **there are enough cells in the population so that many additional cells will arise during a small time interval** (Assumption 3). In other words, random fluctuations can be neglected, and the specific growth rate will be a definite value and not vary by chance from time to time.

This growth process is depicted in Fig. 3.1. The circular arrow represents the first-order, pseudo-monomolecular reaction of a cell forming an additional cell, i.e., giving rise by division to two daughter cells (Case A). Because the concentration of bacteria increases, the rate constant for the process is called the growth rate constant (sometimes specific growth rate), and is usually designated by μ; it is the probability of division per cell per unit time.

A first-order process is one in which the rate depends only on the concentration of a single reactant that enters only once in the process (see Fig. 3.2, top line). An example of a first-order process is the decay of radioactive atoms. One atom of ^{32}P decays totally independently of anything else in the universe, and therefore

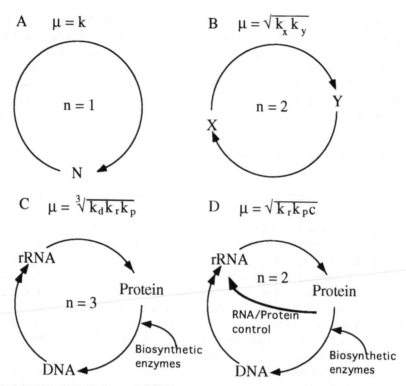

Figure 3.1. Schemes for autocatalytic growth. See text. A, true or simple autocatalytic growth. B, two-component growth, in which one component determines the rate of synthesis of the other and, reciprocally, the other determines the synthesis of the first. C, a mechanistic scheme in which DNA manufactures rRNA which manufactures miscellaneous proteins, some of which in turn determine the rate of DNA synthesis. D, the same scheme, but with the addition of the fuller specification that proteins are part of the protein-synthesizing machinery, and in addition, that the cell has a control link that determines when chromosome replication starts. This regulatory system functions in such a way that the ratio of DNA to protein remains constant independent of the growth rate of the culture.

if the number of atoms in a sample is doubled, twice as many electrons will be emitted and twice as many atoms of ^{32}S will be formed per unit time. Second-order reactions in which two molecules must interact are common, but if one component is in excess or remains constant, then the reaction is pseudo-monomolecular or pseudo-first order (see Fig. 3.2). Finally, if the reaction causes the conversion of A and B to form more B, then there is positive feedback, a plus instead of a minus sign in the rate equation, and the reaction is called, in addition to the above terms, "autocatalytic."

If ΔN is the increase in the number of cells, then we can write the relationship in three equivalent ways:

First Order $\qquad$ B $\longrightarrow$ X : $\quad$ -d[B]/dt = k_1 [B]

Second Order $\qquad$ B + A $\longrightarrow$ X + Y : -d[B]/dt = k_2 [B][A]

Pseudo First Order $\quad$ B + A $\longrightarrow$ X + Y : -d[B]/dt = k_1 [B]

$$k_1 = k_2 [A]$$

Autocatalytic $\qquad$ B + A $\longrightarrow$ 2B: $\quad$ +d[B]/dt = k_1 [B]

$$k_1 = k_2 [A]$$

Figure 3.2. Rate equations for various types of reactions. These definitions and terms are taken from the field of chemical kinetics. Most of the reactions of importance in biology are second order or pseudo-first order. The simplest example of autocatalysis is that in which B is trypsin and A is trypsinogen because trypsin converts trypsinogen into trypsin catalytically.

$$\Delta N_0 = \mu \Delta t N_0 \text{ or } \Delta N_0/N_0 = \mu \Delta t \text{ or}$$
$$\Delta N_0/(N_0 \Delta t) = \mu. \qquad \textit{Autocatalytic rate equation}$$

Using the first form, the total population after the time Δt will be:

$$N_1 = N_0 + \Delta N_0 = N_0(1 + \mu \Delta t).$$

Then in the next Δt:

$$N_2 = N_1(1 + \mu \Delta t) = N_0(1 + \mu \Delta t)^2.$$

Therefore, after the *n*th time interval:

$$N_n = N_0(1 + \mu \Delta t)^n.$$

This is the answer, and with it we could calculate how the number of organisms would increase with time and compare these results to an actual experiment. The equation, however, is not in a very useful form. The following manipulation will help. Define $n\Delta t = t$, the elapsed clock time; and $\mu \Delta t = 1/X$. (The latter is done so that we can play a mathematical trick later.) Then we can replace the exponent n with $t/\Delta t$ and eliminate Δt by substituting $1/\mu X$ for it. As a result the exponent becomes $X\mu t$ and the penultimate rewritten expression in our derivation is:

$$N_t = N_0[(1 + 1/X)^X]^{\mu t}.$$

Now consider the quantity within the square brackets, $[(1 + 1/X)^X]$. It is devoid of biological meaning, it is just a mathematical expression. Take your calculator and evaluate it for $X = 10$, the result is 2.59374; try $X = 100$, and the value is 2.70401; try $X = 1,000$, and the value is 2.71692. If you could try $X = $ infinity, you would get 2.71828. This number is called e and is the base of natural logarithms. As such, it applies to physics, simple compound interest, and high finance just as much as biology. What we have really assumed, then, is that even though we know that a bacterium is born, grows, and finally divides as discrete events, we can treat the growth of the entire culture as if the bacteria were so many radioactive atoms, which on decay became two radioactive atoms, which in turn would further multiply. A better analogy is taken from investments: a dollar invested at compound interest will eventually become two dollars and in a like additional time will become four dollars. Any situation in which Assumptions 1, 2, and 3 hold will yield the same formulae. These will hold when dealing with large populations of asynchronous cells (i.e., cultures with cells in all phases of growth). Consequently, as long as the population of bacteria throughout our period of examination is growing asynchronously and is very large so that it is reasonable to treat the culture as if the cells divide at random, the logic works. In that situation, μ is the probability of cell division per unit time averaged over all the phases of the cell cycle and all the random variation in the cell growth process itself.

We can choose Δt to be very small and consequently treat $1/X = \mu \Delta t$ as small so that X is large, and then $[(1 + 1/X)^X]$ becomes essentially e. Now we can substitute, simplify, and write the familiar equation for exponential growth:

$$N_t = N_0 e^{\mu t}. \qquad\qquad \textit{Growth equation I}$$

The conclusion from this derivation is that this equation is generally applicable to growth with only three rather weak, almost trivial, assumptions. To repeat, if the average properties of the bacteria remain the same throughout growth of the culture, lots of cells are involved, and the biology is such that each cell grows independently of the others, then this equation holds exactly.

This basic equation can be written in many forms; e.g., by taking the natural logarithms:

$$\ln N_t = \ln N_0 + \mu t \qquad\qquad \textit{Growth equation II}$$
$$\ln (N_t/N_0) = \mu t. \qquad\qquad \textit{Growth equation III}$$

Of more microbiological interest is to relate it to the doubling time T_2, by choosing $t = T_2$. We must simultaneously set $N_t = 2N_0$. After all, if the equation applies throughout all growth phases, it applies just as well when the culture has doubled. When these values are substituted in any of the forms of the equation, it is found that $T_2 = \ln 2/\mu$, where $\ln 2 = 0.693147$. Substituting $T_2 = \ln 2/\mu$ back into this logarithmic equation and thereby eliminating μ, taking antilogarithms, and remembering that the antilogarithm of $\ln 2$ is 2; i.e., $e^{\ln 2} = 2$, we can also rewrite the growth equation as:

$$N_t = N_0 2^{t/T_2}. \qquad \text{\textit{Growth equation IV}}$$

This is the logical form when considering a case of binary fission. Or if we repeat the process with the number 8 instead of 2, we can write the growth equation as:

$$N_t = N_0 8^{t/T_8}. \qquad \text{\textit{Growth equation V}}$$

This form would be the appropriate form if we were considering the growth of certain algae that grow by the mechanism of forming eight cells within an enclosed sac and then the sac suddenly opens to liberate simultaneously eight daughter cells. Even populations with very complicated growth patterns can obey the *Growth equations I–V* in any of these five forms. The equation applies as well to human populations where individuals engage in biparental replication, are not destroyed in the act of reproduction, and where some individuals produce many offspring and others none. The law fails most commonly because the availability of resources becomes limiting for one or another reason, or when some catastrophe intervenes.

Development of Growth Equations with Calculus

With the mathematical tools from a beginning course in calculus available, the above treatment can be made easier, which means we will be able to handle more complex cases. In the calculus, what is done is to replace the Δ's with d's. This change is done under the notion that we are now considering a much smaller interval. The middle form of the *Autocatalytic rate equation* then becomes:

$$dN/N = \mu dt. \qquad \text{\textit{Differential growth equation}}$$

Our problem is now to integrate both sides of this equation to see its long-term, large-scale consequences. This equation directs us to look up in a table of integrals the integral of dN/N and of dt. One finds that the integrals are in $\ln N$ and t, therefore:

$$\ln N = \mu t + C,$$

where C is a constant of integration that must be the same throughout the growth of an individual culture. (C will have several meanings within this chapter and elsewhere in the book; it is a catchall term and will be disposed of quickly each time it is used.) Our current C has the same value when time is zero and $N = N_0$ and when time is t and $N = N_t$. Then we have:

$$\ln N_0 = C \qquad \text{and} \qquad \ln N_t = \mu t + C.$$

Now by eliminating C between the two equations again we obtain:

$$\ln N_t = \ln N_0 + \mu t. \qquad \qquad \textit{Growth equation II}$$

This derivation is a paradigm for more difficult cases and allows us to approach bacterial growth in more detail and to handle cases that are not "balanced".

Cells with Only Two Key Components

We should like to build upon this beginning to consider some details of cellular processes. To begin with we will solve a problem that is only a little more difficult than the one just considered. For a moment let us use meaningless symbols and then later make biological sense out of them. Imagine a culture of cells, each composed of only two components, X and Y. Further imagine that X makes Y and Y makes X and that both the kinetic rate equations are first order; i.e., the rate of formation of one component is proportional to the first power of the concentration of the other component (see Fig. 3.1B). Then we can write:

$$dX/dt = k_x Y \qquad \text{and} \qquad dY/dt = k_y X, \qquad \textit{2-Component growth rates}$$

where each rate constant is designated by k with a subscript indicating what is made. The complete solution of these two equations is:

$$X = [(\mu X_0 + k_x Y_0)e^{+\mu t} + (\mu X_0 - k_x Y_0)e^{-\mu t}]/2\mu \qquad \textit{2-Component growth}$$

and

$$Y = [(\mu Y_0 + k_y X_0)e^{+\mu t} + (\mu Y_0 - k_y X_0)e^{-\mu t}]/2\mu, \qquad \textit{2-Component growth}$$

where X_0 and Y_0 are the initial values at $t = 0$. The solution could be derived from what has been presented above with no new concepts from calculus. In fact, with every step explicitly written out, it would take only a page.

An important use for this pair of equations is to justify the concept of balanced growth. Let us assign the roles of two key cellular components to X and Y. Because these equations work just as well no matter what proportion of X and Y was present to start with, calculations based on this pair of equations are presented in Fig. 3.3 for the cases of a sudden increase and a decrease in k_Y of a culture that has previously achieved balanced growth. (When X has the significance of protein and Y corresponds to ribosomal RNA, then Fig. 3.3 corresponds to a nutritional shift-up and a shift-down experiment; these concepts will be developed below.) As the system starts to grow under the new conditions, both X and Y increase, but at different rates, and their ratio alters. After a while the composition of the cells approaches that of cells in balanced growth in this new medium. That a new balanced state will arise can be seen from the *2-Component*

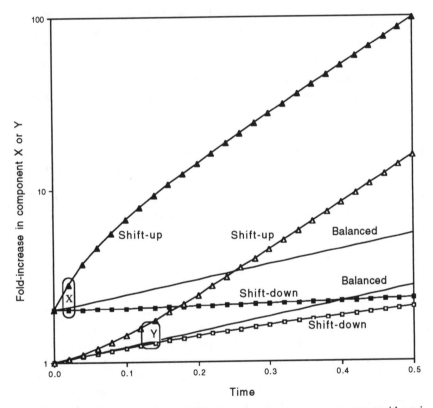

Figure 3.3. Theoretical shift-up and shift-down for a two-component system with positive feedback. The hypothetical cell has two components, X and Y, and a population that at time $= 0$ is in balanced growth. The continuation of this growth is shown by the solid lines; the amount of X remains twice as much as that of Y indefinitely. The curves shown with triangles are for a shift-up in which the rate of X production from Y was increased 4-fold. There is an initial overshoot in X and a lag before Y synthesis is increased. The squares show a shift-down in which the rate of production of X from Y was decreased 4-fold. These changes change the rate of growth by $\sqrt{4} = 2$-fold. These theoretical curves mirror the results with $X = $ RNA and $Y = $ protein in the old experimental studies of *E. coli* and *S. typhimurium* shift-up and shift-down.

growth equations by considering the case when t becomes large. Then the second quantity within the square brackets of both equations will become negligible because each is the number e raised to a large negative power. (Any number greater than 1 raised to a large and negative number is very small.) When time becomes very large, both X and Y increase exponentially with the growth rate constant μ characteristic of the new medium. Progressively the ratio approaches:

$$X/Y = (\mu X_0 + k_X Y_0)/(\mu Y_0 + k_Y X_0) \qquad \textit{Balanced ratio I}$$

which is a constant independent of time because all the quantities on the right-hand side are independent of t. Therefore, any culture as it continues to grow under constant conditions will achieve and maintain a new set of proportions.

In the new state of balanced growth the specific growth rate constant is $\mu = \sqrt{k_X k_Y}$. We can show this directly by multiplying together the two components of the *2-Component growth rates* equation and then rearranging factors:

$$(dX/Xdt)(dY/Ydt) = k_X k_Y.$$

If C (this is a new use for C) is the ratio of X/Y once the steady state of balanced growth is achieved, then we can replace Y with X/C and replace dY with dX/C in the second factor on the left-hand side. C cancels from the numerator and denominator, the second factor becomes equal to the first factor, and we can take the square root of both sides and be back to the *Differential growth equations* with just different symbols for the specific growth rate:

$$\frac{dX}{X} = \sqrt{k_X k_Y}\, dt \text{ or } \frac{dY}{Y} = \sqrt{k_X k_Y}\, dt.$$

Actually $\sqrt{k_X k_Y}$ is equal to μ, but, to be formal, because we have not shown this yet, we will set $\sqrt{k_X k_Y}$ equal to μ' simply because these equations look like the *Differential growth equation* given above. Therefore, the mathematical treatment of integration will work as before with μ replaced by μ'. Consequently, if a hypothetical kind of bacteria composed of two components in balanced growth were subcultured into the same medium when the amounts were X_0 and Y_0, and allowed to undergo further balanced growth, then growth would start without lag and the amount of both components and the total mass, $M = X + Y$, would be given by:

$$X_t = X_0 e^{\mu't} \text{ and } Y_t = Y_0 e^{\mu't} \text{ and } M_t = X_0 e^{\mu't} + Y_0 e^{\mu't} = M_0 e^{\mu't}.$$

Thus, X and Y will increase exponentially with the same specific growth rate as does the entire system composed of X plus Y. Further, because during balanced growth the average cell size is constant, we can now relate mass to numbers $M = CN$ (this is a third use for C) by writing:

$$N_t = N_0 e^{\mu' t},$$

which is identical to *Growth equation 1* except for having μ' instead of μ. Consequently, we have shown that the growth of numbers, mass, or cell components will, for the two-membered case, increase with the same growth rate constant that depends on the underlying rate constants according to the square root law:

$$\mu = \sqrt{k_X k_Y}. \qquad\qquad \textit{Square root law}$$

A number of other relationships about the ratio of X/Y now easily follow. You may be surprised that the ratio does not depend on the proportions that we started with. To see this, multiply both top and bottom of the *Balanced ratio I* equation by k_X/μ to obtain:

$$X/Y = k_X/\mu(\mu X_0 + k_X Y_0)/(k_X Y_0 + k_Y k_X X_0/\mu).$$

By substituting $\mu = k_X k_Y/\mu$ one can also write:

$$X/Y = k_X/\mu. \qquad\qquad \textit{Balanced ratio II}$$

Similarly, we can write:

$$X/Y = \sqrt{k_X/k_Y} \qquad\qquad \textit{Balanced ratio III}$$

and

$$X/Y = \mu/k_Y. \qquad\qquad \textit{Balanced ratio IV}$$

These are useful equations for a number of reasons. First, they show that the specific growth rate and the ratio of components during balanced growth are entirely independent of the nature (and origin) of the inoculum. Second, they allow us to use measurements of the proportions observed in balanced cultures together with μ to calculate the rates of synthesis that would be measured per unit mass of cells or particular cell components. Third, they explicitly demonstrate that during balanced growth the ratio of any two extensive properties must remain constant as required by the original definition of balanced growth proposed by Campbell (1957).

Now we can delve deeper into the kinetics of growth by making other sets of choices for X and Y. We will consider two examples. For the first example, let X be the growing cells and Y be the dividing cells (and in this case, X and Y are not even individual chemical components at all, but cell phases). The mathematical results then imply that the proportion of dividing cells in the population will be constant during steady-state growth and that the entire population, like each component, will increase exponentially with a specific growth rate μ equal to the square root of the product of the two first-order constants. In this case, one first-order constant is the probability per unit time that the cells in the Y class divide to form immature cells, and the other is the probability per unit time that the cells in the X class grow into mature "dividing" cells.

This new treatment is more realistic than the earlier treatment, which assumed that the probability of division was proportional to the total number of cells. This probability is actually proportional to the number of cells that are about to divide. But what we have just shown is that if growth is balanced, the subpopulation of dividing cells is a constant proportion of the total, so that the original crude assumption is justified. With more mathematics we could treat the cell cycle as a succession of stages or phases and describe quite accurately the distribution of doubling times with the suitable choice of parameters. This was first done by Kendall in 1948 and 1952.

Let us make another assignment for X and Y, remembering that μ will always be given by $\mu = \sqrt{k_x k_y}$ no matter how we change the significance of X and Y. Now let X be the number of flagella in the culture and Y be the total biomass; we could then argue that the more flagella, the faster the rate of finding a resource; and that the more biomass, the faster the resource is converted into flagella. In this assignment, ecology and behavioral biology are thus seen as also involved in balanced growth. Below we will make still another assignment to illustrate an additional aspect of microbiology.

The Chemical Kinetics of the Bacterial Cell

The revolutionary ideas of Sir Cyril Hinshelwood allowed the development of quantitative microbiology. He was far from a microbiologist. His most important work consisted of the analysis of simple photochemical processes in the gas phase reaction of molecular hydrogen plus elemental iodine to form hydroiodic acid. For this work he was awarded the Nobel prize. He thought naively that the kinetic approach he had used in these studies would apply as well in microbiology. As it turned out he was right in the part pertinent to this chapter, but he was also dead wrong about the relevance of genetics to bacteria, although the latter gaffe is what he is mainly remembered for now. But let us discuss what he got right.

Hinshelwood (1946) generalized the approach developed in the last section. He considered the case where there were several, say n, first-order reactions in an autocatalytic system so that the amount of each species determined the rate at which the next component in the sequence was formed. The reactions were arranged in a circular fashion, i.e., the last component was essential for the formation of the first (see Fig. 3.4). He showed that no matter what proportions of the various species were initially present, after a short time the system would increase in a balanced autocatalytic way in which all the species that made up

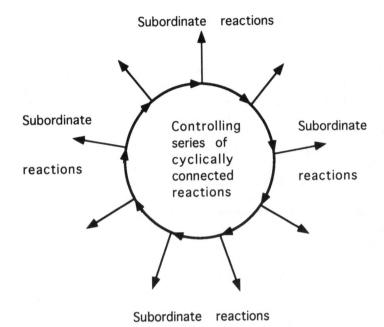

Subordinate   reactions

Figure 3.4.   Hinshelwood's cyclical autocatalytic system. A series of chemical species is depicted in the central circle, each species is made under the control of the previous one. Because the series connects back on itself, this constitutes a positive feedback loop. The arrows emanating from the circle represent side reaction paths or appendages. Hinshelwood showed that the whole system, no matter how complicated, after enough time in constant conditions would grow exponentially. The specific growth rate μ would be determined by the nth root of the product of all the n pseudo first-order rate constants involved in the circle and be independent of all of the other rate constants.

the loop increased exponentially with the same growth rate constant. In this case, μ was given by the nth root of the product of all the pseudo first-order rate constants for the individual steps. Moreover, Hinshelwood showed that other parts of the system, namely other cellular products that were not part of the autocatalytic loop, would also increase exponentially with the same value of μ, although the specific growth rate of the autocatalytic exponential growth would have nothing to do with the rate constants of the steps in the side paths. Of course, the value of the rate constants would control the steady-state proportion of immediate side product and the side products further downstream.

How Many Steps in the Basic Cycle?

By examining the rate of synthesis of every cell component per unit amount of every other component, one could, in principle, dissect the growth mechanism and establish which components are in the master loop and which are in its appendages. The rate constants could be computed by relationships similar to those presented earlier, from the composition of cells in balanced growth—if the number of steps, n, were known. Lacking this knowledge, one could measure the constants directly by appropriate experiments with radioactive precursors. But even if one knew the precise composition of all cell components during balanced growth, and knew n, this would not be an easy task, because there would be many nth roots in the formulae.

One might imagine that n is 1 as in Fig. 3.1A. This could be the case if some one substance (say DNA) engaged in autocatalytic replication and every other substance in the cell were dependent on the amount of that substance. On the other hand, it could be imagined that a very large number of compounds is involved because so many biochemical reactions are involved in the cell's manifold and many-stepped pathways and because there are many interactions between metabolic sequences during cell growth. At the end of the 1950s, if I had been asked to bet, I would have placed my money on n being 3 (see Fig. 3.1C). After all, the central dogma had just been handed down and the trinity of DNA, RNA, and protein was beginning its reign. I was surprised, then, by the results that came from the several laboratories of Schaechter, Maaløe, and Kjeldgaard (1958) in Copenhagen and of Neidhardt and Magasanik (1960) in Boston. The data of both groups suggested that $n = 2$; moreover, they suggested that the control cycle dealt with stable RNA and global protein (see Fig. 3.1D). This idea was supported soon thereafter by Ecker and Schaechter (1963). In these laboratories, experiments had been carried out to measure cell components in balanced growth. They compared the concentrations and found that the more rapidly growing cultures had much more RNA per unit cell mass. The ratio of RNA to dry weight (or to protein), in fact, followed very closely the assumption that the ratio of RNA to protein was proportional to μ. This result suggested that one of the two controlling processes responded as if the rate of ribosome production were dependent on the content of total protein in the cell and that the other responded as if the rate of making protein were dependent on the total content of ribosomes. The results corresponded to the sum of all proteins controlling ribosome synthesis, but also would be consistent with control being exerted by a certain protein or a group of proteins (if their amount is representative of proteins in general). The other controlling process was the effective rate at which ribosomes could function in the synthesis of protein. In this view, DNA is not limiting, nor is messenger RNA limiting, nor are a thousand and one other specific proteins, nor indeed are any other cell components, they are all appendages. These conclusions seemed to run counter to what we knew about the cellular biochemical processes. Of course, an absolute block in any essential step whatsoever would bring growth to a halt. But no matter how media were concocted or compounded, if they allowed growth, then from the specific growth rate the observed ratio of RNA to protein could be predicted as expected from the assumption that the loop of rRNA to protein and back to rRNA was the master circuit and nutritional details and other metabolic pathways did not regulate global growth.

Three conclusions came from this. One was that the cell arranges its other matters so that the costly item is used to the fullest and most efficient extent possible. The second was that the ribosome factories in which all cellular proteins are made are the cell's major capital equipment and are the expensive item for the cell to manufacture (Maaløe and Kjeldgaard 1966; Koch 1971). Third, the process of DNA coding for rRNA has been adjusted through evolutionary time so that the appropriate number of rRNA operons are present in the genome and the regulation of their action is seldom limited by the amount of the corresponding DNA.

The implication is that the cell has mechanisms to adjust the value of k_r (the pseudo first-order rate constant for the production of ribosomes per unit amount of cellular protein) and to a much lesser degree the value of k (the pseudo first-order rate constant for the production of protein per unit amount of ribosomes). Additionally, DNA synthesis is controlled secondarily in such a way that there is always enough DNA to transcribe enough mRNA and the tRNAs, and each tRNA must be charged adequately with its cognate amino acid. This balance of rate process could happen only if the bacteria, through evolutionary processes, had developed regulatory mechanisms to make the rate constant k_r responsive to the cell's nutritional environment. These general concepts were tested by physiological studies in a number of laboratories, and the results supported the importance of quick readjustments of the rate of ribosomal RNA synthesis when the environment was altered.

Then the mechanism for growth rate regulation is composed of two parts. First, there is passive "default" regulation leading to the achievement of the same state of balanced growth no matter what the initial composition of the cells at the time of growth shift. This works because during evolution mechanisms to adjust k_r in a major way and adjust k in a minor way in response to shifting conditions have been developed and the number of rRNA genes in the chromosome have been optimized for the range of conditions encountered in the past. If the value of each k was set during balanced growth and some cellular regulatory element(s) responded abruptly to environmental changes to alter some of the pseudo first-order rate constants, then the kinetics of the shift are given above for the *2-Component growth* equation as shown in Fig. 3.3. On the other hand, if there is purposeful sophisticated regulation in which the cell has control of the rate of its processes and actively and continuously adjusts certain k-values, for example, causing them to overshoot, the adaptation to change will be more complex. Such mechanisms do exist and do exert a limited action (see Koch 1970, 1979). These speed up or slow the adaptation relative to the default response in which when given a new medium the rate constants become discontinuously changed to new constant values. Of course, even for the actual, more complex case when the regulatory process is completed, the rate constants become constant, and the simple mathematics derived above for the *Balance ratio IV* equation apply.

With ribosomes and global protein as the key players, this $n = 2$ interpretation of the effect of cellular regulation is a good first approximation, but it has important limitations. I was the first to propose this two-membered cycle, which soon became dogma under the umbrella of the "constant efficiency hypothesis" of Maaløe and Kjeldgaard (1966). Later I had second thoughts, and our group spent a good deal of time showing that although this rule was found to be a good approximation in the middle range of growth (for doubling times between 40 min and 100 min for *Escherichia coli*), it failed both for very slow growth and for very fast growth. Our group showed that slowly growing cells have protein-synthesizing machinery that they are not using (Koch and Deppe 1971; Alton and Koch 1974); this machinery, however, is capable of functioning very rapidly after a nutritional enrichment. It was also found that very rapidly growing cells in a very rich medium use their ribosomes at a higher efficiency than do cells growing in a minimal medium with doubling times between 40 and 70 min (Koch 1980b). In fact, we now have more detailed information on how the regulatory circuit functions to tune ribosomal RNA synthesis and degradation to the current availability of resources for growth (Jensen and Pedersen 1990). The current evidence at the more detailed biochemical level is in harmony with the earlier conclusion drawn from our studies of the global growth process.

The assumption of an rRNA-protein reciprocal control downgrades the role of chromosomal DNA synthesis and cell division to a subordinate position in the cell's biochemical success. On reflection, this is reasonable: DNA synthesis and cell division, although important in the extreme for the genetics, ecology, and evolution of the bacteria, either when viewed as individuals or as species, are cheap processes in terms of the environmental resources needed when compared with those needed for making ribosomes, proteins, and cell walls.

How Cellular Protein Content Controls the Rate of Ribosome Synthesis

Mechanistically, it is clear that synthesis of DNA is a very important player in the biosynthetic processes of the cell, and yet it does not seem to be a factor in the overall growth kinetics. This independence is due to a regulatory system to adjust DNA replication. A sufficient model for moderate cellular growth rates is that initiation of chromosome replication is dependent on the amount of some protein whose amount is very nearly proportional to the cell's global protein content. It is a fact that the ratio of DNA to protein remains nearly constant at all growth rates (Donachie 1968; Koch 1970, 1988e). Although the experimental fact is clear, the mechanism of this regulation has not been clearly elucidated. The existence of such a regulatory system implies that the scheme of Fig. 3.1C can be replaced by Fig. 3.1D in which regulation at the DNA level is effectively bypassed. Both Cases C and D still have DNA, RNA, and protein arranged in a circle. The difference is that the latter figure has extra arrows; one emphasizes that proteins are needed to make the ribosomal proteins and other proteins needed directly for protein synthesis; and the second additional arrow indicates a controlling link between success in protein synthesis and the triggering of chromosomal synthesis. The effect is that the growth cycle is again formally like Fig. 3.1B, except that X is rRNA and Y is protein, but k_x is composed of the product of k_d and another constant, C. C in this case is the ratio of RNA to protein. Consequently, for our understanding of global regulation of growing Gram-negative organisms only two rules are needed: One is that $n = 2$ and the other is that somehow cell biomass controls the initiation of chromosomal replication. This two-prong control is no doubt the coarse control of the growing cell, and there is evidence for the existence of many refinements. Our profound ignorance, unfortunately, is that we know little about mechanisms whereby the DNA-to-protein ratio is regulated to a particular value.

Growth of Single Idealized Cells

If we were to follow the growth of a single cell, or if we had many cells that were all in the same phase of growth at the same time (a perfectly synchronized or synchronous culture) and all the cells had the same doubling time, then the growth curve would be composed of discrete steps. Then *Growth equation IV*: $N_t = N_0 2^{t/T_2}$, would apply with a special proviso: the numerical value of t to be used in the formula would have to be constant throughout a given generation and then change to the next higher integral multiple of T_2. This would cause the growth curve to have discontinuous steps with each step corresponding to a 2-fold increase. Such synchronous discrete behavior has never been observed (However, partial synchrony sometimes has been observed), and we can conclude that even though the cells have precisely the same genes, they vary in the length of time between when they arose by division of their parent cell and the time at which they in turn divide. (We can ignore mutations in this connection as being quite rare.) Physiologically, this implies that in order to divide each cell must take a variable period of time to grow through individual phases of the cell cycle. If the times for completion of some, or all, of the stages varied from cell to cell, then the doubling times would also vary from cell to cell in a complex way depending on why and how the variation occurred and how the variation in one stage is related to the others. The culture would gradually become asynchronous as it grows, if the variations were random. If the fluctuations had any random character at all then eventually a balanced asynchronous culture would be produced.

By simply following growing cells under the microscope, microbiologists have found that many microorganisms grow in such a way that the coefficient of variation (CV = standard deviation divided by the mean) of values of age-at-division (individual values of the generation or doubling time) has a value in the neighborhood of 20%. In contrast, the CVs for the values of the size-at-division are usually of the order of 10%. This 2-fold difference was first established by Schaechter et al. (1962) and has been supported by later work (see Tyson 1985). This implies that the cell's division is a response to the cellular achievement at some point of a "critical mass" and not to a "critical timing event" (Koch and Schaechter 1962). The speed with which synchrony is lost can tell us much about the underlying control or lack of control in the cell growth process. If the CV for the age-at-division was zero, growth conditions uniform, and the size of the two daughters exactly the same, then a synchronized culture would grow and remain synchronized indefinitely.

If cells did vary in the size-at-division, but divided precisely in two, then if the cell mass of each daughter grew exponentially, a perfectly synchronized culture under constant conditions would lose some synchrony at the first division and then remain partially synchronous indefinitely (Koch and Schaechter 1962). Such retention of semisynchrony has been observed in a number of studies (see Koch 1986b) and results simply because if division takes place at random around a mean size, then a cell whose mother had divided prematurely will take longer to grow to the mean size and vice versa. Fluctuations in division size, then, will cancel out after the first division event of a previously synchronized culture. Of course, synchrony is lost more quickly if the culture is not well synchronized in the first place, if growth conditions vary from spot to spot during culturing, or if the process of cell division does not produce equal-sized cells.

Approach to Exponential Growth of a DNA $\Rightarrow$ RNA $\Rightarrow$ Protein Model

Consider the simplest possible consequences of a model for bacterial growth in which DNA codes for mRNA that is then translated into proteins utilizing ribosomes assembled from both rRNA and proteins. To begin with, let us assume (for simplicity) that DNA is synthesized linearly throughout the cell cycle. This is not a bad approximation for *E. coli* cells doubling every 60 min. Then continuous DNA synthesis would occur because a new round would start just as the old cell cycle finished. The amount of DNA within a cell would increase linearly with time within the cell cycle. Let us assume that rRNA is made proportionally to the DNA concentration. This would occur if there were many ribosomal gene clusters spread uniformly around the chromosome (a situation not far from the truth). Then rRNA will rise in a parabolic fashion with time. Also, assume that protein is made proportionally to the amount of rRNA. Consequently, protein will increase as a third-order parabola. Lastly, assume that the concentration of all components exactly doubles in one cell generation.

Using just the mathematical methods we have discussed above, this extremely simple model can be solved analytically and evaluated for this idealized cell cycle. Alternatively, one can simply set up a computer program to simulate this case (see Fig. 3.5). The results of this simulation are graphed as a ratio of amounts at any time to those that would be expected for the components if there were balanced exponential growth. The numerical results of this model are the same to four significant figures as those predicted by an exponential law for the rise in protein content. For this model the DNA deviates by a maximum of 6% from exponentiality, the RNA deviates by less than 1%, and the protein deviates almost not at all. If we started to improve the model and make it more realistic, it would come even closer to pure exponentiality. Such improvements would include the modified assumption that DNA synthesis was bilinear, that chromo-

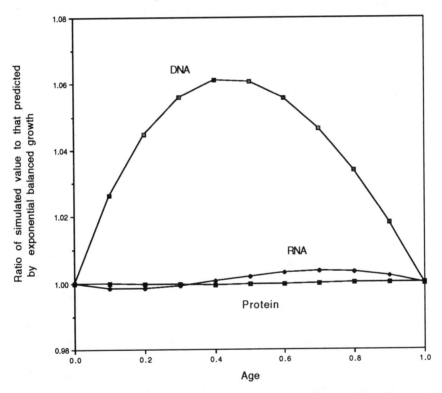

Figure 3.5. DNA, RNA, and protein cellular content for a simple model. The graph shows the results of a simulation in which the DNA of the cell increases linearly, the rate of RNA production is proportional to the amount of DNA, and the rate of protein production is proportional to the amount of RNA. Shown is the ratio of the amount of these cell components relative to those that would be there if all species grew exponentially with the same specific growth rate constant. It is clear that if this were, indeed, the growth process, then much greater experimental accuracy than is now possible in measuring biomass growth would be necessary to show the deviations from exponential growth.

somal replication is dichotomous, and that DNA chain elongation increases during the cell cycle as the level of proteins such as ligases and polymerase III, etc., increases. It would be a little worse if the model were altered to include gaps in the rate of DNA synthesis and not particularly different if turnover of mRNA were included. Because protein is the largest cellular constituent, one may conclude that the entire cell biomass (neglecting the envelope) would increase nearly exponentially in any of the cases and that it would be difficult to measure any deviation from exponentiality in the sum of proteins, RNA, and DNA components.

Synthesis of the envelope would have additional deviations, however. These effects would be greater at low growth rates where some kinds of cells become small and the surface-to-volume ratio is great. The cell components, wall, and phospholipids make up a small portion of the total biomass, but they also increase through the cell cycle in a fashion to provide a covering for the cell cytoplasm. Although not strictly exponential, the contribution of the wall component causes little deviation of the exponential increase in the total biomass (Cooper 1988c). Consequently, we shall expect an exponential increase in the cell's biomass and in most of the components of that biomass.

Precision of Cellular Processes

Even though the age-at-division varies a good deal, there is reason to believe that many cellular processes are quite precise. Consider the variability in the time needed to synthesize a protein. Even though a bacterial cell is small, it contains and deals with a large number of molecules. Therefore, it is reasonable to surmise that energy generation, intermediary metabolism, and protein biosynthesis all take place substantially continuously at the same rate, and that any cubic micron of protoplasm behaves like any other. In addition, even though the step time (the time needed to add an amino acid to a growing polypeptide chain) depends on the appropriate charged tRNA diffusing up to the right portion of the messenger bound to a ribosome at the right time, the synthesis time of a protein with a number of amino acids will be essentially the same for any two copies of the same protein because the fluctuations in time for the addition of each amino acid will nearly balance out for the entire protein. The mean time to string the amino acids together is proportional to the number of amino acids, and the variation in the total time necessary to string all the amino acids of a protein together is only somewhat longer than the variation in time to add one amino acid. The standard deviation of the total time increases, in fact, with the square root of the number of amino acids in the chain. (This is the central limit theorem of statistics.) Similarly, the same argument can be made with respect to all the other contributing processes of protein chain elongation.

Relevant Statistics

This section is inserted here to explain, and to be a reference to, statistical concepts needed in the following sections and throughout the book. It can be skipped if you are well founded in the subject. A little more detail can be found in Koch (1981, 1994a).

The binomial distribution describes the chance of occurrence of two alternative events. For example, it provides the answer to the question, "What is the chance of having 5 boys in a family of 9, assuming that births of boys represent 0.56 of the total live births?" The numerical answer is $P_5 = 0.2601$. The relevant formula can be derived from consideration of the number of ways different combinations can occur; it is as follows:

$$P_r = (n!/\{(n - r)!r!\})\, p^r\, (1 - p)^{n-r},$$

where p is the chance of a specified response on a single try, n is the total number of trials, and r is the number of specified responses, which would vary from 0 to n. The symbol ! means factorial; i.e., $n! = n(n-1)(n-2) \ldots 1$. In different families of 9 children, the formula shows that more families would have 5 boys than any other number. For example $P_4 = 0.2044$, $P_5 = 0.2600$, and $P_6 = 0.2207$. The plot of P against r is an example of a distribution histogram. The binomial distribution provides a way to estimate the mean and standard deviation of p from experimental data. Assuming that there are data on only a single family with n children and that there are by chance r boys in that family, statistical theory shows by calculating the values of P_r for different values of r, as we have just done, that the best estimate of p is r/n. Numerically, for the example where $r = 5$ and $n = 9$, p is $5/9 = 0.5556$ and its standard deviation is:

$$\sqrt{\frac{p(1 - p)}{n}} = \sqrt{\frac{5 \times 4}{9 \times 9 \times 9}} = 0.1656.$$

We would therefore estimate the chance of having a boy on a single birth as 0.5556 ± 0.1656 and would multiply both of these numbers by 9 for the total number of boys in families of 9 children.

This result is not very reliable because the coefficient of variation (*CV*), i.e., the ratio of the standard deviation to the mean value, is nearly 30% ($CV = 0.1656 \times 100\%/0.5556 = 29.81\%$). The precision in estimating p could be improved by looking for a family that had many more children and applying the same formula. It would not only be easier, but better, to pool the census data of a number of families. Suppose 50,000 boys are counted out of 90,000 children in a large pool of families. Then the same formula yields $p = 0.5556$, but now the *CV* will be 0.2981%. Not only is this more precise because large numbers are counted, but also it will average over other sources of variation. Some families, for genetic and sociological reasons, may have different values of p. It may, or may not, be desirable to estimate the value of p that applies generally to the entire population.

The formula for the distribution is cumbersome to calculate when the numbers are large. An important contribution of Gauss was to rewrite the binomial distribution for this large-number case. Thus, the Gaussian distribution applies as a generalization of the binomial distribution for the case in which the numbers involved are so large that they can be treated as a continuous distribution instead of one with discrete variables. The variables of the Gaussian distribution replacing n and p are the population mean m and standard deviation σ. The formula, whose derivation requires a little calculus beyond that reviewed above, then becomes:

$$P_x = (1/\sqrt{2}\sigma)\, e^{(-(x-m)2/2\sigma)}.$$

In this formula the continuous quantity x replaces the positive integer variable r as the measurement of response. This distribution, like the binomial, also can be mathematically manipulated so that one can go from data to estimations of these two parameters. The Gaussian distribution is also called the normal distribution, partly because it is symmetrical about the mean and partly because it is so frequently observed.

For data that follow a Gaussian distribution, the estimate of the mean is called m and is given by $m = \Sigma x/n$, where Σx means the sum of all the individual values of x. The estimate of the standard deviation is called s; i.e., the roman letter corresponding to σ, the value relevant to an infinitely large population, and is given by:

$$s = \frac{\sqrt{\Sigma x^2 - (\Sigma x)^2/n}}{n-1}.$$

The coefficient of variation CV, as mentioned, is given by s/m.

The other limiting distribution of the binomial is the Poisson distribution. It applies for the case where n is very large and p is very small, but the product $n{\cdot}p$ is finite. The best estimate of $n{\cdot}p$ is N, the observed number of total responses of a specified kind. This distribution would be useful, for example, if boys occurred very rarely (say, 1 in 10,000) but families were large (say, 100,000 children). Then an average family would contain a mean of $N = 10$ boys out of the 100,000 with a standard deviation of:

$$N\sqrt{p(1-p)/n} = \sqrt{Np(N/n)(1-N/n)} = 3.1621.$$

Poisson's insightful simplification of the binomial distribution was to assume $Nn \ll 1$. Then the formula for the binomial distribution simplifies to a one-parameter distribution:

$$P_r = e^{-m} \, m^r/m!.$$

The best estimate of the mean is $m = N$, again because P_r is maximal at this value of r and the best estimate of the standard deviation is also N. Note that this is not much different than the binomial distribution, because $\sqrt{10} = 3.1623$ is not much different than 3.1621 for the more complex formula. Note that again n and p are replaced by different symbols, in this case by a single one, m. The important point is that the count of the number of discrete objects provides not only the best estimate of the mean value (i.e., $N = m$), but also an estimate of the precision of the estimate ($\sqrt{N} = s = \sigma$).

The standard deviation has been defined above. It is frequently confused with another term, the standard error, also called the standard deviation of the mean. The standard deviation measures the typical deviation of an individual measurement from the mean of many measurements. The standard error is the standard deviation divided by the square root of the number of measurements. It measures the deviation of the mean of all the data actually observed and calculated from a particular sample from the mean of a hypothetical data base containing an infinite number of observations, and is a measure of how close the actual average is from the "true" mean value.

The accuracy of an estimate depends on the accuracy of its component measurements. For typical use in microbiology in enumerating a viable count, the Poisson error of a colony count and the error of the dilution procedure both contribute to the error in the estimated concentration of organisms of the original undiluted suspension. Combination of uncorrelated errors can only further blur the results or make them less precise. Even though errors in one part may compensate for errors in another part of an estimate, on the average random errors will make errors larger. When errors in one measurement are independent of (uncorrelated with) errors of another measurement, the overall error can be calculated by two rules. If two quantities (x and y) are to be added or subtracted, then the standard deviation s of the combined quantities is:

$$s_{x+y} = s_{x-y} = \sqrt{s_x^2 + s_y^2}.$$

If two quantities are to be multiplied (as in the estimation of the viable count) or divided, then the coefficient of variation (*CV*) of the combined quantities is

$$CV_{x \cdot y} = CV_{x/y} = \sqrt{CV_x^2 + CV_y^2}.$$

Constancy of Rate of Macromolecular Synthesis

Consider a protein of 360 amino acids being synthesized at a typical rate of 16/s in a cell with adequate amounts of enzymatic factors, tRNAs, and ribosomes loaded with mRNA. Therefore, the mean step time to add any one amino acid is 0.0625 s, and it will take 22.5 s to make the protein. Let us assume that the time for amino acid addition has a CV of 10%. Then the standard deviation will be 0.00625 s. But if the variation in the time to add any one amino acid has no influence on the variation in time for any other amino acid's addition, then the variation in time to make the entire protein is 0.00625 $\sqrt{360}$ = 0.119 s. The CV would then be (0.119/22.5)100% = 0.527%; this is a small quantity. If the cell makes 100 copies, the standard deviation of the mean value of all of them is 0.119 s/ $\sqrt{100}$ = 0.0119 s or a CV of 0.0527%, which is even less. Consequently, we ought to be able to treat protein synthesis as if it occurred, under a specific set of conditions, as almost a precise process. The same argument could be made for the process of RNA synthesis. Global protein synthesis and global formation of ribosomes, therefore, would seem not to be subject to much statistical random fluctuation.

Let us take another example, that of partition of substances at cell division. Consider a bacterial cell with a volume of 2 μm^3 that has 10,000 ribosomes. What is the variation in rate of protein synthesis in each of two 1-μm^3 halves of the cell? This is the same statistical question as the error in a colony count on a plate that has an expected mean of 5,000. The answer is $\sqrt{5,000}$ = 70.7; the CV thus is (70.7/5,000)100 = 1.4%. This suggests that chance fluctuation in the partition of ribosomes at division would be small. Of course, as we have seen, any fluctuation would be damped out by further growth, if the rate constants are indeed constant. For cells growing at slower rates the number of ribosomes per cell is less and the percentage variation would thus be greater, but even slowly growing cells have quite a large number of functional ribosomes, although few may be working at any time. Consequently, the fluctuation from cell to cell and from moment to moment could be significant. However, the degree of this variation remains to be measured.

Continuous Versus Discrete Cellular Processes

As long as we deal with the total amount of most chemical species in a sample of a balanced culture, we need not consider how the components are partitioned among the individual cells, be they small metabolites or macromolecules. The classical example is to consider the number of protons in a 2-μm^3 bacterium. If the cytoplasm has a pH of 7, then the average number of protons is 10^{-7} · $2 \cdot 10^{-15} \cdot 6.023 \cdot 10^{23}$ = 120. So even though pH 7 corresponds to a low concentration, there is still an average of 120 so that two-thirds of the time (one standard deviation), the number of protons will be between 120 − $\sqrt{120}$ = 109 and 120 + $\sqrt{120}$ = 131. Although this does not seem like a large fluctuation, the same arithmetic applied to a volume the size of a protein or even a virus would imply that the pH was either plus or minus a large enough quantity so that essentially the pH was never in a biological range. The answer to this seeming paradox is that dissociation and reassociation of water to form protons and hydroxyl groups occurs so quickly that the relevant time average does not depend on the instant-to-instant variation.

This deterministic concept must certainly have been the expectation of Sir Cyril Hinshelwood for the idealistic principle that the same rules should apply to bacteria as those that held for his favorite system of small molecules such as I_2 and H_2 in the gas phase of low pressures.

Now consider cellular events that depend on small numbers of particles and discrete events. Bacteria have one component present in very small numbers: chromosomes. Variation in the timing of the initiation of replication would cause large changes in the properties of individual cells because the variation is not ameliorated, in this case, by a large number of events. Variation in the partition of chromosomes at division could cause even more fluctuations. When a cell with two chromosomes divides in half, what is the random chance that both chromosomes go to one daughter and the other gets none? This is the same type of problem as calculating the chance that in a family of two children both will have the same sex. 1/4 will be two-girl families, 1/4 two-boy families, and 1/2 one-boy and one-girl families, so the answer is 50%. Unequal partitioning of the chromosomes in bacteria occurs, but only rarely. It is only observed in 0.3% of the cell division events. That means there is some accurate mechanism segregating the chromosomes and that a default random segregation is not relevant to properly functioning bacteria. Although some small copy number plasmids have elegant mechanisms for segregating at least one copy to each viable daughter cell, random processes do apply to the segregation for many other plasmids.

There are other cell components, such as repressors, that are represented by very few copies within an individual cell. There are individual kinds of repressor molecules that may number as few as only 10 copies per cell. The *CV* in their partition (according to the binomial theorem) between daughter cells can be large, $CV = \sqrt{0.5 \times 0.5/10} = 15.8\%$.

As Max Delbrück first pointed out (1949), we can also imagine metabolic systems serving roles as all-or-none switches that could cause the cell to behave discontinuously for certain design features of regulatory systems. This would happen if the product of one branch inhibited a second branch and vice versa. Switches constructed of permease systems and feedback repression can also function in an all-or-none fashion.

To add one more item to the list of discontinuous circumstances, the chromosome in its usual configuration has no ribosomes inside the nucleoid region; they are all constrained somehow to remain outside this region. The chromosome would be expected to be nontranscribable unless the DNA for a particular gene happened to be on the outside of the DNA mass. While this is an *a priori* possibility, it can be argued that there are enough clefts in the nucleoid so that all regions of the chromosome are rapidly accessible and that the growing mRNA chain may thread out of the nucleoid.

I have listed more potential problems than actually seem to occur. Fortunately in fact, for the problem of deciphering bacterial growth, few of these are of significance to the overall process. There does appear to be pausing or stuttering; i.e., intermittent starting and stopping at particular regions of the genome during the synthesis of nucleic acids. But as for the rest of the possibilities, the cell's controls appear to be adequate to make all the reactions serve as slave reactions to the master rRNA $\Rightarrow$ protein and protein $\Rightarrow$ rRNA cycle of Fig. 3.1D and to swamp out or overcome the statistical fluctuations expected from chance events. Of course, these control mechanisms are part of establishing the values of the apparent first-order rate constants for various processes appropriately for the conditions defined by the internal and external environment.

The Forever-chemostat and Forever-turbidostat

Up to this point only bacteria growing for a relatively few generations in batch culture have been considered. We have presumed that these cultures did, however, reach a physiological state of balanced growth. Now let us imagine a culture that has been growing in a constant environment for eons of time so that many evolutionary changes have occurred. In a closed vessel this is a physical impossibility, because exponential growth would quickly use up any conceivable supply of nutrients. We need to imagine some form of continuous culture. The effect of evolutionary selection would be quite different depending on how the culture was maintained. There are two limiting ways to carry out continuous culture.

The Chemostat

In the chemostat the culture is continuously partially harvested at a constant rate and the loss in volume made up by replenishment with fresh medium. Although the replacement medium has a moderately high level of the single limiting low-molecular-weight nutrient, the well-stirred, steady-state culture has only a low level, because of consumption of this one material. All other components are assumed to be in excess. This ambient level is self-adjusting because the level of bacteria will increase or decrease until the residual concentration is just sufficient to give a specific growth rate equal to the mechanical dilution rate of the pumping system.

The derivation of the *Chemostat equation* is as follows. If N is the concentration of bacteria within the growth chamber, if v is the volume of the chemostat culture vessel, and, as usual, if $dN/dt = \mu N$, then the production of bacteria by the apparatus is:

$$v(dN/dt) = v\mu N.$$

The rate at which bacteria emanate from the chemostat is equal to Nf, where f is the flow rate. Consequently,

$$\mu N v = N f.$$

Notice that the units of both sides of this equation are the number of cells formed in, or released from, the growth chamber per unit time. It follows that

$$\mu = f/v = D,$$

where D is the dilution rate, defined as f/v, the ratio of the flow rate divided by the culture volume. Another way to derive this result is to augment the *Differential growth equation*: $dN/dt = \mu N$ to include the loss of organisms from the system. Because organisms are eliminated from the culture vessel, an additional "washout" term has to be added to the equation. The equation then becomes:

$$(dN/dt)_n = (dN/dt)_g - (dN/dt)_w,$$

where the subscripts are defined as follows: n = net, g = growth, and w = washout. In that case,

$$dN/dt = \mu N - DN, \qquad \textit{Chemostat equation}$$

where now these subscripts from $(dN/dt)_n$ have been dropped. Now at the steady state where N is constant, dN/dt is zero and $\mu = D$. D, like μ, has the units of "per unit time." Thus, μ is forced to become equal to the dilution rate established by the experimenter, who chooses the flow rate and the volume of the culture vessel and, therefore, sets D. The growing culture must and will adjust its growth accordingly, if it is physiologically possible. Usually the input concentration of the limiting substrate S_0 is chosen to produce a suitable crop of bacteria. After inoculation the organisms grow rapidly, then the organisms deplete the substrate sufficiently to slow their growth, then a steady state develops where μ has changed to satisfy $\mu N - DN = 0$. Thus, when balanced growth is achieved, the condition $\mu = D$ will ensue.

Although this equation for chemostat growth is given in most microbiology texts, and in Kubitschek (1970), it concerns a fictional value of μ. The μ in this equation would be equal to the μ that would be observed with an extremely low-density batch culture that had the same concentration of the limiting nutrient in the medium as established within the chemostat. Following growth under these dilute culture conditions would require the use of a very sensitive method like carrying out viable counts because the experiment would have to be set up so that the bacterial concentration is low in order for consumption to be negligible. Because there have been unpublished claims to the contrary about this point (Jeanne Poindexter, personal communication), I will restate the case differently. If a chemostat were instantly stopped and simultaneously the vast majority of the bacteria instantly removed, the remainder, now growing as a batch culture, would grow exponentially according to *Growth equation I* but with the μ determined by the value of D dependent on the previous flow rate f and culture volume v.

Now turning to evolutionary biology (Koch 1985c, 1987a, 1995c), we might ask, What would happen if such a machine were to run for a very long time (perhaps a billion years) under absolutely constant conditions? Let us assume that the bacterial cells cannot stick to the walls of the vessel nor to the nutrient inlet and that mixing is instantaneous and perfect. As the cells adapted to this habitat, they would become more effective at foraging the limiting resource, and the ambient concentration of this nutrient would become smaller and smaller. It is self-evident that very small cells would evolve so that the surface-to-volume ratio would be as great as possible. They would ideally adopt the habit of long thin cells or possibly the irregular shape of prosthecate bacteria (appendaged bacteria) in order to have a larger surface-to-volume ratio. Additionally, they would evolve very efficient pumps and have many copies of the permease system per cell to scavenge the limiting nutrient as completely as possible. Finally, their genome would contain no spare baggage, and induction and repression would be missing or at a minimum because with conditions always constant, the best strategy for bacteria is to maintain constantly the optimum physiology. When further evolutionary changes became very rare, the eventual complement of cell components and their size and shape would depend on compromises between these factors and other nonnegotiable aspects of the cell biology. Additionally, the evolved strain would probably would not be motile because diffusion of a small nutrient molecule is faster than flagellar motility, and they wold not grow in clumps, chains, or filaments because that would impede diffusion. They would truly be "monads"; i.e., single cells, and after cell division the two daughter cells would quickly break ties with each other and separate completely.

The Turbidostat

The second form of continuous culture is the turbidostat. In such a machine, a portion of culture is also removed and replaced with an equal volume of fresh medium, but this is done by laboratory machinery that responds to the growth of the organisms. Usually the concentration of organisms is measured photometrically, but there also other ways to maintain cell density. If the medium is very rich, growth for a very long time under the constant conditions would lead to evolution of cells capable of very rapid growth. These might be large and quite nearly spherical because that would minimize the cost of wall manufacture and a high surface-to-volume ratio is not as important in this case in which the environment is always sufficient and resources and energy metabolism adequate. These cells should have very many ribosomes. Also like the cells in the chemostat case, they should eliminate excess genetic baggage and many of the special features that allow bacteria to grow and survive in a variable natural environment (Koch 1987a). In neither of these fanciful cases wold spore or resistant forms be useful. For the limiting case, if ribosomes could work only at the efficiency at which they actually function in the laboratory today and ribosome function was only needed to make ribosomal proteins, associated protein factors and very little else, calculations made by Robert Harvey (personal communication) and Stephen Cooper (personal communication) show that the doubling time would be about 8 or 9 min. I have tried to find out how fast bacteria could grow. I chose a strain that had not been subject to mutagens or mutagenic agents in the laboratory, and allowed it to grow on a very rich medium; I observed doubling times of 15–16 min at 37° C (Koch 1980b). Usually, a 20-min doubling time is the maximum growth rate reported at 37° C in a rich medium; probably I was able to achieve this faster rate because of the strain chosen and because the rich medium, besides amino acids and vitamins, also contained extra glucose and had a higher than usual osmotic pressure.

Life in a Fluctuating Environment

Because of the vicissitudes of life, bacteria in nature must apparently waste a good deal of their resources and carry a large number of genes in order to be able to respond to changes in their environment. These expenditures are necessary even if the environmental changes are mandated as the results of their own depletions or pollution of their environment. Among the responses are shape changes, formation of spores and resistant forms, and the differentiation of some daughters to different strategies in order to hedge bets for long-term survival of the culture (Koch 1985c, 1987b, 1991b). Indeed a great deal, possibly most, of the bacterial genome is composed of unique sequence DNA that is transcribed very rarely indeed. Perhaps some has been inactivated mutationally and may need to be reactivated under some circumstances, some may function under extreme conditions, and some may be just plain junk. But a good case can be made for the cell's having a "catastrophe kit" of genetic tricks to survive some extreme challenges (Koch 1993a, 1995c; Kolter, *et al.* 1993).

Phases of the Culture Cycle

The topic of this section is one that is covered in all beginning microbiology texts. Even if deified, the culture cycle is simply a laboratory artifact. If one dilutes a stationary culture, say of *Escherichia coli*, there is usually a lag, then exponential growth that approximates balanced growth to some degree. Cells in late lag grow to their largest size, but the cells are still quite large during exponential growth. The exponential phase is followed by a slowing to a stationary state. During this transition the cells continue to divide but produce mass more slowly so that the cells become smaller. That phase may be followed by one in which the cells gradually die and may lyse. Of course, the outcome of observations on the growth of the culture depends on the age of the parental culture, the degree of dilution, the medium, the kind of measurement used to assess growth, and the particular strain used. The biology that can be gleaned from experiments of this type is important, but the concepts can be more critically established by experiments that attack only one of these changes at a time.

Consider the response of the forever-chemostat and the forever-turbidostat of evolved bacteria to dilution into a large amount of both kinds of medium used for the continuous cultures but now grown as batch cultures. Neither evolved strain would do very well. One's first thought is that they would both die very rapidly in the subsequent starved state because they would have lost mechanisms such as starvation resistance (Matin *et al.* 1989; Kolter *et al.*, 1993). Consequently, the so-called stationary state would be short-lived. Upon the depletion of a medium component, an abrupt cessation of increase in cell numbers and their rapid destruction would be expected instead of a gradual slowing of biomass production and the still more gradual slowing of cell division shown by normal cells when approaching the stationary phase carried out to modify their own physiology. On redilution there would be an extensive, possibly infinite, lag because of the large proportion of dead cells.

The organisms from the chemostat might use the rich medium less efficiently than those from the turbidostat because the former were allowed to utilize only one substrate for carbon and energy and had the opportunity to specialize on it. On the other hand, the turbidostat culture might be incapable of meeting all its needs in the chemostat medium. Actual utilization of a rich medium would depend on the composition of the medium of the chemostat and turbidostat versus the medium used for study of the proposed culture cycle experiment. Of course, each organism would have difficulty in the medium to which it was not adapted. In addition, the bacteria adapted to the chemostat would have difficulty in their own growth medium, because now the level of the previously limiting substrate would be much higher. They would pump much more of the previously limiting medium, leading to metabolic unbalance and severe osmotic problems.

Actual bacteria, prototypically *E. coli*, respond to dilution in fresh rich medium by forming many ribosomes and becoming larger (as in a shift-up); they respond to degradation of their environment by becoming smaller and reducing the ratio of RNA to DNA (as in a shift-down) and by the development of more resistant forms. They respond to a surfeit of a nutrient by feedback inhibition and repression of the uptake system. Without the ability to wait a billion years, the shift-up and shift-down types of experiment with an organism from a natural habitat will permit more certain interpretation than can be deduced from the study of the phases of growth of the culture cycle when it is the organism that is the agent that controls the changes in the medium.

Two Contrasting Aspects of Growth: Mass Growth and Number Growth

Above we focused on the closed loop in which protein begot RNA and RNA begot protein; it was argued that DNA synthesis and cell division are subservient to the process of mass growth as long as there are other regulatory processes maintaining the ratio of protein to DNA. Although this is a good argument when conditions are continuously satisfactory, it fails when conditions become poor. With only a low concentration of resources a large surface-to-volume ratio is needed, and cells optimally should become small and acquire as nonspherical a shape as possible. Under sudden starvation or shift-down conditions, ribosomes would suddenly be present in excess of ability to use them. In response, the cell do stop ribosomal synthesis and do increase ribosomal turnover.

The culture cycle, when experimentally followed, reveals a number of features showing that organisms have developed mechanisms to cope with fluctuations in the natural environment, in the face of an often unpredictable alternation of unfavorable and favorable growth conditions. These responses include size and shape changes in which the guiding principle is that when conditions become poor, the cells partition their residual resources and their biomass so as to maximize the number and survivorship of cells. On the other hand, when conditions are good, the cells increase the rate at which resources can be turned into biomass. Cells can then be greater in size and behave independently of surface-to-volume considerations, and consequently the cells may become not only bigger but more nearly spherical.

The "Growth Law" at the Cellular Level

Over the last half-century, there has been considerable discussion in the microbial literature about the "growth law" at the cellular level, but much of it, I feel, misses the real point. A chemist such as Hinshelwood, upon entering the field of biology fifty years ago, would expect from the kinetic considerations that every component should eventually increase exponentially with the same specific growth rate as every other component. In addition, I imagine that he would think when balanced growth is achieved, exponential growth should apply as well at the level of the single cell; this follows from the simple argument that an enzyme or a ribosome cannot know how big a cell it happens to be in at the moment.

The statistician would argue that the cell cycle is made up of many parts and that the kinetics of all aspects of growth should be subject to statistical fluctuations that depend very much on the numbers of the parts and on the circumstances. For substances present in a large number of copies the deterministic laws of chemistry would be followed, but the laws of small number statistics should apply when there are few copies. The microbiologist knows that there are at least two discrete stages of the cell cycle: chromosome replication and cell division, where the numbers elements are small (one or two). These events would not be described by the kind of equations we have been dealing with (although their average behavior at the level of the culture would). These discontinuous processes would be of less importance when conditions permit rapid growth, where the cells are large with low surface-to-volume ratios.

It has been shown that when X and Y were associated with different phases of the cell cycle (Case B of Fig. 3.1 where $X \Rightarrow Y$ and $Y \Rightarrow X$) during balanced growth, discontinuous events would not cause deviations from the exponential increase of all aspects of cellular life. Although we have not discussed the effects of variability in the division time of individual cells, it can be shown these effects would average out when one considers an entire growing population. The actual argument is even stronger than this statement, because no matter what the kinetics and variability of those kinetics, either within the growth of a single cell or among cells in the population, all parameters—that is, all the k's and μ's, when averaged over the population—would serve in the appropriate equations as developed above. The only exception arises if there are countermanding regulatory mechanisms serving to create periodic fluctuations.

Ribosome Function in a Perfectly Evolved Organism

I hope the reader has convinced himself or herself that either of the two hypothetical strains that should arise after very long-term continuous culture would have evolved to maximize their growth by acquiring the ability to work at the highest achievable efficiency for all aspects of macromolecular synthesis. Because of the importance of protein synthesis it would follow that ribosomal components and factors would become as physically small as possible consonant with rapid and accurate catalytic activity. Because ribosome efficiency would be paramount in both types of continuous culture and during growth in nature, it can be presumed that evolution to quite high efficiency has already occurred. Even so, ribosome construction is a very major expense for the cell. This leads to the concept that the cell has evolved to arrange its physiology so that the ribosomes are always the limiting factor. Moreover, as each new ribosome is made it would be put to work and not allowed to idle away its time.

In addition, we could assume that the turbidostat-evolved organism would have ribosomal cistrons in great number distributed over the genome so that the rate of production would increase throughout the cell cycle, and that the more slowly growing chemostat-limited organisms would have fewer rDNA cistrons. If they had only one, they would control initiation so that the output of ribosomes increased exponentially throughout the cycle. Then RNA and protein would exactly increase exponentially and in parallel. The sum of all cellular components would increase almost exponentially; there would be a few quantitatively minor deviations caused by the necessarily discrete aspects of cellular growth (chromosome initiation, chromosomal replication, cell division), and effects due to the shape and the amount of wall material needed.

How Can Steady-State Growth be Anything Other Than Exponential at the Single Cell Level?

The concept of the most efficient mode of ribosome formation and utilization does not necessarily apply to the details of the life cycles of real bacteria, because of their need to adapt to environmental fluctuations. So much of the effort and genome of actual bacteria is devoted to long-term survival that they apparently very often take a short-term loss and are less efficient than ideally possible. Still, I can think of no *a priori* reason or mechanism to allow efficiency to vary throughout the cell cycle. In the next several sections, I will discuss the efficiency with which bacteria conduct protein synthesis and the formation of the protein-synthesizing machinery throughout the cell cycle.

There are two ways that any deviation from constant efficiency could be expressed in actual bacteria. First, the cell may have a mechanism to shut down global protein and/or rRNA synthesis either as part of the cell cycle or as a strategy that only some of the cells in the culture elect. Little evidence for this has been published, although a few examples have been found (Koch 1979, 1987a,b). For example, if cells from a chemostat in which the bacteria grew with a 10-h doubling time are shifted into the same medium with an excess of the limiting factor, glucose, they take 6 h before achieving balanced growth with a doubling time of 45 min. If, however, they are shifted up into a very rich medium, their rate of growth increases in much less than a minute to the definitive balanced growth rate (under these experimental conditions, the doubling time was 20 min). In this first case the cells placed in glucose-containing medium adjust only slowly the rate of ribosome formation k_r, and this means that those cells have elected to decrease their potential growth below their maximum metabolic capabilities until they "feel assured" that glucose is, indeed, no longer limiting.

The second way that apparent inefficiency could be produced is by turnover. It is evident that protein turnover is contrary to optimum efficiency. Some turnover does occur and may be unavoidable as a result of wear and tear, as part of the function of control mechanisms for the discrete aspects of the cell cycle, as excretion or secretion of proteins (exoenzymes), as loss of outer membrane components, or as a result of defective protein synthesis. The only available estimates show that extremely little turnover takes place for the bulk of the cellular proteins under conditions of balanced growth (see Koch 1991b).

Consequently, if growth of dry biomass at the level of the individual cell cycle were linear or bilinear or any relationship other than a very good approximation of pure exponential growth, one should be able to demonstrate either cell cycle-dependent turnover or cell cycle-dependent inhibition of macromolecular synthesis. In principle, this should be easier to detect than deviations from exponentiality in the increase of cell constituents. After all, there is only a 6% discrepancy between the linear and the exponential models that assume no turnover or variation in specific rates of synthesis.

Kubitschek's Idea of Linear Growth

Kubitschek was a physicist who entered the field of biology at the end of World War II. He made and analyzed time-lapse films of *E. coli* growing on slide cultures and then developed a more accurate form of the Coulter counter to measure the distribution of cell sizes within cultures. Both endeavors aided him in studying the cell cycle of single cells. In the late 1960s he followed the growth of a thymine-requiring culture after the thymine was removed (Kubitschek 1971). He observed that at the time of starvation, the cell-size distribution had a positively skewed shape that was quite broad. Cells continued to divide for a brief period, and then the cell number remained constant while all of them grew bigger (longer), i.e., the whole distribution moved to the right. That was not surprising in terms of then current knowledge because by stopping DNA synthesis, those cells that were before a certain point in the cycle no longer divided, but instead did filament (i.e., grow as "snakes"). Although the cells stopped DNA synthesis, there would be no reason why they could not make ribosomes and consequently accelerate the rate of protein synthesis. Kubitschek's measurements, however, showed that cells of all phases of the original distribution became longer as a linear function of time instead of an expected exponential increase. He spent the rest of his career focusing on the linearity of growth of the individual cell. He knew that his ideas were quite different from those of other workers and ideas espoused here. In fact, it was workers in his laboratory (Ecker and Kokaisl 1969) who had carried out the crucial experiments with autoradiography (see below) that were often cited as the definitive proof of exponential growth. His interpretation of all the data was that the cell's increase in volume was limited by the number of uptake sites for medium constituents. Over the years, some of the specifics changed, but his final view was as follows: Just before cell division, new transport proteins were made and inserted into the membrane, or at least became functional. Consequently, through most of the cycle, the cells had a fixed complement of permeases that, although working at their maximum rate, still limited cell growth. Thus, the volume of the cell would increase linearly during the cycle, the rate of growth become doubled a moment before division, and then the rate per cell halved when the cell divided. Kubitschek supported his contention in a number of ways, primarily by measuring the rate of uptake of various substances and attempting to measure the amount of materials in the cell's soluble pools.

In some of his papers he assumed that there was an exponential increase in protein content during normal growth. He accommodated this to his own observations that the buoyant density of the cell was constant (Kubitschek 1987), by postulating that the cell adjusted to these contrasting requirements by filling and emptying the acid-soluble pool (consisting of molecules of less than 500 molecular weight).

The Controversy Over the "Growth Law"

Most workers, myself included, focused on the *a priori* exponentiality of cell growth and ignored Kubitschek's ideas. But, because of the debate between

Cooper and Kubitschek (Cooper 1988, 1991; Kubitschek 1990), the issue of growth at the level of the single cell once again became a hot topic. Cooper subscribes to the exponential model, in contrast to Kubitschek's subscription to linear accumulation of total cell components during the cell cycle. In the next several sections I will deal with relevant studies, both experimental and theoretical, and attempt to tie them together (for more details, see Koch 1993b,c). To anticipate, the conclusion will be drawn that the experimental work is more in line with exponential growth. The experiments do exclude the linear model, but none of the published studies has enough resolution or has proper controls to "prove" exponential growth.

Nontracer Work and Elementary Theory

A number of early studies have tried to establish experimentally the rule by which single cells of a bacterial species grow by studying the size, age-at-division, and size-at-division of single cells and populations of cells. Although the emphasis of this section is on how the cell grows, for this purpose cell division is only an obstacle. If all cells had the same age-at-division, then a variety of methods would have been adequate to define the kinetics of the growth process. Questions concerning the control of the division process cannot, in fact, be dissociated from the growth process. The term "growth law" as usually used is the relationship between the rate of increase of the cells dimensions (volume, surface area, length) and the time within each individual cell cycle.

The various experimental approaches that have been used will be illustrated by citing the predictions of the exponential "default" model in which growth of cell volume is exponential. For the initial exposition we will assume that the cell density is constant throughout the cycle and the contributions of DNA and cell envelope are both small. Although cell volume is the parameter of choice, surface area or length would substitute as well when the cells were long and thin, or if corrections were made to calculate equivalent area or length if the cells were right cylinders.

Method 1 is the integral method. If V is the volume of a single cell, the default growth law could be symbolized by:

$$V = V_0 e^{\mu t}: 0 \leq t \leq a, \qquad \text{\textit{Default growth law I}}$$

where t is the age of the cell since birth and a is the age-at-division.

Method 2 is the differential method. The default growth law may be written as:

$$dV/dt = \mu V: \qquad V_0 \leq V \leq 2V_0. \qquad \text{\textit{Default growth law II}}$$

There are a number of ways to test these relationships experimentally. It would be sufficient to measure V as a function of age (Method 1) or to measure the rate of volume increase dV/dt of single cells as a function of V (Method 2). Following individual growing cells in slide culture under the microscope requires little equipment, but the accuracy permitted by the phase light microscope (both because of the wave properties of light and because of the phase halos) is not adequate to resolve linear, bilinear, and exponential models. This is so because the maximum difference between the two extreme models is only 6% during the course of a single cell cycle. If we were to restrict attention to the central part of the cell cycle, the effects of the variability of the cell division process could be assumed to be negligible, but then the limitations resulting from poor spatial resolution would be even more severe. If cells are followed for longer than a doubling time, then the effects of cell division and the *a priori* necessity for doubling the rate every cell cycle would be dominant. As a result, critical tests of possible modes of growth by either of Methods 1 or 2 have not been believable.

Method 3, analysis of steady-state distributions, involves measurements of the distribution of ages or, most often, sizes of many cells drawn at random from a balanced growing population. For example, this could be measured on fixed preparations of numerous cells and then the distribution of cell volumes prepared. Of course, the photography is best done with the electron microscope, but then the drying and fixation artifacts must be considered. In principle, Method 3 could be done with staining with the light microscope, and the effects of the Heisenberg uncertainty principle averaged out by combining a very large number of measurements by computer techniques. Evidently, with video enhancements or the confocal scanning light microscope (CSLM) the accuracy could be improved. Unfortunately, so far these instruments have not been used for this purpose.

Method 4 measures the kinetics of permease incorporation. It has been little used (Koch 1973), so it will not be discussed.

The expected distributions for the case when all cells behave precisely alike are well known, the age distribution being first derived in the seventeenth century and the size distribution thirty years ago (MacClean and Munson 1961; Koch and Schaechter 1962). The formulae are shown pictorially in Fig. 3.6, and the equations are:

$$\Phi(a)da = 2\mu e^{-\mu a}\,da: \ 0 \le a \le T_2 \qquad \textit{Canonical exponential age dist.}$$
$$\Phi(V)dV = 2V_0/V^2\,dV: \ V_0 \le V \le 2V_0. \qquad \textit{Canonical exponential size dist.}$$

Here a is the age and V_0 is the volume of the newborn cell. $\Phi(a)da$ and $\Phi(V)dV$ are cell size distributions. Both of these distributions are called canonical because they apply only to this simplest case of growth and division in which cell cycles of all bacteria are identical. More general cases are considered below.

The interesting feature of the canonical exponential age distribution is that the frequency of cells about to divide is half as great as the frequency of newborn cells. This 2-fold factor in the age distribution arises for the reason that one cell divides to become two cells. There is a 4-fold factor when the same comparison

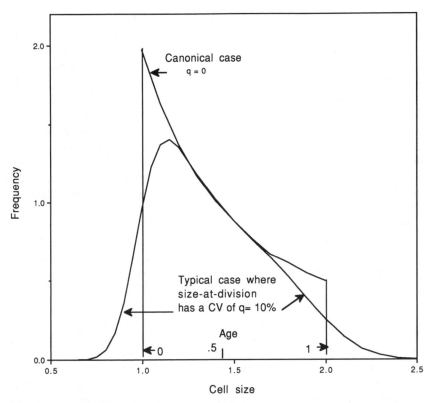

Figure 3.6. Distribution of cell sizes: Exponential law. The population is assumed to be asynchronous and in balanced growth, and biomass synthesis throughout the cell cycle is assumed to be exponential. The curve marked $q = 0$ is for the case where all cells divide when they reach exactly the same size. The curve marked $q = 10\%$ applies to the case where the size of cell in the act of division is normally distributed with a coefficient of variation of 10%. Both distributions are calculated for the case where every cell divides to produce two identical daughter cells. The average size-at-birth of a cell is taken as 1 and the mean size-at-division as 2.

is made for the size distribution; i.e., the frequency of cells that are about to divide is one-fourth as great as the frequency of newborn cells. The extra factor of 2 arises from the general nature of distributions. Distributions specify the fraction of the items that are classified within a constant interval of the abscissal scale. In the latter case, this is the fraction of cells that are between volume V and volume $V + dV$. For exponential growth, the rate of increase with cell size, dV/dt, becomes faster as the cycle progresses. It will cause a cell to pass twice as fast through a fixed range of cell sizes when near division than moments later when it has divided into two smaller, newborn, more slowly growing cells.

Canonical distributions for other candidate "growth laws" can be derived.

Consequently, any of these three methods could be used to establish unequivocally a "growth law" if precise measurements could be made and if each bacterium was born with the same precise size and grew exactly according to some explicit, but initially unknown, law and then divided precisely into two daughters in the same time and at the same size as all other bacteria in the same culture. So far no microorganism has been found that divides at very precise sizes and/or ages. And therefore, not surprisingly, the data obtained with well-studied organisms do not fit the corresponding predictions for exponential, linear, bilinear, or several other chosen forms for the "growth law." This is to say that because cells do not divide at a precise size and age, the variation confounds the establishment of growth kinetics. Clearly, no matter what the growth law, if variation in cell division were large, the size distribution would be broad and nearly the same for any growth law.

In addition to the above problem, there are problems with the very basic concept of a "growth law." First, the rates of the cellular processes could vary from cell to cell or from second to second in either a systematic or a stochastic way. Variations in the biological properties of cells of the same size would be an important part of growth. Consequently, the complete "growth law" should cover statements on the statistics of growth as well. The unabridged "growth law" should thus include the fluctuations as well as the mean rate of growth of the absolute mass, volume, length, or any extensive property of the cell as a function of cell age or, better, cell mass. Better still, the letter of the "growth law" should include the mechanism generating such variations.

A second problem in the study of bacterial growth has to do with the variables used to specify progress through the cycle. I used above the age of the cell since its birth or the cell's current size. For the canonical case, age, mass, volume, and effective length could be used equivalently. But not if there is variability in the duration of the cell cycle. In some circumstances the age from birth is inappropriate. Koch (1977), Koch and Higgins (1982), and Cooper (1991) have argued that time before division ("backwards age") is appropriate in some cases and is particularly relevant in those cases where the time to replicate the chromosome and the time between the end of replication and division are relatively constant compared with fluctuations in the period before chromosome replication and in the age-at-division.

It may be that the size of cells at division varies subject to trivial chance fluctuations that are irrelevant to other aspects of the cell cycle. For example, it takes a time, that may be variable, for two sisters cells finally to become unstuck after they are in fact two physiologically separate individuals. Such variation would not cause a problem, if the cell passed through many observable and identifiable phases and if growth were related to the identifiable phases. In the study of embryonic development of higher organisms, similar problems arise, and workers in these field have established morphological criteria for a series of stages that are more relevant for development than is clock time. A parallel development is essential to our further understanding of the cell cycle process in prokaryotes. Then the data would be organized not in terms of age or size, but in terms of observable stages or phases. Such distinctions may be possible in the future with autoradiography and/or gene expression markers and flow cytofluorography, but they are not available now. Alternatively, it is likely that the initiation of chromosome replication is the important event in the cell cycle and time should be measured from that event. Without this ability to identify phases, there is an inherent blurring in experiments that use either forward or backward age. If it can be shown that if cell mass is a precise measure of progress through the cell cycle, then the phase of the cycle should be reflected by cell mass. Certainly current thinking is that cell biomass is much more closely related to the trigger of initiation of chromosome replication (Donachie 1968) and of cell division (Koch and Schaechter 1962) in *E. coli* and in other organisms (Tyson 1985). Consequently, for the present, experiments based on cell size are to be preferred over those based on time, but better measures may be forthcoming.

History of the Theory

Although the theory and the experiments in this field were often parts of the same papers, it is convenient to divide the two. Further historical insight can be gained from Koch and Schaechter (1962), Burdett and Kirkwood (1983), Cooper (1988, 1991), Tyson and Diekmann (1986), and Koch (1993b).

Statistical theories were proposed by Rahn (1931–32) and by Kendall (1948, 1952), who imagined that genes replicated at random and that cells divided only when the cell had replicated a complete set of parts. Rahn pictured independent replication of each cellular "gene," whereas Kendall's modification was to imagine that the "genes" had to replicate in a particular order and that for a particular "gene," its opportunity to replicate did not start until its predecessor had replicated. Neither scheme makes much sense from our modern knowledge of biology; both genes and processes are sufficiently numerous to make small-sample statistics irrelevant. These early attempts to account for the observed variability of the age-at-division of cells were influential, however, in leading to concepts in statistics and to the study of the cell cycle.

Powell (1955, 1958) was the first to consider the mathematics of growing populations from a microbiological perspective, but it was Collins and Richmond (1962) who adapted the mathematics of population growth, originally developed for actuarial practice, to the case of balanced growth of microorganisms. They derived a relationship between the rate of growth of cells of any given size and three observable distributions of cells: those engaged in division, those in the act of being born, and those in every phase of the cell cycle. This last distribution of the entire observed population is called "the size distribution of the extant population." Their calculation combines integrals of these three distributions (the extant population, λ; the momentary distribution of dividing cells, ψ; and the momentary distribution of newborn cells, ɸ) to calculate the growth rate V_x of cells of size l_x. The *Collins-Richmond* equation, using their symbols (except using μ for the growth rate constant), is:

$$V_x = \mu\{2\int_0^{l_x} \Psi(l)\, dl - \int_0^{l_x} \phi(l)\, dl - \int_0^{l_x} \lambda(l)\, dl\}/\lambda(l_x). \qquad \textit{Collins-Richmond}$$

It we are interested in cells of a given size l_x, then each of the three distributions must be integrated from size zero up to this size. This equation indicates that when the algebraic sum of the integrals is multiplied by the growth rate constant and divided by the frequency of cells of the class of interest, the mean rate of growth of cells of that cell size class is obtained.

The *Collins-Richmond* equation applies to any parameter of growth (length, volume, dry mass, DNA, etc.). Of course, the simplest relationship is obtained if the observer chooses to measure the variable most important for the organism's physiology. The mathematics was extended quickly, and Painter and Marr (1968) and Marr, Painter, and Nilson (1969) gave limitations, generalizations, and extensions. Work from Marr's laboratory served the role of disseminating the "growth law" problem and the power of the Collins-Richmond procedure. A very important feature of this procedure is that it generates the mean growth rate of cells of particular sizes no matter what variations and fluctuations occur. So if cells of a particular size class are divided, by their physiology, into classes with different growth rates, the Collins-Richmond algorithm will produce the weighed average of the growth rates. This is an advantage for some purposes and a disadvantage for others.

Almost simultaneously Koch and Schaechter (1962) published a paper that was less elegant but more powerful: it simply reversed their strategy. It started with the assumption that the statistics of the cell cycle could only be defined given specific assumptions about four different aspects of the cell cycle; these qualitatively different aspects are listed on the left-hand side of Table 3.1. Each of the four general assumptions of the cell cycle has to be assigned a specific form in order to calculate the properties of growing cells that reproduce by binary fission. The specific forms that seem to apply to some rod-shaped enteric bacteria are shown on the right-hand side of Table 3.1. Thirty years after this formulation, this set of specific aspects still seems applicable to Gram-negative rod-shaped bacteria. A key point was the idea that the significant form of the "growth law" for *E. coli* was the exponential default applied identically to every cell with no variability from cell to cell. With the experimentally observed values for the CV of the size at division and the CV of the evenness at division, the default "growth law" generated the distribution of sizes of the cells in the growing population quite exactly. The mathematical limitations in the original paper were discussed and some of them removed by Powell (1964), whereas other (mathematical) limitations remain that require analysis by computer simulations. The original

Table 3.1. Postulates needed to model the population size structure of an organism by the Koch-Schaechter procedure

General and Specific Growth and Division Laws	
Areas for Which a Postulate is Needed	Specific Form of Postulate for *E. Coli*[a]
Biomass growth	Exponential growth[b]
Division control	Critical mass at division[c]
Variation of division size	$CV \leq 10\%$[d]
Variation of the evenness of division	$CV \leq 10\%$[e]

[a]This is the "default" model; it assumes that identically the same value of μ applies to all cells and that it does not vary from second to second.

[b]Growth is exponential with the exception of replication of the chromosome and the formation of cell envelope. The synthesis of DNA is discontinuous and chromosomes may be replicating at multiple forks. The envelope is replicated in a manner to contain the newly synthesized cytoplasm. The rate of envelope formation is dependent on the synthesis of biomass and on the shape and proportion of side wall versus pole being formed.

[c]Because the *CV* of the size-at-division is usually about half as big as the *CV* of the age-at-division, the default model assumes that the cell controls the size-at-division and not the age-at-division. There is good reason to believe that the cell measures some cell component that is representative of the total biomass and that the cell had initiated at an earlier time in the cell cycle the global process leading to division; perhaps this is the initiation of chromosome replication.

[d]The actual size-at-division has a *CV* for many strains of *E. coli* that is less than 10%. This follows from extensive measurements available from many studies carried out in Nanninga's laboratory at the University of Amsterdam. These results are consistent with other less comprehensive studies. Even these reported values may be overestimates because of difficulties in extrapolating from the size of cells at a stage near division and because slightly thinner cells may be longer as they divide (as pointed out by Cooper).

[e]Division of the cytoplasm is quite even; for a "well-behaved" strain; e.g., B/rA, it has been measured (Trueba 1981) to be 4%. Again, this may be slightly high for reasons noted in footnote d.

theory for Gram-negative enterics, however, is fairly generally applicable because the size distribution of many kinds of prokaryotic cells at division is quite narrow (see Tyson 1985), division in many strains occurs very near the middle of the mother cell, and cellular growth appears to be close to exponential. Of course, for those that divide asymmetrically and have poor control of the size-at-division, the predictions based on the right side of Table 3.1 are irrelevant.

The Koch-Schaechter approach can be applied to any set of specific assumptions, at least by computer simulations. It is considerably more versatile than the Collins-Richmond approach in that it can be used to predict not only the extant distributions of cell sizes, ages and backward ages, and life remaining, but also the correlation coefficients of the age-at-division values of mothers with their daughters and of sisters with their sisters. The predictions of the canonical exponential model generated by the assumptions on the right side of Table 3.1 are that the correlation coefficient of the age-at-division times of mother with daughter should be −0.5 and that of sister with sister should be +0.5. Experimental values are usually somewhat more positive than these values. Such values can be taken as evidence against this specific model, and therefore as evidence against one or more of the four component postulates; alternatively, they can be taken as evidence for experimental difficulties, such as the failure to maintain uniform conditions.

Smith and Martin (1973) published a paper that stimulated a flurry of interest in the cell cycles of both eukaryotes and prokaryotes. Like the Koch-Schaechter formulation for *E. coli*, their model contained a deterministic and a stochastic element. They suggested that part of the cycle was determined by the length of time required to carry out certain essential processes of the cell cycle. The total duration of the deterministic parts was fixed and applied precisely to all cells of the population. They assumed that the remaining portion was distributed in a Poissonian manner and followed a negative exponential distribution. Later this was amended to include two stochastic steps (Brooks *et al.* 1980). although much work was stimulated by the Smith and Martin model, it no longer appears to be valid and has had little permanent impact upon the study of prokaryotes or eukaryotes other than reawakening interest in the subject.

An important but very simple insight came from Donachie (1968) who concluded that cells had the same mass at a key phase of the cell cycle no matter what the growth rate was. This was deduced by considering the mean size of bacteria growing in different media. Assuming exponential growth at the cell level, he concluded that *E. coli* cells, no mater how fast they were growing, when they achieved a certain size, triggered processes (probably the initiation of chromosome replication) that resulted later in cell division.

History of Measurements

Almost as soon as cinematography through the phase microscope became possible, Baynes-Jones and Adolf (1933) were at work measuring the kinetics of the growth of length of bacterial cells. Others without time-lapse movie cameras watched cells growing on microscope slides for endless hours. Rahn (1931–1932), Kelly and Rahn (1932), and Powell and his coworkers made measurements of the age-at-division. Following this, Schaechter *et al.* (1962) prepared high-contrast microphotographs to follow both the division of individual cells together with the division of the nucleoids within the cells and to establish the distribution of the timing of cell division. This work employed a trick devised by Mason and Powelson (1956): if the surrounding medium has an appropriately high index of refraction, then the nuclear mass, which has a higher index of refraction, can be seen and distinguished from the cell cytoplasm. Shortly after that Errington, Powell, and Thompson (1965) used the Dyson eyepiece to improve measurements on single living cells and followed the growth of individual cells. As mentioned, although the age-at-division measurements generated in these ways are generally valid, attempts to measure the "growth law" directly by measuring cell sizes are not valid because of the resolution of the phase microscopy. As this was realized, alternative approaches were sought.

Deduction from Size Distribution of Balance Population

Because of the theoretical and experimental work of Collins and Richmond (1962) described above, and the experimental work of Harvey *et al.* (1967) that followed, interest in the field was stimulated. Both groups found growth to be roughly proportional to size for the central bulk of the cells but not for the large and small cells. This is very weak and limited support for the default model. The Collins and Richmond results are limited because they counted fewer than 1,300 cells, and very many organism are needed for sufficient accuracy. In addition, the resolution of the light microscope and the choice of length rather than volume or mass as a parameter further limited their relevance to cell growth processes. The work by Harvey *et al.* (1967) was limited by the precision of their specially modified Coulter counter.

Measurement of the length of individual cells has played an important role because lengths can be measured, if only roughly, in living cells photographed under the phase microscope, whereas widths and precise shapes cannot. Lengths and widths can be determined more accurately in samples of dead cells taken from living, balanced cultures and fixed (with incumbent artifacts) for the electron microscope. Using only length measurements leads to significant limitations; measurement of the width in addition allows the computation of cell surface areas or volumes. Although volumes are the quantities relevant to the growth of bacteria according to the default model, estimation of volume increases the uncertainty because of the large contribution of the width measurement to the overall error. Moreover, in many studies in which the width is measured, the actual shape of individual cells has been neglected. Accurate assessment of cell volumes is relevant to the determination of the growth rate throughout the cell cycle in order to test hypotheses whether the cell is primarily controlled by length, surface area, or volume increase.

As a good approximation, in the extensive studies by the group at the University of Amsterdam, it was assumed that the "rod-shaped" bacteria are right cylinders of constant width and are capped with hemispherical poles. For many rod-shaped organisms this is not a bad first approximation, but certainly not an accurate second approximation. Even though it is considerably more work, it is possible to digitize the electron microphotographs and more completely and accurately estimate the cell surface areas and volumes as has been done in work emanating from the laboratories of Higgins and of Burdett. Of course, there is an inherent limitation in that the photographs do not contain information about the depth measurements normal to the plane of the photograph. Consequently, the results are limited by the validity of the assumption that the cells have cylindrical symmetry. Therefore, some mathematical idealization has been needed in order to calculate the surface area or the volume of the cell from the two-dimensional photographs.

It is a long leap from the linear dimensions of a living cell to the dry mass in that cell. Even assuming that the volume of a cell has been accurately measured, a knowledge of the dry weight concentration of that cell is needed. However, this inference can be made for Gram-negative rod-shaped bacteria because the density of the living cells remains constant throughout the cell cycle. Demonstration of this was an important contribution of Kubitschek (1987). Constancy is inferred from the narrowness of the distribution of cell buoyant densities from growing cultures. Buoyant densities are measured by centrifuging a population of bacteria through a gradient produced with some dense material, often Percol®. Percol® is manufactured from very fine particles of silica (sand) coated with polyvinyl pyrrolidone to prevent them from adsorbing to the wall of the bacteria. Because the buoyant density of the cell is determined by the density of cell components weighted in proportion to their abundance, constant density implies that the water content remains constant. (Cells are usually 70% water.) Most cell constituents composing the dry mass have a higher density than water. The observed constant buoyancy is sufficient to justify a proportionality between volume and biomass. (The constancy of density throughout the cell cycle also implies that the proportion of small molecular weight compounds remains constant, because they determine the internal osmotic pressure—see Chapter 5—and therefore the cell's water content.)

The data accumulated by the University of Amsterdam group have been analyzed by Grover et al. (1987) and by Koppes et al. (1987) employing refinements and extensions of the Collins-Richmond theory. The Collins-Richmond approach has also been generalized by Kirkwood and Burdett (1988) and employed by Burdett, Kirkwood, and Whalley (1986) to study the cell cycle of *B. subtilis*. Unfortunately, neither group manipulated its data in terms of cell volumes to test the validity of the exponential versus linear model.

Methods for Measurement by Pulse Labeling

The general approach of considering size only or some aspect of it was superseded by tracer techniques. Radioactive precursors of high specific activity became available in the 1960s; these provided probes to study cell growth more precisely through pulse-labeling of cells. Not only could the formation of major components such as proteins and RNA be studied, but, particularly with thymidine, the replication of the chromosome within bacteria could be correlated with growth and division. However, the use of tracers required some analytical way of separating cells in different phases of the cell cycle. Above, this approach was called the differential method, and given the designation Method 3. Such experiments are very much more powerful because they are differential; i.e., they directly measure the rate of growth so that there should be a continuous increase, a 2-fold increase altogether, over the cell cycle for the exponential model and no increase for the linear model over the majority of the cell cycle, but possibly with an abrupt and temporary 2-fold increase a short time before division.

The "Baby Machine" and the Chromosome Replication Cycle

Before considering experiments defining the "growth law," let us consider the technology and results concerning the regulation of DNA synthesis obtained with the membrane elution apparatus. This apparatus is familiarly called the "baby machine." With it a culture of cells in balanced growth is pulse-labeled and then quickly filtered through a large membrane filter. Immediately the filter is inverted and conditioned medium passed through the membrane at a constant rate and fractions collected. All these steps are carried out at constant temperature with aerated medium. Adequate and uniform flow through the various parts of the filter is presumed. Some bacteria stick to the filter and remain and grow normally. When they divide and the daughter cells separate, ideally one remains attached to the filter while the other is carried away by the flowing medium as a "baby." The fractions are examined for cell titer and radioactivity. This type of procedure is referred to as the "backward" method; it is useful in estimating the time to the next division rather than the time from the previous division. The first cells to elute are descendants of the cells in the last phases of the cycle at the time of labeling.

Helmstetter and Cooper (1986) were able to show clearly that there were discontinuities in the release of radioactive material after the cells had been labeled with tritiated thymidine. They found that the time of the division measured back to the time of the cell initiated chromosome replication was independent of growth rate; for an *E. coli* strain, this time was about 60 min. They found that chromosome replication took about 40 min and was followed by a 20-min period before division took place. This meant that at rapid growth rates the cells had multiple replication forks so that the bacteria could divide in less time than they could replicate the chromosome. The position of chromosome replication within the cell cycle, counting birth as age zero, is often specified in terms of B, C, and D, where B is the gap before chromosome replication (the equivalent of the G_1 period in the eukaryotic cell cycle; it can be positive or negative and appears to be very variable), C is the chromosome replication time (the equivalent of the S period), and D is the gap before division (almost the equivalent of the G_2 period) and T_2 and I are symbols used for the doubling time; i.e., $T_2 = I = B + C + D$.

Resolution of the Separation Techniques

In addition to the membrane elution technique, a variety of other techniques have been used for the "growth law" problem; these are listed in Table 3.2. Each

Table 3.2. Published differential experiments to determine the law of biomass increase[a]

Label	Separation Technique	Size Parameter	Conclusion	Author	Date
Uracil	Vel. Grad.	Position	Continuous	Manor Haselkorn	1967
Amino acids, glucose, etc.	Vel. Grad.	Volume	Linear	Kubitschek	1968[b]
Leucine	Autoradio.	Length	Exponential	Ecker Kokaisl	1969
Uracil	Memb. Elut.	Backwards age	Exponential	Dennis	1972[c]
IPTG	Vel. Grad.	Position	Exponential	Koch	1973[d]
Leucine	Memb. Elut.	Backwards age	Exponential	Cooper	1988

[a]Experiments are listed chronologically in which a culture, guaranteed to be a balanced growth, was labeled for a short period of time, usually much less than 10% of the cell doubling time. The bulk radioactivity (or inducer) was removed and the cells fractionated by a variety of techniques into different phases of the cell cycle and the radioactivity activity determined.

[b]Kubitschek used many radioactive amino acids and other precursors for pulse labeling. Then he treated the cultures with formaldehyde and washed the cells well before separating them by the Mitchison-Vincent velocity gradient procedure. Although he thought that he was measuring rate of uptake into pools and hence uptake capacity, it is certain that in many, possibly most, cases the measures more nearly reflected the rate of macromolecular synthesis.

[c]The uracil used in this ecperiment also became converted into deoxycytosine and thymidine derivatives and therefore labeled DNA as well as RNA. This complicates the analysis greatly, and although Dennis argues that RNA accumulation must have been continuous and exponential, the force of this conclusion is weakened because of these metabolic conversions. On the other hand, his experiments are very much strengthened because of the controls he ran and because he could use the DNA data to establish the degree of synchrony of the cells because DNA synthesis is all-or-none.

[d]This experiment is actually not the Method 2, but the Method 4 kind. If the linear model were right, the experimental results should have been: no permease inserted in any size class except that one containing large cells just about to divide. Actually, the permease levels per unit biomass were constant in all size fractions consistent with exponential insertion of the permease into the medium.

method, however, has strong limitations and, in fact, all are quite poor because the size-*CV* of a synchronized culture should be zero and instead the experimental values approach those expected for a totally nonsynchronized culture (see Table 3.3). The membrane elution procedure does not, in fact, produce a population of cells that are purely "babies" because the size distribution of the freshly eluted cells is much broader than expected from the range of sizes observed in the electron microscope for dividing cells. In spite of the blurring, important conclusions about chromosome replication were drawn; these were proved to be correct by later studies (Skarstad *et al.* 1985). It is worth commenting at this point that discriminating the all-or-none synthesis of the chromosome requires less resolution than does the problem of establishing the kinetics of cell growth.

Turning to the other techniques, in the sucrose-density velocity-gradient centrifugation technique, labeled cells are concentrated and overlaid on a sucrose-

Table 3.3. Comparison of the separation techniques used in differential experiments with those expected from the size-at-division measurements[a]

Separation Technique	CV (%)	Skewness	Author
Sucrose density gradient— Mitchison-Vincent[b]	17		Kubitschek
Membrane elution— baby machine[c]	22.7 22.0	0.852 1.57	Helmstetter Cooper
	Exp *Linear*	**Exp** *Linear*	
Theory for balanced extant populations[d]	22.6 22.0	.568 .367	Koch
Experimental results for dividing cells[e]	~8		Nanninga

[a]Ideally, a separation technique should yield cells of a single size or age or (for the case of the membrane elution technique) the cells released at any time should have been in a single phase at the time of labeling and thus have a CV of zero. However, this is far from the case, as deduced from the experimental results of the various authors. In fact the reported CVs were much closer to those expected from an unsynchronized culture in balanced growth, which, as shown in the bottom part of the table, are in the range of 22 to 23%. The skewness is a measure of the asymmetry of the population. This should be zero for a Gaussian population.

[b]The separation technique is the centrifugation of a concentrated sample of cells through a sucrose gradient in a bucket rotor. It is limited by mixing during acceleration and deceleration and in the sampling procedure. In many experiments only the small cells on the top of the gradient are sampled. These should have a small CV as sizes (see the rest of the table); however they have almost, but not quite, the same CV of the original unseparated populations of cells.

[c]The membrane experiments were carried out by filtering pulse-labeled cells and then inverting and flushing. Various laboratories have measured the size distribution of cells coming from the membrane and have uniformly reported that there appear to be two populations. One has the narrow distribution expected from the sizes of dividing cells observed in the electron microscope (see footnote e). The other half has the distribution expected for cells of the range found in the original asynchronous population.

[d]The theoretical distributions were calculated by computer using the Koch-Schaechter method for the "default" system of assumption shown on the right-hand side of Table 3.1. The calculations were for the CV of the size-at-division = q = 10%, but for division into two equal-sized daughters. The value q = 10% is high, it should be about 8%, and actually the value of the CV of the evenness of division is about 4%. However the 4% and the 8% should combine to make the values reported here quite applicable.

[e]The experimental values that appear to be most precise are those obtained in the laboratory of Nanninga at the University of Amsterdam and reported in a series of papers. They were obtained by fixing and photographing cells in the electron microscope and examining the photographs by computer.

gradient pre-formed in a centrifuge tube. The tube is briefly centrifuged in a centrifuge that rotates very smoothly. After gradually stopping the centrifuge, samples at different depths are removed. The radioactivity is counted and the cell numbers determined and their volumes analyzed in the Coulter counter. The technique is limited by stirring during the acceleration and deceleration of the rotor and by mixing during the mechanical process of separating fractions.

In the autoradiography technique, labeled cells are washed to remove radioactivity that is not associated with the macromolecules. The cells are then covered with photographic emulsion and stored in a cool, dry, dark place. Later the film is developed, the silver grains overlaying the cells are counted, and the length of the cells is measured. Autoradiography of pulse-labeled cells is limited by the variable efficiency of the yield of silver grain above different regions of the cell. In the autoradiography experiment listed in Table 3.2, length instead of volume was used as the measurement of progression through the cell cycle.

Critique of Differential Pulse-Labeling Studies

The reason for reviewing the membrane elution work on chromosome synthesis at this point is that it is this technique that Cooper (1988) has used with a pulse of tritium-labeled leucine to label cellular protein. He has concluded that protein synthesis is unequivocally exponential during the cell cycle of *E. coli*. Thus, he confirms what Ecker and Kokaisl (1969) had found with the same label and contradicts what Kubitschek had found with labeled leucine and other labeled cell metabolites. Most other early studies listed in Table 3.2 also concluded that growth was exponential.

Cooper and Kubitschek each raised a number of criticisms against the work of the other. Some of the criticisms may be valid, but I feel they are not as severe as the ones now to be leveled at work from both laboratories. I do not believe that the different conclusions by the two workers depend on their choice of different separatory and analytical techniques; each of the available analytical techniques has poor resolution that blurs phases of the cell cycle. Without these difficulties both techniques would be adequate to find the difference expected between the exponential and linear models.

In studying the papers and in conversation with Kubitschek, I uncovered a flaw in his experiments, which is discussed more extensively in Koch (1993b). Kubitschek had used the predictions of the linear model to estimate the average birth size when testing the linear model against experiments, but then had used the same estimate of the average birth size when testing the exponential model, and he should have used a different estimate for the next model he was testing.

In studying the papers and in conversation with Cooper, I have uncovered a serious flaw in his experiments as well as all the other differential experiments (Table 3.2). It too is discussed in more detail in Koch (1993b). All the differential experiments so far reported in the literature can be faulted because of lack of knowledge concerning the pulse-labeling kinetics. All experiments listed in Table 3.2 were conducted by adding a small amount of labeled compound to the growing batch culture. The cells before this time were manufacturing their own metabolites. The isotopic metabolite had to be pumped through the cell membrane, act as feedback inhibitor of the *do novo* pathway, mix with and replace the endogenous pool, and be incorporated into macromolecular form. Consequently, the distribution of radioactivity from short pulses with the cells of the population would be a measure of the rate of protein synthesis only if the transport were very fast, if inhibition of the *de novo* pathway were immediate and complete, and the content of the cellular pool infinitesimal. It would, in fact, not be unexpected if cells with a higher surface-to-volume ratio were able to block the *de novo* pathway more quickly or extensively. No controls concerning these possibilities were part of any of these experiments. As most of the experiments cited in Table 3.2 concluded that growth was exponential, we are at a strange impasse: the experimental evidence for exponentiality of protein synthesis appears less solid now than before. The rate of RNA accumulation was studied by Dennis; his work included controls about the pool kinetics and, although there are complications (see Koch 1993b), his studies are quite convincing that total RNA accumulates exponentially.

Tests of Linear Uptake

Another approach was to measure the rate of uptake of labeled precursors into cells in various phases of the cell cycle. Kubitschek (1968) took the smallest cells separated by velocity-gradient centrifugation, allowed them to grow, and measured the uptake of many labeled substances with time. There are many problems with this approach; e.g., cells may be thrown into lag and the fraction of small cells may preferentially contain dead or dying cells. Moreover, there are severe limitations to the interpretation that measured uptake kinetics reflect the number of uptake sites.

One experiment listed in Table 3.2 belongs to yet a different category; it really is a Method 4 experiment and of the class of differential experiments of Method 2. In this experiment the *lac* operon was pulse-induced with isopropyl-B-D-thiogalactoside (IPTG). The cells were separated by velocity centrifugation. All fractions were found to have the same level of galactoside permease per unit dry weight. This experiment was performed to provide a test of Kubitschek's linear model, which predicts that transport sites will be introduced only during a very narrow portion of the cell cycle. Consequently, the resolving power of this experiment is much larger than the other differential experiments because the specific linear model predicts zero incorporation in most of the fractions. This experiment cannot be taken as critical evidence for the exponential model, however, because the size distributions of the individual fractions were not measured. It is, however, very strong evidence against the linear growth hypothesis, with the proviso that galactoside permease is typical of all membrane transport proteins, a proviso for which there is no evidence.

Conclusions About the Growth Law

Discussion of the growth law has been examined in some detail here and in more detail in Koch (1993b); the results of considering all the published experiments is discouraging. A definite knowledge of the kinetics of processes during the cell cycle should have been obtained. In fact any of the four methods with current advances in their fields are presently capable of critical exploitation on this issue.

First, single cells had been followed during growth under the phase microscope in the past. In the future, with the emergence of the confocal scanning light microscope (CSLM) and computer-linked video microcinematography the resolving power and accuracy has been sufficiently increased to resolve the "growth law" experimentally by Method 1. This would require computer manipulation to calculate the volume of the cell from the length. Differential isotope techniques (Method 2) can be refined and controls run to establish the kinetics of pool turnover to allow valid conclusions, especially if refinements of the separation techniques are devised, and implemented, and size distributions included in the experimental report. Recently a breakthrough has occurred in the development of the inverted microscope flow cytometer (Skarstad et al. 1985). With the development of the technology to measure bacteria and their content of DNA, etc., much greater accuracy can now be obtained and very large samples of individual cells can be examined. Because light scattering can be used to measure cell biomass, Method 3 can be used to analyze the cell size distribution utilizing the Collins-Richmond technique and its various mathematical improvements. Various fluorescent probes can be used and it is hoped that they can be used to distinguish sister cells that remain attached from cells that have not divided. Method 4 could be refined and run in a well-controlled manner. If the number of transport proteins functioning in the cell membrane were known for cells in various phases of the cell cycle, this too could be very strong evidence for a growth law. Any of these four methods, consequently, could be used and thereby avoid the artifacts of older experiments. These could definitely answer one of the more important questions in bacterial physiology: the "growth law" question.

Hope For Future Studies

A bacterial cell is an organism; it is very complex and sophisticated, as is any organism. We can begin to understand it by studying the "balanced growth" of cultures with large numbers of individuals. The study of large numbers of cells from a growing culture washes out the variability among cells and the steady-state growth conditions lead to constant composition and rates. A simple exponential growth law results and can be used by the microbiologist to define the doubling time and specific growth rate, μ. An extension of this logic leads to the finding that of the thousands of genes, their gene products, and the rates of functions of these products, only two reciprocal processes are controlling (RNA begetting protein and protein begetting RNA) and all the rest are but appendages.

For most prokaryotic organisms, macromolecular growth is basically exponential (or such a close approximation that it cannot be differentiated). A clear exception is the process of DNA synthesis during chromosome replication. The DNA itself contributes little to the cell's mass or volume. Although the processes of chromosome replication, nucleoid separation, and cell division are discontinuous, the precise details do not matter for the question of biomass increase. No matter what the duration of the B, C, D, and T_2 times, growth would be effectively exponential both at the populational and at the single cell level. Similarly, the details of the attachment of nascent peptidoglycan and the enlargement of the cell envelope are almost irrelevant to the kinetics of increase of the total mass or volume, because wall constitutes only a minor fraction of the cell's dry weight.

Possibly for all free-living organisms that do not have a circadian rhythm, the cellular algorithm is to grow as smoothly (and efficiently) as possible and not halt the major chemical work of protein synthesis by arranging that there is no unnecessary limitation in the steps of the cell cycle (the constant efficiency hypothesis). This rule should apply only while the cell is in balanced growth. Of course, the organism should be wary of changes in the nutritional environment and have strategies to cope with either an improvement or deterioration of the ambient conditions. Probably genes for such precautions occupy a considerable amount of the genome.

As one tries to ferret out the rules of growth in adequate and constant conditions, the major problem results from the variability in the sizes of dividing cells, although it is actually very small. These fluctuations affect all kinds of processes. They may not be significant to the cells, but they certainly are for the investigator. In addition to variability in the normal processes of growth, there is the added problem of cells dying, dead, or resting, of cells that have failed to divide at the appropriate stages and are now too big to grow as efficiently, or of cells that divide to become minicells or their metabolic equivalent. Although these represent a very small minority of the cells, their influence on the analysis of certain types of experiments described above can be large.

It can be hoped that in the future, the cell cycle in bacteria will reveal further details beyond the default exponential "growth law." Alternatively, convincing proof of the generality of Hinshelwood's logic may emerge. For this reason I feel the analysis of the population distribution, whether by the Collins-Richmond approach with the novel mathematical extensions or the Koch-Schaechter approach, will be important in dissecting more deeply the cell cycle of prokaryotes. The problem of experimentally establishing the "growth law" turned out to be harder than imagined, but now there are many viable approaches, any of which is capable of giving an unequivocal answer.

4

The Synthesis of Functional Bacterial Wall

KEY IDEAS

Murein, or peptidoglycan, is the covalently linked fabric forming the exoskeleton, or sacculus, of most eubacteria.
Murein makes it possible for bacteria to resist osmotic challenges.
The wall fabric is strong and elastic.
Murein has an exotic sugar and unusual amino acids permitting crosslinks.
Murein is formed by multiple linking of disaccharide pentapeptide units through bonds between hexoses and between peptide groups.

Peptide bonds are made with specific synthases—by a process not akin to protein
   synthesis.
Precursor units are pre-energized so that they can be crosslinked after externalization.
Passage through the cytoplasmic membrane requires a lipid carrier.
Autolysins are necessary for enlargement of the sacculus.
Before stress-bearing covalent bonds are cleaved, new covalent bonds are formed to
   accept the stress.
Turgor leads to stresses, which in turn favor cleavages and enlargement.
Murein addition and cleavage patterns establish cellular morphology.
Turnover is essential for the elongation of the side wall of the Gram-positive rod.
Turnover may be essential for Gram-negative rod side wall elongation.
Turnover of poles is much slower than that of side walls.

The Crosslinked Fabric-Like Structure of Bacterial Wall

The chemical nature of the wall allows it to be a two-dimensional stress-resistant
fabric forming the exoskeleton of most eubacteria. The key feature is that it is
composed of functional units containing both carbohydrate chains and peptide
chains such that a basic unit is capable of being connected to other units in at
least three places. The formula for the peptidoglycan unit structure of *Escherichia
coli* and *Bacillus subtilis* is shown in Fig. 4.1. This specific formula represents
the most common structure in eubacteria, and is similar to the basic units used
by all eubacteria, except for the wall-less mycoplasma. Many of these units are
linked together to form the sacculus surrounding the growing bacterium. Thus,
carbohydrate chains of many disaccharide units are formed via $\beta(1\rightarrow4)$ glycosidic
bonds; such chains are chemically similar to the cellulose of plants or more
similar to the chitin of insects. These substances are the major mechanical element
in the external skeleton of those organisms. Murein contains a very special sugar
(muramic acid) that has an appended D-lactyl group. The attached peptide,
for example, L-alanyl-D-isoglutamyl-*meso*-diaminopimelyl-D-alanyl-D-alanine
shown in Fig. 4.1, is quite different from peptides in proteins. This is due to
the D-amino acids, the linkage of glutamic acid through its γ-carboxyl group,
and the exotic diaminopimelic acid residue. However, most of the amino acids
are connected via their α-amino group to the α-carboxyl group of the adjacent
amino acid, as in normal proteins. The peptides of murein have properties
similar to those found in the fibrous proteins of mammalian skin or hair. But
peptidoglycans are profoundly different in their mechanical properties than either
fibrous carbohydrates or proteins. The fundamental chemical difference of having
both carbohydrate and protein features in the basic unit allows the development
of multiple covalent crosslinks to form two- and three-dimensional structures.
Both ends of the disaccharide are used to extend the carbohydrate chain and
there are three carboxyl groups and one amino group in the peptide through
which linkages can be formed. In the peptidoglycan unit structure of *Escherichia*

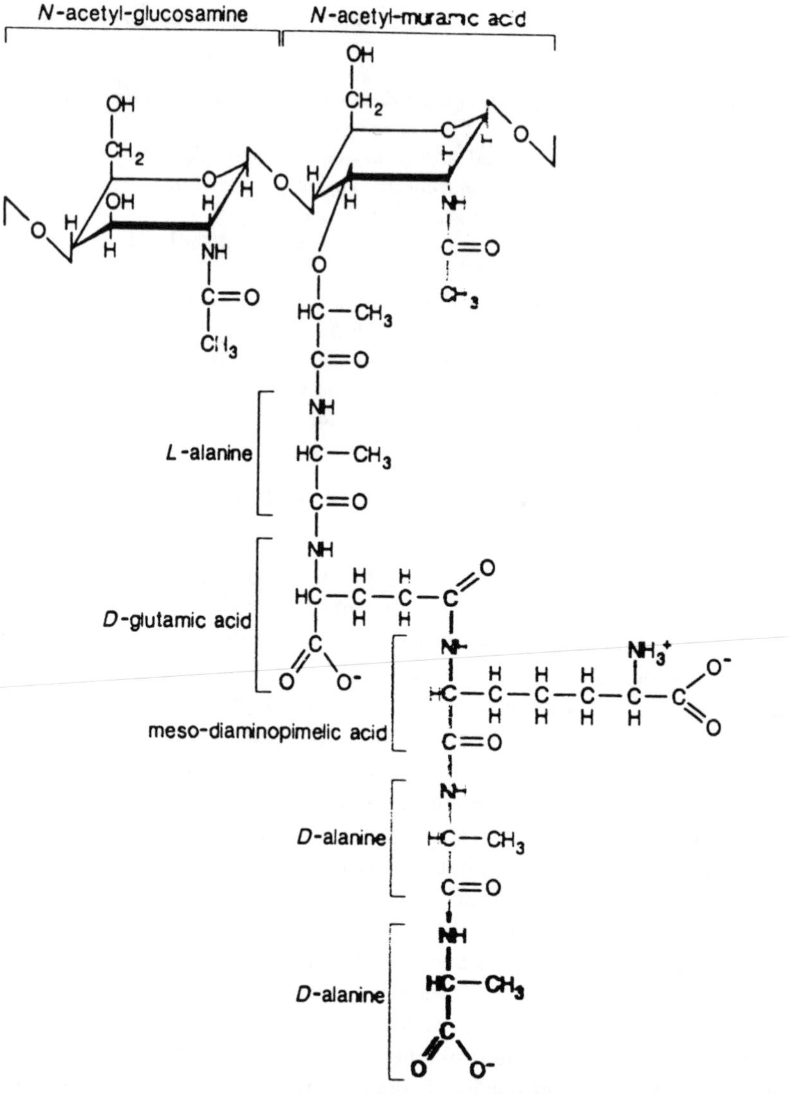

Figure 4.1. The structure of *B. subtilis* and *E. coli* peptidoglycan. The stress-resistant wall contains both carbohydrates and amino acid moieties. The carbohydrate *N*-acetyl-glucosamine occurs very widely; however, the carbohydrate *N*-acetyl-muramic acid occurs only in bacterial walls. It is unusual because it has a substituent that contains a carboxyl group, making crosslinking by amide bond formation possible. The peptide portion has some amino acids in the D optical form not found in proteins; one amino acid (glutamic acid) forms a peptide bond from its γ carboxyl group instead of the usual α carboxyl group; and one amino acid is the unusual compound, *meso*-diaminopimelic acid. This compound does not occur in proteins, but is an intermediate in the biosynthetic pathway for lysine in some organisms. Although only a pentapeptide disaccharide is shown here, the sugar units are joined end-to-end to form long, nearly straight glycan chains. The peptides create crosslinks in which the amino group of a *meso*-diaminopimelic acid in one chain replaces the terminal alanine of another chain that extends the fabric in the perpendicular direction.

coli and *Bacillus subtilis,* shown in Fig. 4.1, only the free amino group of the diaminopimelic acid and the carboxyl of the first D-alanine (after the second is removed) are used. The polymerized structure should be thought of as both a stress-resistant fabric or a fisherman's net. The two dimensions of the fabric covering the surface of the bacterium "feel" the stresses expressed due to the cell's turgor pressure, and the strength of the fabric is of critical importance.

The ability to crosslink new material after it is secreted through the cell membrane depends on two novel kinds of biochemistry in the precursor structure. The first is the attached peptide forming a branch off the saccharide chain. The carbohydrate chains are made of alternating units of $\beta(1\rightarrow4)$ linked residues of 2-acetamido-2-deoxy-D-glucopyranose (more commonly called *N*-acetyl-D-glucosamine and abbreviated GlcNAc) and 3-O-(D-1-carboxyethyl)-2-acetamido-2-deoxy-D-glucopyranose (more commonly called *N*-acetylmuramic acid and abbreviated MurNAc). GlcNAc is also found in the chitin of insects and fungi and is the second most abundant hexose on earth, but the MurNAc is found nowhere else in biological systems except in bacterial walls. Muramic acid is simply *N*-acetylglucosamine with an ether-linked lactic acid residue on the third carbon atom (3-O-D-lactyl derivative). During synthesis, the carboxyl group of the lactyl grouping allows the crosslinking to an amino group, usually L-alanine, although it may be serine or glycine. Then, by subsequent additions of other amino acids, it is elongated to become a peptide of five amino acids.

The second universal feature is that the peptide is able to form a crossbridge with another such peptide by a tail-to-tail arrangement. Usually this is accomplished by the presence of a special polyfunctional amino acid with one more amino group than is present in most naturally occurring amino acids. Typically, the diamino acid is *meso*-diaminopimelic acid, although it can be LL-diaminopimelic acid, L-ornithine, L-lysine, L-diaminobutyrate, or *N*-acetyl-L-diaminobutyrate. Any of these diamino acids provides an extra functional amino group to crosslink the head of one peptide belonging to one glycan chain to the tail of another peptide belonging to another glycan chain to form a linkage holding the fabric together. In some bacterial species, the third amino acid in the chain instead of the diaminopimelic acid shown in Fig. 4.1 can be L-glutamate, L-homoserine, or L-alanine. In these three cases, the tail-to-tail linkage is made, but depends on the extra α-carboxyl group in the D-glutamic acid, which in all species is the second amino acid in the peptide chain, with the formation of a diamino acid-containing crossbridge. In some cases a short bridging peptide is also inserted.

In every case, a membrane-bound transpeptidase attaches the amino group of one glycan chain to the carboxyl group of the penultimate D-alanine of a nearby peptide in exchange for the terminal D-alanine. The inclusion of an extra terminal D-alanine during synthesis is the feature that allows the peptide bond to form outside the cytoplasmic membrane. Inside the cell peptide links are formed using the energy of the high-energy phosphate bonds of ATP or GTP. These substances are not available outside the cell and it is necessary to store the energy in the peptidoglycan precursor when it is secreted through the cytoplasmic membrane. Thus, this transpeptidation step is critical to the formation of a bacterium's exoskeleton; it is the Achilles' heel of the wall growth mechanism and is the site of action of many useful antibiotics (see below).

Let us summarize the special chemical features of the bacterial wall. In the peptide portion, there are some D-amino acids instead of the L-amino acids commonly found in proteins. D-glutamic acid occurs in an uncommon linkage, i.e., through its γ-carboxyl instead of the α-carboxyl group. There are usually diamino acids that do not occur naturally in proteins. These features not only serve in wall formation, but also make the peptidoglycan resistant to many proteolytic enzymes that are widespread in nature.

Wall Strength Comes from Wall Structure

A unit thickness of bacterial wall is much stronger than the walls of higher plants. This extreme strength is due to its covalently bonded crosslinked structure, quite similar to the structure of vulcanized rubber, Lucite, or Bakelite. Thus, all parts of the bacterial wall are connected together by a network of covalent bonds forming an extremely large macromolecule. The integrity due to the covalent bond structure leads to a mechanical strength that maintains cellular shape (like the exoskeleton of insects). This type of structure is much stronger than cellulose or chitin, which are formed of one-dimensional chains. Those materials should be thought of as cables that do give great strength under tension, but in the side-to-side direction they are held to other chains only by many weak hydrogen bonds. These can therefore rupture and the chains can slide, and creep quite readily; these polymers are much less resistant to any tension not applied along their main axis.

Consider tearing a piece of paper. The paper may resist a certain amount of tension without giving much, but then forms a single tear when the tension becomes too great. If the paper is dampened, it stretches first and then tears more readily. Of course the cellulose chains are just as strong wet or dry, but paper is formed of cellulose fibrils oriented in many directions and matted together. The fibrils in the paper are held together by hydrogen bonds. When a hydrogen bond breaks and tension is present, the chains move relative to each other and a tear starts. When wet, this process occurs more readily because as stress-bearing hydrogen bonds are broken, new non-stress-bearing bonds form briefly with water molecules so that the mechanical elements can move relative to each other more readily, permitting creeping. Hydrogen bonds reform, linking a saccharide or peptide of one chain with that of another, but now in different places so that the strands have slid relative to each other. This latter process is part of the creeping and viscoelastic extension that then reaches a limit after the creeping process has allowed many cellulose chains to reach the end of many others, and a tear commences.

In the example of the piece of paper, only one tear develops. This fact, i.e., that stress favors a single fracture of materials, is central to material sciences and is a restatement of the Griffith (1920) "propagation of cracks" theory. The theory proposed that in a material that is subject to an even force, the stresses will be more intense in some regions than in others because of imperfections. Bonds in the vicinity of regions where there are flaws are those that will rupture first. When they fail and rupture occurs, the stresses become redistributed and now some bonds will feel less stress while other nearby bonds are subject to intense stress and then, in turn, quickly fail. This leads to the formation of cracks that relieve the stress in the bulk of material, stops other cracks that were incipient, and causes the fragmentation of the solid by "brittle fracture."

From measurements of cellulosic materials containing oriented fibrils stretched until they break, it is possible to demonstrate that wood samples, even when dry, fail much more readily than expected from the properties of diamonds, which have a high degree of perfection (Mark 1943). In principle, a cellulose chain should be as strong as a chain of C-C bonds in a diamond. In fact, there is propagation of cracks starting at weak points that are part of the biological structure of the wood. These overstressed points fail and then, like a series of dominos, the whole structure fails. Electron microscopy suggests that covalent bonds seldom fail. In fact even the weaker hydrogen bonds holding chains together often remain intact. It is the imperfections due to the detailed anatomy of the wood that determine just where the ruptures take place.

The bacterial wall material, like the diamond, has "wet strength" because of its reliance on covalent bonds. Of course, there are also hydrogen bonds linking chains together, but only a few because of the paucity of hydroxyl functions on the saccharide chains. Hydrogen bonds can form between the peptide structure and the saccharide chains, other peptide chains, and other wall components. However, many of these would be broken and reform with water as a partner when the wall is under mechanical stress. As our understanding increases, these hydrogen bonds no doubt will be important in quantitating the elasticity of the wall. The major point is that the crosslinked nature of the bacterial wall explains how the rupturing effects of turgor pressure are resisted in objects with such a thin wall, but it creates the conundrum of how the bacterium can possibly grow without rupturing.

Wall Synthesis and the Make-Before-Break Strategy

Wall enlargement involves two processes: the crosslinking of nascent oligopeptidoglycans and the cleavage of old wall. This concept has been appreciated since Weidel and Pelzer's seminal paper of 1964. For the safety of the cell these two processes must be well coordinated. The formation and crosslinking is an interesting example of biochemistry and will occupy most of this chapter, but the cleavage is no less interesting and very important from the point of view of controlling bacterial growth. Although the wall must be cleaved to be able to insert new material for enlargement, when weakened it is in danger of ripping or tearing, leading to the destruction of the organism. Although Weidel and Pelzer assumed that cleavage occurred first, we (Koch *et al.* 1981b) have taken the reverse position. We feel it is axiomatic that in order to keep the wall strong during growth, new wall be incorporated and crosslinked before existing stress-bearing wall is cleaved. We call this the "make-before-break" strategy. Different bacteria carry out this strategy in different ways, however, as will be delineated in later chapters. Safe enlargement requires that the strategy be essentially fool-proof, and for the one understood best, i.e., the strategy of growth of the Gram-positive rod-shaped bacterium, it is very simple and very clever indeed. Even so, the mechanism can be fooled, however, and the cell can be made to destroy itself.

Biosynthesis of Peptidoglycan

From the structure and function of murein, it must be clear that nothing about the process of its formation can be trivial; this is made evident through a study of the biosynthetic process. The engineering design is clear in the fabrication of components and their assembly. The unit processes that will be itemized below remind one of the prefabrication or assembly line methods where parts made elsewhere must be brought together that fit perfectly, and then are joined precisely. Yet other parts of the process remind one of the random diffusion and collisions of molecules involved in ordinary chemical reactions.

In some sense, the job of wall growth is more difficult than chromosomal replication, RNA manufacture, or protein synthesis because the integrity of a stressed structure must be maintained during each step. It has many of the general properties of these other processes, but the feature that stands out is that the final assembly must take place at sites remote from where normal cellular regulation can be expressed. Another special property is that the installation of peptide crosslinks occurs not by a templating procedure, but in a way more closely akin to the way that a nascent polypeptide chain spontaneously folds itself into a native functional form with only information from its own primary sequence. This auto-control feature is necessary for wall formation, as for protein folding, because there is no alternative source of information. This brings us back to the major point that the last steps are biosynthetic processes that must take place outside the cell in a region in which the cell can maintain, at best, only very limited control.

There is one more general feature. This is that the wall seems to be made by general rules and not specific algorithms or molded to existing templates. There are many ways, in principle, to build a stress-resistant wall made of a covalently closed fabric. There are many naturally occurring options for the details of the chemistry, and a dozen different crosslinking types are known. Distances are not critical because the walls of some species have no extra peptide inserted in the bridge, some have pentapeptide polyglycine inserts, and some have yet other short amino acid chains. Even in the walls of species that have pentapolyglycine inserts, all the links do not have to be glycine and not all chains have to have exactly five glycine residues. Consequently, the structure is very likely not a regular, precise, semicrystalline one. There are many additional arguments against a regular structure for peptidoglycan (see Koch 1988b). So we should not be looking for the kind of templating characteristic of protein or nucleic acid replication, but probably something closer to the process of glycogen and starch biosynthesis, in which enzymes work at random at elongating or branching the chains, but in unison yield the desired product, but a product in which no two molecules are exactly alike.

The biosynthetic scheme has been divided into three phases in earlier reviews (Schliefer and Kandler 1972; Tipper and Wright 1979; Rogers *et al.* 1980; and Höltje and Schwarz 1985). The first phase is the prefabrication of the disaccharide unit with its pentapeptide chain (and sometimes additional peptide chains). Below, I will separate the carbohydrate chemistry from the peptide chemistry. The second is the process of moving these units across the hydrophobic cytoplasmic membrane and depositing (inserting) them at the base of growing oligopeptidoglycan chains. The third phase is the polymerization (crosslinking) process. This is the stage in which the fabric that covers the cell is reinforced with new murein crosslinks, but with units that do not yet bear stress. But these stages are only half the process. To these stages we must add a fourth phase concerning the autolysins and the cleavage process that permit expansion growth, and a fifth phase concerning the general elements of wall turnover. Determination of cell morphology will be left to later chapters.

Outline of the Assembly of the Basic Unit

The basic "building block," or repeating unit of most eubacteria is a pentapeptide linked to —β-D-GlcNAc(1→4)-β-D-MurNAc-(1→4)—. Each muramic acid residue contains as an integral part a D-lactyl group that is linked to L-alanyl-D-isoglutamyl-*meso*-diaminopimelyl-D-alanyl-D-alanine for the typical case of *E. coli* and *B. subtilis* (Fig. 4.1). Variations in the composition of the peptide chain occur in other organisms and can be taxonomically important. The most extensive discussion of these themes and variations is to be found in Schleifer and Kandler (1972), Tipper and Wright (1979), and Rogers *et al.* (1980). In the scheme of Schleifer and Kandler (1972), the glycan shown in Fig. 4.1 is A1γ. The A indicates that the third residue is *meso*-diaminopimelyl (diaminopimelic acid, abbreviated DAP or A_2PM). Its ω-NH_2 is used to form the crossbridge. The 1 indicates that there is no intervening peptide sequence, and the γ indicates that the third amino acid is *meso* and not LL diaminopimelic acid. The Schleifer and Kandler scheme does not cover all cases because further substitution of the chain can occur, in particular, the amidation of the α-carboxyl group of the D-glutamic acid. In the B-type peptidoglycans, the polyfunctional third residue is not used for crosslinking; instead the second residue (glutamic acid) is used. In these cases there is always a polyfunctional bridge diamino acid. This means that invariably the crossbridge peptide is formed in the same chemical way; i.e., a terminal D-ala-D-ala bond on the donor peptide is subject to transpeptidation combining the carboxyl group of the penultimate D-ala to form a new peptide bond with an amino group now present in the bridge amino acid. It can be presumed that the bridge amino acids in all these cases was added before the externalization of the unit.

There are some common structural features among the various eubacterial peptidoglycans. One is that the glycan chain is fairly nonpolar by virtue of extensive derivatization of the hexoses. The only free hydroxyl group remaining are those on the 6-position of both sugars and the one at the 3-position of the N-acetylglucosamine moiety. In some species there is a reduced level of acetylation on the glycan chain, which would make the chains considerably more polar, but this deacylation presumably takes place after transport through the wall. For the structure shown in Fig. 4.1, the peptide chain possesses three negative charges and one positive charge at neutral pH, but in some species, for example *Enterococcus hirae,* formerly *Streptococcus faecium,* there are no remaining charges.

Another common feature is that the peptide chain always has some D amino acids instead of the usual L configuration. One possible reason for the D configurations, suggested above, is to prevent hydrolysis by the ubiquitous proteases, but there probably is a deeper reason. This is suggested by the alternation -DLDLD- of configurations, starting with the D configuration of the lactyl group. Such a structure certainly will not assume the forms typical of proteins: α-helices, β-pleated sheets, etc. For example in the usual α-helix, each residue points outward from the helix; consequently, with a regular alternation of configurations in the muropeptide, every other R-group would try, and fail to be able to protrude inside the helix. The design consideration seems to be that these peptides are the springs of the bacterial armor plate. When the pentapeptide or tetrapeptide is not crosslinked, the peptides do roll up in compact balls, possibly making it easier to transport the units through the lipid membrane, but when linked they must obviously largely straighten out. These considerations suggest that the design criteria of the peptidoglycan might be associated with bacterial shape and morphology, so that form can lead to function. For example, different organisms may need different amounts of springiness in their walls, and this in turn may depend on the detailed structure and on the charges in the crosslinked chain. It is worth keeping in mind when thinking about the evolution of the eubacterial wall that the sugars in all cases have been found to have the structure of D-glucose and that the last two amino acids of functional pentapeptides are always D-ala-D-ala.

Formation of the Disaccharide

Starting from glucose a number of steps are needed to get to the beginning of the unique part of the pathway for murein formation. Glucose is converted to glucose-6-phosphate with the phosphate donated by ATP. Isomerization produces fructose-6-phosphate. This product is aminated with the amino group of glutamine to form glucosamine-6-phosphate. Next, the phosphate is moved to the more energetic 1-position and then the amino group is acetylated to yield GluNAc-1-phosphate. The last two steps sometimes take place in the opposite order. In some organisms (*Nocardia* and *Mycobacteria*) the acetyl groups are then oxidized to *N*-glycolyl groups and in other organisms, e.g., in some bacilli, streptococci, and lactobacilli, the externalized wall is deacetylated. This latter strategy leads to resistance to lysozyme.

We may consider that the starting material for the unique branch of peptidoglycan synthesis is *N*-acetylglucosamine-1-phosphate. This is a modified glucose that already has in place two needed groupings and is primed with a supply of free energy in the form of the phosphate group. But apparently the glycosyl phosphate hemiacetal bond is not reactive enough to form the glycoside bond of the disaccharide, so the cell has recourse to one of the oldest of its biochemical tricks: it reacts the compound with a triphosphate reagent. The product, UDP-GlcNAc, has two phosphates and retains one high-energy bond needed later to form a glucosyl hemiacetal. The other product is pyrophosphate; it is believed that its removal via pyrophosphatase then couples more energy into the process and pulls the reaction further forward. Thus, in the coupled reaction, there has been the net consumption of one high-energy phosphate bond. Note that the reagent employed is the one usually reserved for carbohydrates, uridine triphosphate, UTP.

Up to this point in the biosynthesis the biochemistry is standard, simply variations of steps found in many metabolic pathways. The rest of the pathway is quite different. In the next step, UDP-GlcNAc then acquires the functional grouping needed for later crosslinking. The reactant in this case is phosphoenolpyruvate, which is a very high-energy phosphate compound because in most biochemical reactions when the phosphate is removed the molecule spontaneously isomerizes to form *keto* pyruvate. By conversion into the stabler keto form the reaction is pulled forward. But keto tautomerization cannot take place in this case because the phosphate is not replaced with a hydrogen; instead the phosphate ester forms an ether linkage with the 3-position hydroxyl group of the *N*-acetylglucosamine as inorganic phosphate is liberated. This step is made irreversible by the reduction of the double bond with NADP + H$^+$. This stabilizes the product, fixes the ether linkage, and fixes the stereochemistry as D. Now we have a carbohydrate with an attached carboxyl group similar to the chemical reactivity of the carboxylate groups present in proteins.

Formation of the Peptide Portion

The peptide chain is formed and elongated with a series of amino acids, not in the serial way that proteins are made, but rather in a way closer to the synthesis of peptide antibiotics and some small hormones. The normal mechanism for protein synthesis would not suffice because it could not cope with D-configurations nor function in the formation of the isoglutamic acid (γ-carboxyl bond). In most instances, L-alanine is added to the lactyl group; in a few species this first amino acid is glycine or serine. The coupling is done by stereospecific enzymes that are called synthases because they couple the free energy of the terminal phosphate bond in ATP to force the covalent bond formation (with the formation of ADP and inorganic phosphate). Then, in the same way, the amino group from D-glutamic acid is incorporated. The D-glutamic acid is made from L-glutamic acid by L-glutamic acid racemase. The next peptide bond formed is via addition to the γ-carboxyl group of the D-glutamic acid. This also is unusual biochemistry. The amino group comes from *meso* diaminopimelic acid, for the specific example under consideration. This unusual amino acid is an intermediate in the biosynthesis of lysine. Because one end of a *meso* compound has the D configuration and the other L, it is not surprising that it is the L end that reacts to the D glutamic acid. Then finally, D-alanyl-D-alanine (prefabricated by alanine racemase and D-ala-D-ala ligase with the concomitant hydrolysis of ATP) is coupled to the carboxyl end. From energy minimization calculations, this pentapeptide forms into a compact, quite hydrophobic mass when viewed from the outside. This may be essential in the later process of transport of the completed basic unit through the membrane (Labischinski *et al.* 1979, 1983; Koch, in preparation).

To recapitulate: (i) It can be seen why individual enzymes are required for the amino acid additions. Each is specific for its amino acid substrate, its configuration, and the nature of the linkage. The energetic cost is paid with the phosphate bond energy of ATP. (ii) The peptide structure is special. The alternation of D and L configurations probably serves the functional role of creating a compact form to pass through the membrane, and yet allowing an open extended configuration when subsequently bearing stress. (iii) The universal aspect of the chain is that, as formed, it has five amino acids terminating, with no exceptions, in the D-ala-D-ala grouping that supplies the energy for crosslink formation.

Transport and Linking

After the basic unit has been prefabricated, it is delivered to the building site. The problem is that this site is on the other side of an obstacle: the cytoplasmic membrane. Lipid-soluble compounds can pass through the bilayer, whereas polar molecules of the size of the basic unit (molecular weight ca. 1,000) cannot. A similar problem arises in the synthesis of periplasmic proteins, outer membrane proteins, and exoproteins. The usual solution for transport of these proteins is now well known. The coding regions for these proteins lead to a product with an initial hydrophobic stretch of amino acids, the leader sequence, which can pass through the membrane. Once threaded through, the rest of the chain with an assortment of hydrophilic and charged residues can be pulled through because as one hydrophilic group enters, another hydrophilic group is likely to be leaving. The leader sequence is cleaved off by an enzyme, the signal peptidase. For the wall unit, the strategy is similar: the muramic acid pentapeptide is attached to a nonpolar membrane-bound carrier. Then the other saccharide, the UDP derivative of N-acetyl-glucosamine is added to the MurNAc derivative to form the completed basic unit in a state already linked to the carrier molecule. At this point in some species, extra bridges such as a pentapeptide of glycine residues may be added. These are donated by charged tRNA molecules. Because charged tRNA molecules could be formed only on the cytoplasmic side, it is clear that at this point the basic unit is still on the cytoplasmic side of the membrane. Then the hydrophobic molecule functions to aid the penetration of the disaccharide-pentapeptide. Finally the unit is dissociated from the "carrier" and attached, i.e., inserted between another carrier and the growing oligopeptidoglycan chain, composed of a number of peptidoglycan units. After crosslinking through the peptides, the chain of oligopeptidoglycan is cleaved from its hydrophobic handle. Fig. 4.2 shows transport, insertion, crosslinking, and cleavage from a bactoprenol handle.

The lipid-soluble carrier is undecaprenol (also called bactoprenol). It is a C55 compound containing two *trans* double bonds and nine *cis* double bonds. It is one of a family of isoprenoid lipids that occur throughout living cells. From the structure of lipid bilayers with two fatty acid chains apposed tail to tail it can be estimated that generally it takes a span of about 36 carbon atoms to reach across a bilayer. (Think of a bilayer with phospholipids each with fatty acids of 16 to 18 carbon atoms attached to the glycerol moieties.) So these C55 molecules could go across and curl back through the phospholipid membrane. Other than the double bonds on this molecule, there is only one functional group, a single terminal hydroxyl group. This hydroxyl group is esterified to inorganic phosphate, which imparts a negative charge to the molecule. In the transport cycle it is bound via a pyrophosphate bond with three negative charges at neutral pH to the glycan unit. After the still-mysterious transport across the membrane, the unit is almost as mysteriously inserted into the growing peptidoglycan chain at its base.

Chemically, such basal insertion is similar to the way that other extracellular polymers are made, such as dextrans and levans. These are formed by enzymes that have two binding sites; one holds the growing chain and the other accepts the new unit, then a transfer occurs that moves the growing chain to the end of

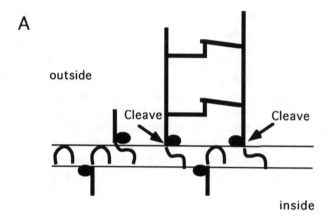

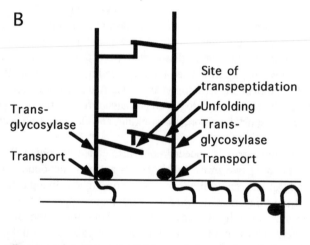

Figure 4.2. Transport, insertion, crosslinking, and cleavage from the bactoprenol handle. A. Shown on the left-hand side: a bactoprenol, a bactoprenol attached to a pentapeptide disaccharide facing the cytoplasm, and a bactoprenol attached to a pentapeptide disaccharide facing the environment. Towards the right a pair of growing glycan chains crosslinked at two places is shown. Sites of cleavages for the transglycosylase steps are indicated. B. The transport of the disaccharide-peptidoglycan unit is shown moving through the phospholipid membrane. Additionally, the growing chains have been cleaved by a membrane-bound enzyme transglycosylase and linked to the new disaccharide units. On the right the return of the bactoprenol to face the cytoplasm, its recharging, and its return to transport another unit to the outside are shown. In the peptide crosslinking the two rolled-up pentapeptides unroll enough to get together and become attached via the agency of a membrane-bound transpeptidase with the release of D-alanine.

the new unit. Then the binding site must alter its chemical function, or more likely, there must be a translocation step, equivalent to the translocation step in protein synthesis. The details have not been elucidated yet, so we must speculate on another possibility. This is that the relevant chain-elongating enzyme must react with the pyrophosphate-linked growing chain to split the carrier pyrophosphate and a glycosyl-enzyme complex, which then donates the chain to a new isoprenoid-linked unit.

Because some of the penicillin-binding proteins (PBPs) have transglycosylase activity as well as transpeptidase activity, they are probably involved at this stage. If the scenario just described applies, the carrier is at this point bound to a singly esterified pyrophosphate (which in an aqueous environment would have three negative charges). An unresolved question is: On which side of the membrane is the terminal phosphate group removed? The conventional wisdom is that it is on the inside in order to conserve phosphate. There is no evidence on this point, but I would suspect that hydrolysis takes place on the outside so that a less polar derivative of undecaprenol with only one negatively charged phosphate has to be transported across the membrane to the inside. The freed phosphate can then be taken up by energy-consuming systems that the bacteria have on hand for their normal need for phosphate. To start the whole cycle anew, a fresh MurNAc with both a pentapeptide and the UDP group attached reacts to form pyrophosphate and liberate UMP and attach the MurNAc-polypeptide to the carrier.

The Assembly

How can oligopeptidoglycan chains be functionally linked together? First principles tell us that they cannot simultaneously be linked and made to support stress. Something equivalent to having a friend put a finger on the knot while one is tying up a package is needed. Even the stretching of a bond by 0.05 nm is enough to favor irreversible hydrolysis of the peptidoglycan, or equivalently two molecules must come together very exactly for a bond to form. This will be discussed in Chapter 6. Therefore, to form the bond it is necessary to position the two groups to be linked very close to each other and hold them in precise orientation. In many chemical reactions, diffusion is the rate-limiting process, and it is necessary to wait until the appropriate conjunction of reactants spontaneously takes place. In other cases, such as some enzymatic processes, the enzyme binds one substrate rapidly and the rate-limiting reaction may be the diffusion of a second substrate to the complex. The use of a binary complex greatly speeds the process compared to the case when all reactants have to collide simultaneously to form a ternary complex. But what if the reacting molecules are tied onto tethers that will not let the reactive group come to apposition? Then nothing will happen. If the limit of overlap of the reacting molecules is small, then the reaction will be very slow. Studies in physical organic chemistry have shown that these effects can alter reaction rates a great deal. So how can the D-ala-D-ala bond at the tail of the pentapeptide be brought to an enzyme (possibly PBP 1a, 1b, or 2) and transferred to an acceptor amino group of another oligopeptide? In the process the D-ala-D-ala bond must be cleaved, D-ala released, the enzyme-bound peptide must find the acceptor group, say diaminopimelic acid, from another chain, interact with the enzyme to form the crosslinked tail-to-tail pair of peptides, and free the enzyme. As an important complication, the two chains to be linked are constrained by the position and linkage of the remainder of the chains and the enzyme is bound to the membrane and not free to move. It does seem pretty impossible! Recently, however, it has been found that a transpeptidase, although membrane-bound, has a catalytic domain with a flexible connection to the membrane-bound domain (Spratt et al. 1988). Thus, the magnitude of the seeming impossibility is reduced because the donor-transpeptidase complex would be capable of diffusing to interact with the acceptor peptide. Such flexibility would seem to be an absolutely essential feature early in the evolution of eubacteria in order to be able to make use of the exoskeleton strategy.

There are two possible classes of strategies for linking oligopeptidoglycan chains together. In one scenario the entire procedure is very mechanized. It requires that the groups to be reacted and the enzyme to be pre-arranged very precisely. The membrane must be arranged in a systematic way so that paired donor and acceptor units of both parts of the chains are held next to the membrane with the isoprenoid handles. The chains must already be positioned in such a way that the peptide can crosslink in a very precise way. A number of systematic ways have been proposed, ranging from glycan chains (possibly with interruptions) going all around the cell, indefinitely long helical chains spiraling around the cell, to pairs of nascent chains that have been linked together, etc. (see Koch 1988b for more details). For any of these systems to function, the membrane and the extant structure must be organized and remain organized during growth. There are very many reasons for doubting that such a regular system exists. Let me review the arguments. First for rigid systems, the wall growth templating process would determine the cell shape and growth pattern. The new chains would need to be laid down in precise relationship to old chains and the latter incapable of stretching. Because the turgor pressure is positive, it would then follow that any error would result in an enlargement in the diameter and there would be no way of correcting this deviation. Other objections are that the glycan chains are not long enough to reach around the cell; the composition of hydrolysis products is very complex, suggesting that the wall is not a precise structure; not all crossbridges are formed; and some crossbridges appear to be formed between two peptides on the same chain. Consequently, the wall is not a well-organized system of struts, braces, and guy wires; a haphazard structure also follows from the point made above concerning the variability of the bridge structure within the wall of the same organism.

The alternative to a methodical mode of wall growth is to assume that the cytoplasmic membrane is fluid and that oligopeptidoglycan units are inserted through the membrane at random in growing parts of the wall. Additionally, the chains are elongated at random sites on the cytoplasmic membrane; the membrane-bound enzymes are randomly associated with the isoprenoids; a half reaction takes place when a donor collides with the enzyme; and the other half reaction takes place when, by chance, an acceptor bumps into the donor-enzyme complex. To say that the processes are random is not to deny that different regions of the cytoplasmic membrane might affect the process. Control by the cell has to come from its directing enzymes to different regions, like sidewalls or septa, but without precisely arraying them on a template of existing wall. Then at random the crosslinking takes place, linking nascent chains to nascent chains and nascent chains to non-stress-bearing parts of stress-bearing chains. Stress-bearing chains might be linked to other stress-bearing chains but only to form non-stress-bearing bonds, because such chains would have very limited ability to move at random.

This view considers that polymerization is akin to the crosslinking of rubber during vulcanization. It is different in that the crosslinking happens on or in the functioning cell wall. The product in this case might be a random material in which the glycan chains are distributed like the fibers in paper or felt or it might have some partial structure. In this scenario, the shape and detailed structure and growth depend on other factors. These possibilities will be outlined in later chapters.

Conversion of Nascent Wall to Structurally Competent Wall

The process of wall synthesis and crosslinking as described so far is only the prelude to a structural role. As the units are being incorporated they cannot support stress. If this turgor pressure was lost from a cell it might have a structure as shown in Fig. 4.3. In some cases stress may be supplied only slowly, whereas in others the glycan must accept stress very quickly. While using Portland cement, epoxy, or glue, time is allowed before stress is applied. During such "curing" more crosslinking bonds form so that the strength of the material becomes greater. If stress is applied too soon, bonds will tear, masses will separate, the joint will rupture, and the structure will fail. Something equivalent to curing takes place in bacterial wall formation. This can be concluded because of experiments that demonstrate that label for wall enters a sodium dodecyl sulfate (SDS)-soluble state before it is linked to the fully insoluble sacculus (Mirelman 1979). There are also measurements that demonstrate that changes in the murein occur with the phases of the culture cycle; thus, exponential phase wall has longer glycan chains and fewer crosslinks than does stationary phase wall. These may be related to curing or "aging" as the terms would be used in the building industry.

The practical view is the question of how quickly stress can be applied after a unit is crosslinked into the wall of a growing cell. The unlinked oligopeptide

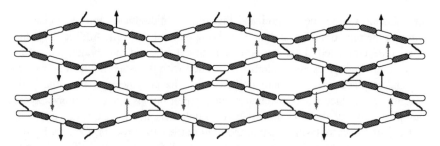

Figure 4.3. Diagrammatic structure of a segment of fully crosslinked peptidoglycan of a monomolecular wall fabric in a planar, but unstressed, conformation. Four oligopeptidoglycan chains, with the glycan running horizontally, are crosslinked via the peptides as shown. Although there are peptides at every other saccharide, it is only at every eighth saccharide that a pair of glycan chains is interconnected. This results in pores in the monolayer fabric that are eight hexoses long and two relatively compact tetrapeptides wide. The glycan chains tend to be fairly straight and nonelastic, but the pore is allowed to be somewhat smaller than when stressed. The peptides are flexible and extensible, but their relaxed state is compact. The drawing is highly schematic and applies to a nonstretched wall. Actually, because the glycan chains are of finite length and not all peptides are formed into crossbridges, parts of the fabric cannot be stress-bearing. Moreover, because some oligopeptide disaccharide units are being continuously inserted within the fabric in the case of *E. coli* and some are being removed, it would be better to depict a three-dimensional structure. In the case of *B. subtilis,* the wall as shown would also occur, but because of the thicker nature of the wall, there would be crosslinks to the more centripetal and peripheral wall.

is not under any stress due to turgor pressure. Only when crosslinked in two places is it capable of accepting stress from the cytoplasmic membrane (CM). When selective cleavage of existing, stressed peptidoglycan takes place, additional stress becomes transferred. This stress will be experienced not only by the newest peptide bond, but also by the intervening portion of the saccharide chain as well as at least one other crosslinking peptide. In later chapters, we will dissect the growth strategies for different parts of the wall of different organisms and show that the time before stress is applied varies a good deal. Of the organisms that have been sufficiently studied, the slowest is the septum/pole of the Gram-positive rod. Portions of the invaginating septum appear to take a considerable length of time (36 min) (Burdett *et al.* 1986) before they are fully formed, have split down the middle, and can serve as the stress-resistant part of the pressurized cell. On the other hand, the very thin wall of the Gram-negative cell must need only seconds between inserting into the extant wall, crosslinking at multiple points, making cleavages that transfer stress to the new chain, and enlarging. After stress is applied, a portion of the fully crosslinked wall might look as depicted in Fig. 4.4.

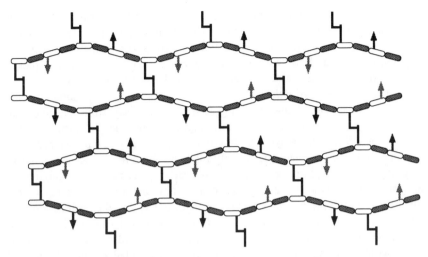

Figure 4.4. Diagrammatic structure of a segment of fully crosslinked peptidoglycan in a planar, stressed conformation. When the wall is serving its function of resisting the cell's turgor pressure, then the conformation of Fig. 4.3 would alter to allow the wall fabric to enlarge (stretch elastically). The structure depicted shows the conformation on being stretched in all directions in the plane so as to cover the largest possible surface area. Note that the glycan chains are almost nonextensible and that the area contained by a section of the wall increases about 2-fold as does the area of a pore formed in this model. See Chapter 10 for experimental results on the extent of stretch.

The Necessary Cleavage of Cell Wall for Growth

For the enlargement of the bacterial cell wall and for cell division, cleavages of previously formed peptidoglycan are essential. There is a spectrum of enzymes that cleave the wall and a spectrum of strategies used by different species for safe growth.

Reworking of the Wall Substance

Reworking of portions of the wall probably does go on after the initial insertion, at least in the Gram-negative wall. Because the wall appears to be synthesized with only partial order at best, one can imagine that some of the newly formed bonds when first stressed would receive much more stress than other newly formed bonds. The more stressed bonds would be much more likely to be cleaved by hydrolysis (see below). Cleavage could be catalyzed by an autolysin or could even arise nonenzymatically. After the new fabric is linked, comes to bear stress, and is pulled taut, then there would be realignment of the structure in the same way any fabric is altered when stretched—some bonds that had been quite loose before may now have only very limited play or may support stress. A number of annealing processes can be imagined. The surviving pentapeptides may serve to form new crosslinks. The transpeptidases may also aid in this process by splitting the peptide bond with the reformation of enzyme substrate complexes or even reform pentapeptides, if any D-ala molecules remain in the vicinity. Also, because there are *anhydro* forms of the terminal N-acetylmuramic acids (MurNAc) of glycan chains, this suggests that these may store the glycoside bond energy in such a way that longer glycan chains may be formed at the expense of splitting the anhydro groups. The mechanism would be that *anhydro*-muramic acids are formed, presumably made by the transglycosylase activity, resulting from the splitting of longer stress-bearing glycan chains. Although the enzyme activity would split a glycan chain, it would store the energy of the glycoside bond in the *anhydro*-glucose linkage. Later, this chemical action may be reversed to reform a more highly crosslinked and tighter fabric. It would be tighter because in this process the synthetic bond would be favored between reactants that can reach each other. Evidently, uncontrolled activity of this kind would lead to creep and an inability to resist long-term stress. However, if the enzyme acted only on nonstressed bonds (and, of course, produced only different nonstressed bonds), then with stress and strain equilibration, a better-cured, stronger product could be formed that would be more resistant to stretching. Clearly this suggestion is hypothetical because there are no measurements of enzyme activities of any kind on paired stressed and unstressed substrates. But one can assume from general knowledge of enzyme activity that an enzyme could exist that would not bind a stressed substrate in particular conformations.

The Difference Between Hydrolases and Transferases

The way many enzymes act is by binding to a sensitive bond if the bond is surrounded by suitable groups. Then bond splitting occurs, liberating one portion to the medium and forming a new bond to a functional group of the enzyme. The thermodynamic energy of the bond is not dissipated, but merely transferred in this process. If this were all that happened the enzyme would be inactivated. In the usual case, a second half reaction occurs, regenerating the enzyme and linking the other half of the substrate with some new group. In the cases in which we are interested, this can be a nucleophilic atom, such as oxygen or nitrogen. Thus, if the enzyme attacks a glycoside or peptide bond the liberated OH or NH_2 group is replaced by the OH or NH_2 moiety of the enzyme (or equivalent group). Then, in the second step, transfer occurs to an appropriate group in a new substrate. Different enzymes have different degrees of specificity for this acceptor group. If there is virtually no specificity, then water is likely to be the acceptor and the result is hydrolysis simply because water is so abundant in biological systems. Such enzymes are called hydrolases. However, this does not prohibit another acceptor from serving. Consequently, hydrolases can behave as transferases and yield a new product and this, of necessity, must also be a substrate for the enzyme. Other trades can take place, each with a chemical bond similar to the original one. Eventually, of course, the system must reach thermodynamic equilibrium, which for hydrolases would result in the original substrate being hydrolyzed to constituents. This will less likely be the case, however, if the system contains very little water, if certain specific compounds are involved, and if the enzyme operates in such a way that the surrounding water is not permitted access to the reactive site. In the last case the enzyme is not classified as a hydrolase, but as a transferase, e.g., a transpeptidase or transglycosylase. It has been shown that hydrolytic enzymes like trypsin, β-galactosidase, or invertase, do have limited transferase activity. It is not yet clear whether the various autolytic and transpeptidase/transglycosylase enzymes of wall biochemistry use other acceptors than water and peptidoglycan respectively.

Autolysins

Enzymes produced by the bacteria themselves cleave a variety of murein bonds. The literature has been reviewed by Höltje and Schwarz (1985) and by Doyle and Koch (1987). The facts about these enzymes will be presented here in a different way. The enzymes will be classified according to what they do to the stress-bearing properties of the wall structure, the biochemical nature of the enzymatic process, and the chemical properties of the chain ends that are produced (see Table 4.1).

The types of cleaving enzymes are classified here into three categories. The

Table 4.1. The types of wall-cleaving enzymes classified by function of the products

I. Enzymes that render both sides unusable for formation of stress-bearing wall (strict autolysins).
   Amidases (*N*-acetyl-muramoyl-L-alanine amidase)
   Glycosaminidases (endo-β-*N*-acetyl-D-glycosaminidase)
   Hexosaminidases
   Muramidases
II. Enzymes that render one side not useful for formation of stress-bearing wall.
   Anhydro-glycosylases or exo-transglycosylases (if the saccharide with the anhydro group is liberated into the medium)[a]
III. Enzymes that render both sides useful for formation of stress-bearing wall that can function as an acceptor.
   Endopeptidases (dissipate the energy of the peptide bond)[b]
   Anhydro-glycosylases or exo-transglycosylases (if the saccharide with the anhydro group is retained as part of the wall structure)[a]

[a]If the *anhydro*-glucose group remains part of the wall, then the energy is effectively stored; if it is part of the fragment that diffuses away, the energy is most probably lost.

[b]Both products can serve as acceptors for new wall formation.

first type comprises the autolysins proper; these are hydrolytic enzymes that render both cleavage products unusable in the formation of new stress-bearing wall. Enzymes in this first category include the amidases, which cleave the L-alanine-muramic acid bond. There is no known way that either the L-alanine or the muramic acid moieties while still linked to other wall components as part of the wall structure, could serve in new bond formation, because there is no known suitable enzymatic activity nor any source of free energy for exergonic bond formation. Also in this category are the glycosaminidases, which hydrolyze the glycan chains, also dissipating the free energy of the bond by hydrolysis. The Gram-negative organisms with their much thinner walls seemingly could use neither type for enlargement of the stressed sacculus.

The second kind are anhydro-glycosylases; these enzymes conserve the bond energy and form one product retaining sufficient energy for the reformation of a crosslinking bond. They cleave the glycan chains and store the energy in the form of an equivalent anhydride bond linking one part of the hexose back upon itself. Such cleavages must be reversible with an equilibrium constant not too much different from one. Consequently, they may be used to reassemble a linkage between different parts of the sacculus. Anhydro-glycosylases are listed in Table 4.1 in two categories. The energy is available for reformation of wall only if the unit remains as part of the wall. Anhydro ends are present in the walls of *E. coli,* but compounds dissociated from the cell with the anhydro linkage of muramyl-containing fragments are found in abundance in products secreted into the culture medium during growth of *Neisseria gonorrhoeae* (Rosenthal and Krueger 1988).

The third category contains endopeptidases. These dissiptate the energy of the peptide bond, but render both products useful for the formation of stress-bearing wall. The products, however, can function only as acceptors. Endopeptidases cleave the same crosslinking tail-to-tail peptide bond whose formation was essential to the formation of a stress-resistant fabric. Presumably, these bonds are formed under the auspices of the PBP 1A, 1B, and 2. These enzymes are those transpeptidases whose activity is inhibited by penicillin. Because cleavage by endopeptidases liberates the energy of the peptide bond, both resultant chains can be used only as acceptors in the formation of new crossbridges.

There are probably other types of enzymatic activities that should be included in a less conservative table. For example, there may be enzymes so far not detected because they function only on stressed substrates. The table probably should include enzymes usually classified as irreversible. For example, there are a number of DD-decarboxylases that free the terminal D-alanine hydrolytically from the pentapeptide, but that may also function *in vivo* as transpeptidases, but possibly only under certain conditions. Such a role would depend on persistence of the enzyme complex and the ability to accept another terminal amino group in place of the D-alanine.

Geometry of Cleavage Action

The strict autolysins are conceptually the simplest cleavage enzymes. They serve the needs for safe growth of cells of organisms of diverse shapes. The simplest and most straightforward case of autolysin function is in cell division of Gram-positive organisms, say either streptococci or bacilli. Synthesis creates a thick septum, but only a modest level of autolysin is needed to split it. However, the split must be done in such a way that the cleavage plane goes right down the middle. Koch *et al.* (1985) indicated ways that physical forces could direct bisection. Evidently for splitting, it does not matter whether glycoside, peptide, or amide bonds, or some combination of these are cleaved, although there may be secondary reasons for one type of cleavage over the other. Hydrolytic enzymes are appropriate in this instance because the externalized wall surface generated by the enzyme does not engage, or need to engage, in further linkage for the synthesis of stress-resistant wall.

Elongation of the cylindrical regions of those Gram-positive, rod-shaped organisms that use the inside-to-outside strategy involves a great deal more hydrolytic activity and is associated with extensive turnover. For Gram-positive organisms, for both the rapidly turning over side walls and the much slower turning over poles, there appears to be no attempt to conserve either bond energy or the previously used wall. During expansion and the cleavages permitting inside-to-outside growth, those seems to be a profligate expenditure of energy and resources. Little of the evidence is in, but turnover may be obligatory in the case of the side wall of Gram-negative organisms as well.

Strategies of Self-Protection Against Autolysin Action

Enzymes march to their own drummer. How can autolysin action be channeled away from a destructive course into a productive one? It appears that a variety of techniques have been employed by prokaryotes. These are listed in Table 4.2 and shown diagrammatically in Figs. 4.5, 4.6, and 4.7. Strategies for the constructive utilization of autolysin may involve control of their synthesis and localization, but do not, in principle, involve cellular control of the actual enzymic activity because, of course, enzymes bind suitable substrates and catalyze reactions without regard to the orientation of the substrate with respect to the cell's geometry or the earth's magnetic field or other external conditions. There are several cases to be previewed here and elaborated on in Chapters 9 and 10.

Fig. 4.5 shows four of the protective factors that have been suggested to serve in the causes of safety during the elongation of the side walls of the Gram-positive rod-shaped bacteria.

(1) The energy metabolism of the cell acts to extrude protons and thus lowers the pH in the wall. This reversibly inactivates native autolysins and so protects

Table 4.2. Strategies for the constructive utilization of autolytic action for enlargement and cell division without compromise of cellular integrity[a]

Gram-positive organisms

Side wall formation by inside-to-outside growth, controlled by
1. Proton extrusion
2. Stress intensity due to turgor pressure of the cell
3. Prevention of folding (naturation) of the secreted peptide chain within the wall
4. Surface rebinding of autolysins leading to high surface concentrations

Pole formation by forming and splitting a septum, controlled by
1. Chemiosmosis
2. Stress intensity due to turgor of the cell
3. Prevention of folding of the secreted peptide chain within the wall
4. Surface rebinding of autolysins
5. Stress intensity due to the existence of a partial split
6. Cracks propagated according to the Griffith principle
7. Murosomes laid down in the middle of the septal wall

Gram-negative organisms

Side wall formation by insertion followed by selective "smart" cleavages[b]
1. Insertion and crosslinking of a glycan strand between two stress-bearing strands followed by cleavages pulling the new chain into the fabric
2. Insertion of a duplex of peptidoglycan chains around a single chain followed by cleavages liberating the old chain
3. The "Three-for-one" mechanism, see Chapter 10

Pole formation
1. Insertion of glycan strand between two stress-bearing strands
2. Insertion of a multiple linked group of peptidoglycan chains around a single chain followed by cleavages pulling the new chain into the fabric

[a]Evidence for mechanisms varies from very clear to nonexistent; i.e., from fact to speculation.

[b]These mechanisms are totally hypothetical and require enzymes with very special properties.

the wall near the cytoplasmic membrane and hence the cell. This safety factor was suggested by Jolliffe, Doyle, and Streips (1981); they studied agents (proton conductors) that shorted out the protonmotive force and expeditiously returned actively extruded protons back into the cytoplasm. These substances caused cell lysis of *B. subtilis*. They suggested that, during normal growth, active proton extrusion causes the accumulation of protons and net positive charge outside the cytoplasmic membrane and this dynamic steady-state lowering of pH within the wall of the actively metabolizing cell normally inhibits the autolytic activity. When pH lowering is prevented, the autolysins act to destroy the stress-bearing part of the cell wall and cause lysis.

(2) Peptidoglycan is not under tension as secreted through the membrane and initially crosslinked. A layer, however, comes to feel tension after still newer layers are added underneath and the cell elongates. Eventually a layer is maximally stretched. Because enzyme cleavage activity depends on the stress in the bond to be cleaved, the wall may then be easily breached. Therefore, autolysin

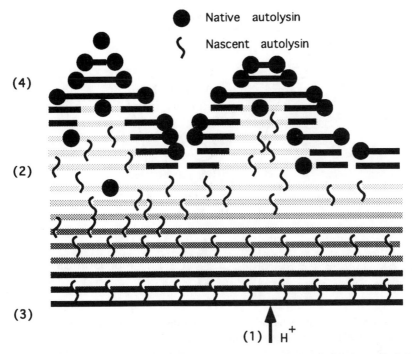

Figure 4.5. Four factors allowing growth and division while protecting the wall of the Gram-positive rod-shaped bacterium from destructive autolysis. The four of the seven factors designated in Table 4.2 are shown. These are described more fully in Chapter 9 for side wall formation by inside-to-outside growth: (1) The chemiosmotic process. The energy metabolism of the cell functions to extrude protons and thus lowers the pH in the wall and inactivates native autolysins and so protects the cell. A circulation is produced because the protons are continuously extruded and return into the cell doing useful work. (2) Stress intensity due to turgor of the cell. Peptidoglycan is not under tension as secreted through the membrane and initially crosslinked. A layer, however, comes to feel tension as new layers are added underneath and the cell elongates. Eventually a layer is maximally stretched. Because enzyme cleavage activity depends on the stress in the bond to be cleaved, the wall may then be easily breached. Therefore, autolysin action is localized to regions at and above intact wall surrounding the cell. (3) Prevention of folding of the secreted peptide chain within the wall. The autolysins are secreted through the cytoplasmic membrane and probably move through the wall with the inside-to-outside growth process. At some point, they fold into a native configuration. If they cannot fold to a native configuration while surrounded by wall, then destructive autolysins will only form near the outer surface of the wall. (4) Surface rebinding of autolysins. On the outer surface, the autolysins act to cleave wall and then are liberated. Only the native autolysin molecules rebounded on the outside of the cell are available to destroy the cell.

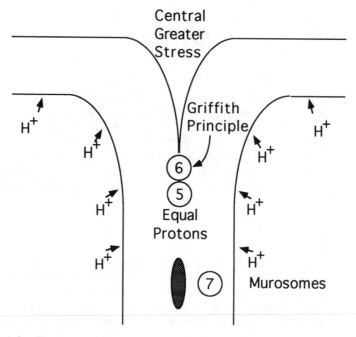

Figure 4.6. Three more factors allowing safe division of Gram-positive rod-shaped bacteria. In addition to the four factors shown in Fig. 4.5, also shown is the role of autolysins formed in previous generations (and to a lesser degree from other cells) that coat the outer surface and lead to a developing crack or crevasse. The side wall cracks in the wall continue to deepen, leading to helical grooves. Over the septum a crack develops and leads to the splitting of the septum to produce two daughter poles. Factors that lead to the precise bisection of the septal wall include: (5) Equal distance from the source of extruded protons, (6) the Griffith principle that leads to propagation of cracks once started, and (7) For *Staphylococcus* the presence of special murasomes in the middle of the wall that are activated when the septal wall is to be split.

action is localized to the most peripheral regions leaving an intact portion of wall surrounding the cell. This concept is fundamental to the surface stress theory (see Chapter 6), which holds that the wall under the most stress is most rapidly cleaved because the stress lowers the activation energy for enzymatic hydrolysis. The stress is greatest in the outermost stress-bearing layer.

(3) The autolysins are secreted through the cytoplasmic membrane and probably move through the wall with the inside-to-outside growth process. At some point, they fold into a native configuration. If they cannot fold into a native configuration while surrounded by wall, then destructive autolysins will only form near the outer surface of the wall where the wall is less dense. The suggestion proposed under this category is that pores within the existing peptidoglycan may be too small to permit folding. Thus, active molecules may be formed only on the wall exterior.

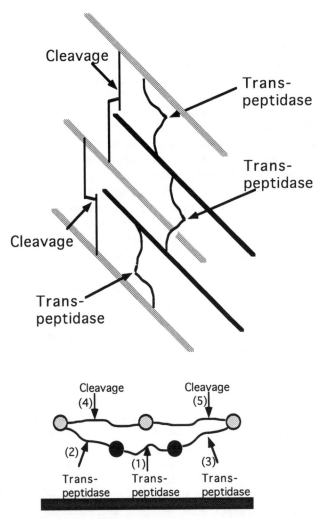

Figure 4.7. The "multienzyme" model for the safe enlargement of the thin Gram-negative wall complex of transpeptidase(s) and autolysin(s). Two views are shown. Upon extrusion of pentapeptide-containing oligopeptidoglycan chains through the cytoplasmic membrane (shown dark), transpeptidases (1) of the enzyme complex link two nascent chains together. Then transpeptidases (possibly a second kind, but part of the complex) link (2) and (3) the raft of the structure of two chains to the existing wall (shown in grey surrounding a stress-bearing strand. The regulation of the parts of the holoenzyme ensures that only after crosslinking can the autolytic component function (4) and (5) to cleave the old strand. Selectivity of the autolysins would depend, in addition, on the completion of the crosslinks, on the stress in the old peptidoglycan, and possibly on the detailed chemistry of the bond to be cleaved. Reproduced from Koch (1990c) by permission.

(4) On the outer surface, the autolysins act to cleave wall and then are liberated, like the classical joke about the man who sits on a limb and saws it off closer to the trunk. However, the native autolysin molecules that rebind on the outside of the cell are available to destroy the cell. The surface rebinding of autolysins is very rapid so that there must be an accumulation of autolysins produced in earlier generations on the cell surface (Koch 1988d). This facilitates hydrolysis of the murein surface immediately adjacent to the growth milieu, including the site of septal splitting as well. Adequate rebinding of autolysins only occurs if the enzyme concentration in the medium, due to past growth, is large.

The safe splitting of the septal wall of Gram-positive organism is a different case, but may occur by a simple, almost default, mechanism. In this case the role of the autolysin is to split the septum evenly in half; i.e., the process must not deviate from splitting right down the middle or weak spots in the wall may form. But what could cause an enzyme to adhere to such a narrow restricted course? Seven factors (Table 4.2) have been suggested including the four factors listed above allowing elongation while protecting the wall of *B. subtilis* from destructive autolysis.

(5) (See Fig. 4.6.) The protons extruded through the membrane would have their least inhibiting action via pH lowering at the midpoint of the septum. Applying this concept to the splitting of septa, it can be seen from geometrical considerations that the septal wall halfway between the two surfaces where separation is occurring would have its pH lowered least and be most at risk for autolysis.

(6) The stress intensity from engineering considerations is greatest where an earlier split ends (the Griffith principle). This is simply the idea that once a crack starts it will continue to extend in preference to new cracks forming. Analysis of the stresses generated when a structure like a partially split septum is under tension show that the tensile strength would be maximum at the center and cause the split to occur halfway between the edges of the septum because there the tension is greatest (Koch *et al.* 1985).

(7) The strategy used by some staphylococci is to form a series of compartments, designated murasomes, in the geometric middle of the septum that contain peptidoglycan-hydrolyzing enzymes in a nonactive form. At a suitable phase of the cell cycle, these become activated and favor cell splitting. Obviously, this is a more advanced strategy, involving both placing the murasomes in the middle of the developing septum, and later activating them (Giesbrecht *et al.* 1985).

Insertion of wall into the existing sidewall of Gram-negative rods is more complex. Some workers, principally Schwarz and Höltje in Tübingen (1988), have argued that the inside-to-outside mechanism functions for the Gram-negative as well for Gram-positive rods, and that there is more than one layer in Gram-negative organisms, although not as many as in Gram-positive organisms (Labischinski *et al.* 1991). They have moved away from this position. It is now generally agreed, as indicated in Chapter 10, that Gram-negative organisms have only one stress-bearing layer, but that the wall is a little thicker because there is newly added wall in some places as well as older wall that is no longer bearing stress but has not yet been detached. A different mechanism must therefore come into play for the Gram-negative wall. Exactly what and how safe wall growth is carried out has not been experimentally determined. Because of its complication and the need for safety, the mechanisms must depend on the postulated, but unproven, existence of "smart" autolysin systems that hydrolyze peptidoglycan only if that act will not compromise the bacterium. There are several possible candidate mechanisms: One of them assumes that the insertion of a glycan strand between two stress-bearing strands and multiply linking it to the stress-bearing strand is a prerequisite for cleavage of the stress-bearing links. This could happen if the autolysin were an allosteric enzyme that was only activated to cleave stress-bearing wall if the regulatory site could bind unstressed but linked wall. Such a mechanism, in principle, could lead to safe elongation (Koch 1990c). A second mechanism had been proposed to explain turnover during growth of Gram-negative bacteria; it assumes the insertion and linkage of a duplex of nascent peptidoglycan chains around a single stress-bearing chain followed by cleavage (see Fig. 4.7) (Koch 1990a, 1990c). Very recently, Höltje (1993) proposed a "Three for one" model in which three linked oligosaccharides are linked into the wall and the templating oligosaccharide cut out (see Chapter 10).

The Role of the Wall in Resisting Turgor Pressure

The cell wall has to overcome two kinds of challenges. One is carrying out the dynamic process of enlarging at the same time that the cell supports a turgor pressure. For this process, a make-before-break strategy is essential. The other difficulty is to support the weak cytoplasmic membrane against the quite large turgor pressure. This problem is overcome by a process akin to that used in filtration technology. In order to separate a suspension into a filtrate and a precipitate, one usually needs two elements to the filtration apparatus. This is because the filter that will let the fluid, but not the solid through, is usually too weak to support the pressure and, if not supported over its area by the apparatus, will burst. The second component is a support (screen) for that filter. The support can have pores that would pass the solid, but it must have the strength to support the pressure. When both function as a sandwich, the filter apparatus can operate effectively. The filter may be mechanically very weak as it only need be strong enough to support itself over the dimensions of the pores of the supporting screen and not over the entire area of the filter.

In the biological case, the cytoplasmic membrane is the filter that is impermeable to certain ions and molecules and the wall is the support. Neither individually can support the pressure, the former because it would burst and the latter because it would pass the elements that create the turgor pressure and thus dissipate the pressure. Measurements have shown that phospholipid membranes can resist very little pressure. But in unpublished calculations I have found that very large pressures would be required to make cytoplasmic membranes either burst or herniate through the spaces between chains in the peptidoglycan.

[Some people believe that the periplasmic space is isosmotic with the cytoplasm—see below. If this were true, then the osmotic pressure is exerted on the outer membrane, which would cause a bleb and rupture unless the Braun lipoprotein molecules (see Braun and Wu 1994) that are covalently linked to the peptidoglycan and embedded in the outer membrane give enough strength due to their very large number to keep the outer membrane in place.]

Evolution of Eubacteria

The details of the chemistry of wall growth indicate many things about the evolution of the eubacteria. From the considerations of Chapter 1 it is clear that with the successful development of organisms with efficient physiological and biochemical processes, the problem of coping with high internal osmotic pressure became paramount. The strategy of having a stress-resistant wall is an obvious one, but the sophisticated structure that we see in modern true bacteria is by no means trivial. Even in its most primitive form it required the simultaneous solution of a number of engineering questions. A polyfunctional unit, basically hydrophilic, must be exported through the hydrophobic membrane. It must be sufficiently nonpolar and compact to get through and it must be pre-energized to become both linked in the carbohydrate direction and crosslinked in the peptide direction. Finally, old functional wall must be cleaved as required by growth, but without hazard to the cell.

These considerations and the commonalities of the peptidoglycans of bacteria define the special properties to be attributed to the first eubacterial descendent of the Last Universal Ancestor. The only exportable energy sources are the pyrophosphate linkages in the carrier-bound form of the peptidoglycan to favor glycan chain elongation and the D-ala-D-ala fourth and fifth amino acids in the pentapeptide that maintain the energy needed for crosslinking. Also required is the secretion into the membrane of transpeptidase and transglycosylase and the autolysins, which need not necessarily be membrane bound. It would appear that these properties would have been present in the first eubacterium. So I would think that the original eubacterium might have had a wall of the Schliefer and Kandler designation of A1γ type. The minor modifications such as are seen in the B family of this classification would have come later under this scenario. Of course, the fundamental thought in this speculation is that much of the regulation for growth and division, then and now, has to come from the biophysical principles themselves.

History

Because the wall of bacteria is different in kind from any other substance, it is not surprising that early workers seemed to feel that something was amiss.

For example, read the recollections of Milton Salton (1964), who first isolated peptidoglycan. The finding of the accumulation of the UDP-nucleotides of muramic acid in cells grown in the presence of penicillin by J. T. (Ted) Park (Park 1949) was without precedence. Just the involvement of uridine nucleotides was novel. Of course, the driving force for the studies of many workers was to understand the way that the penicillins worked. But understanding had to wait until the mode of synthesis of the bacterial exoskeleton was appreciated.

The conceptual breakthrough was presented in the review by Weidel and Pelzer (1964) that brought into focus the idea of a bag-shaped molecule as a structural entity. Then the work of Wise and Park (1965) and of Tipper and Strominger (1965) led to the notion that penicillin reacted to form a covalent bond with an enzyme of wall metabolism whose role is to link peptide chains together in a tail-to-tail manner. This reaction permanently inactivated the enzyme.

From this point the field radiated in many directions. Workers from different backgrounds using different techniques borrowed the results and tools of other specialists to expand our knowledge. In this paragraph I will drop names for the record, but mostly to indicate the variety of fields of expertise that were involved. In one direction was the work of isolation and study of enzymes that attacked peptidoglycan. These provided key tools for the elucidation of the wall structure. Workers in this area included Daneo-Moore, Ghuysen, Glaser, Kandler, Mirelman, Neuhaus, Rogers, Schleifer, Sharon, Shockman, Spratt, and Ward. Important contributors from the field of ultrastructure were Beveridge, Beyer, Burdett, Higgins, and Murray. Physiological studies were pursued by Archibald, Doyle, Goodell, Höltje, Schwarz, and Tomasz.

Ideas sometimes receive immediate acceptance; for example, the double-stranded structure for DNA was welcomed immediately by the biological community. However, sometimes when an idea is fresh, the evidence in support is not accepted as convincing. For example, the idea that the sacculus was the stress-resistant part of the bacterium is one that came full-blown with the Weidel and Pelzer (1964) review. Its corollary, *viz.* that peptidoglycan determines cell shape, is no doubt just as true, but this concept was accepted only gradually and is not a solid fact even yet. Even though many observations of many kinds are consistent with the idea, there were earlier papers and a critical paper by van Heijenoort *et al.* (1975) that reported that the wall of *E. coli* lost its shape after treatment with high concentrations of guanidine HCl. This, apparently, formally and logically excludes the possibility that a purely covalent structure determines the shape of the sacculus. But most workers would not believe this conclusion (although they did not doubt the experiments as van Heijenoort is a very careful and critical scientist), because this claim is only one versus very many contrary observations. It creates a doubt, and one that some day must be reinvestigated.

5

Turgor Pressure of Bacterial Cells

KEY IDEAS

Osmotic pressure is a colligative property depending only on the effective concentration of particles, regardless of size.
Water flows through membranes until its chemical potential is the same on both sides.
Small cations and anions are the major source of osmotic pressure because their numbers are great and they do not generally penetrate the membrane readily.
Hydrostatic pressure is a second cause of fluid flows.
In bacteria, a high turgor pressure compensates for the difference between the osmotic pressure inside and outside the cell.
Most prokaryotes and plants depend on a rigid, stress-resistant covering of the cell.
Eukaryotic cells cannot resist osmotic forces based on the tensile strength of their cytoskeleton alone, but employ additional mechanisms.
The classical way to measure turgor pressure (Mitchell and Moyle 1956) depends on finding a solution that has the same vapor pressure as does the bacterial cell.
The most common method to estimate the cell's normal turgor pressure is to find a

concentration of a nonpermeant molecule that is just capable of causing plasmolysis.

The most accurate, but not generally applicable, method is to choose organisms with gas vesicles and study the effect of external hydrostatic pressure on vesicle collapse.

Turgor pressure is measured by the difference of the collapse-pressure curves with and without enough osmolyte to cause plasmolysis.

The collapse-pressure curves demonstrate that the sacculus in growing cells is normally expanded due to turgor pressure.

Osmotic Pressure

Osmotic pressure depends on the number of particles and is independent of their molecular weight or the mass concentration. Presumably, therefore, a ribosome subunit contributes as much to the osmotic pressure as does a molecule of glucose. This statement is accurate when it is applied to more or less rigid particles, and rigid particles, no matter how big or how small, contribute equally to the colligative properties of the solution. However, a random coil, like the linear chain of glucose residues as found in amylose, would contribute something between one and the number of glucosyl residues in the chain to the particle count. This variation occurs because the real criterion has to do with the number of *independent* particles, and the remote parts of a portion of a flexible chain are essentially independent of each other. On the other hand, long molecules may interfere with each other and lower the osmotic pressure. The low molecular weight cellular solutes control the colligative properties, and in particular the osmotic pressure. The small solutes of the cell include the molecules of intermediary metabolism, largely organic acids and phosphate derivatives, and magnesium and potassium ions needed as cofactors as well as for balancing the charge of nucleic acids. They also include other counterions needed to match the charges on macromolecules.

Although permeable to water, the cell membrane is quite impermeable to salts, many uncharged solutes, proteins, and nucleic acids. Especially important is the fact that many small ions do not easily penetrate the cytoplasmic membrane. The internal accumulation of small ions creates the largest part of the osmotic driving force that causes the uptake of water (in fact, two orders of magnitude greater than the cellular macromolecules). In the absence of other forces, the water content (expressed as water activity, but more correctly the chemical potential) will tend to become equal on both sides of the membrane by a flux causing dilution on one side and concentration on the other. But when hydrostatic pressure is a factor, the situation is different. For a bacterial or plant cell, an equilibrium state is only achieved because the osmotic pressure difference is balanced by the turgor (hydrostatic) pressure after water has entered the cell causing it to swell and to stretch the wall. The strong elastic wall develops tensile stresses that have a normal component that pushes back on the cytoplasmic membrane in the same way that a rubber band stretched around an object pushes the contents that it surrounds inward. Therefore, it is the stretching (or strain) produced in the wall by uptake of water that leads to stresses that then produce (or better, support) the turgor pressure, P_t. Thus, in this equilibrium situation, a different kind of equilibrium is achieved, in which the sum of the turgor pressure and the osmotic pressure on the outside just balances the inside osmotic pressure. The turgor pressure has many roles in plants and eukaryotic algae. Evidently, the turgor pressure has many roles in prokaryotes as well, but these are just beginning to be appreciated.

The Donnan Equilibrium, the Potential and the Osmotic Effect

An interesting situation arises when some molecules cannot pass through a surface or are constrained in some other way. Let us consider a membrane separating two compartments. Imagine that the membrane is permeable to water and to small ions. Imagine that a solution of KCl is initially on one side and there is only distilled water on the other. Initially, water would move towards the salty side to equalize the water activity. Also, because the salt can pass through the membrane, it would travel in the opposite direction, down its gradient, but more slowly. Sooner or later the concentrations of both water and salt will become the same on each side of the membrane, if there is no hydrostatic pressure difference. If the membrane became mechanically stretched from the initial water flow, then at equilibrium the membrane will return to its initial position as a flow of salt solution returns the volumes of the compartments to their original values.

Now consider the case where a large ion that cannot cross the boundary is initially present on only one side of a membrane. Let us call it R^{-n} where n is the number of negative charges on this polyanion. Of course, the molecule must be associated with cations, say n molecules of K^+.

The K^+ and the Cl^- can cross the membrane, but not independently. If one moves slightly ahead of the other, then a voltage develops that prevents further separation of charges. Such voltages, in the steady state, are called diffusion potentials. Because K^+ and Cl^- have nearly the same rate of diffusion, the diffusion potential for this salt is very small. This is why KCl is used in salt-bridges. In the case under consideration, if the $K_n R$ were added to one side of the equilibrium system described above, then the concentration of K^+ would be larger on that side and so K^+ would diffuse faster through the membrane to the opposite side than K^+ would diffuse back towards the side with the impermeable anion. Quickly an electrostatic potential would develop that would speed the movement of the only anion that can move, Cl^-. Electrostatic forces are very powerful; this means to the biologists that the cation must accompany the anion in crossing the membrane with very little imbalance between the two. The slight imbalance is important because it generates diffusion potentials, but the amount of unbalanced (displaced) charge is extremely small in every case. Because the ions must move in concert, transport is a second-order kinetic phenomenon. If we designate k as the rate constant for the second-order rate constant for permeation in one direction, it must necessarily apply in the other direction, and it follows that:

$$k[K^+]_o[Cl^-]_o = k[K^+]_i[Cl^-]_i, \quad \textit{Rate-in equals rate-out}$$

where the subscripts stand for outside and inside. Now at equilibrium, if the concentration of the neutral salt is designated x on the side with the impermeable anion and y on the other side, then:

$$(y)(y) = (x + nR)(x), \quad \textit{Donnan equilibrium condition}$$

where we have canceled the k's. This is the condition for equilibrium of salt diffusing to and fro through the membrane. If the volume of the compartment with salt alone is large, then given nR and the initial value of y, we can calculate x from the *Donnan equilibrium condition* equation and then set the $Cl^- = x$ and the $K^+ = x+nR$. Now we can do several things.

We can calculate the Donnan potential, the gradient potential arising because K^+ (and Cl^-) are present at different concentrations on opposites sides of the membrane. This requires a direct application of the Nernst equation that relates the electromotive force, EMF, between two compartments to the concentrations in the two compartments. Consequently, the potential for the K^+ is:

$$\text{voltage} = \text{EMF} = (RT/zF)\ \ln[(x+nR)/y].$$

For the Cl^-, it is:

$$\text{voltage} = \text{EMF} = (RT/zF)\ \ln[y/x],$$

where z is the number of charges on the particles, and F is the Faraday, 96, 487 coulombs/mole and R is the gas constant. Note that the potential for the cation and anion are exactly opposite; this can also be seen from the *Donnan equilibrium condition* equation. The actual potential that we might measure depends on whether our electrodes respond to the cation or the anion (metal electrodes usually respond to the cation). When the quadratic solution to the *Donnan equilibrium condition* equation is substituted into the equation for K^+, we have:

$$\text{EMF} = (RT/zF)\ \ln[(-nR + \sqrt{n^2R^2 + 4y^2})/2y].$$

If y is very large and you check through all the parentheses, then you can see that the argument of the logarithm function is unity and that the EMF is then zero. We say that there is no potential when the salt concentration is large and that the potential is swamped out by the high concentration of salt.

The potential, however, does become large when there is little salt and x and y are small. To see this you would have to expand the square root by the binomial theorem. First rearrange the last equation to become:

$$\text{EMF} = -(RT/zF) \quad \ln[nR(-1 + \sqrt{1 + 4\,y^2/n^2R^2})/2y\,].$$

Then expand $\sqrt{1 + 4y^2/n^2R^2}$ remembering that the binomial theorem for an exponent with a value of 0.5 is $(a + b)^{0.5} = a^{0.5} + 0.5a^{-0.5}b \ldots$ We need keep only two terms, so this yields:

$$\text{EMF} = -(RT/zF) \quad \ln[nR(-1 + 1 + 2y^2/n^2R^2)/2y]$$

or

$$\text{EMF} = -(RT/zF) \quad \ln[y/nR].$$

The potential depends on the concentration ratio y/nR and in principle it would be infinite if y were zero.

These concepts will be important later on, but here we need to calculate the osmotic consequences. By adding up the concentrations of the three kinds of particles on one side and subtracting those for the two on the other, we readily obtain:

$$\pi = RT[R + (x + nR) + x - y - y]$$

or after using the *Donnan equilibrium condition* equation and doing some algebra:

$$\pi = RT[R + \sqrt{\{n^2R^2 + 4y^2\}} - 2y)].$$

If the salt concentration is high, then the square root becomes $2y$, the last two terms cancel, and the osmotic pressure is as if there were no counterions to the impermeable species and thus:

$$\pi = RT[R]. \qquad\qquad \textit{High salt}$$

On the other hand if the salt concentration is very small, when the same kinds of rearrangements and expansions by the binomial theorem are done, the square root becomes $2nR \sqrt{\{1 + 4y^2/n^2R^2\}}$, and we obtain:

$$\pi = RT[R + nR], \qquad\qquad \textit{Low salt}$$

and since $K^+ = nR$ the osmotic pressure is proportional to $[R]+[K^+]$ and thus under low salt conditions is responsive to the counterions as well.

This is relevant to the action of MDOs (membrane-derived oligosaccharides). These are only formed and accumulate in the periplasm of Gram-negative bacteria when they are grown at low external osmotic pressure. These polysaccharide molecules typically bear five negative charges and are so large that they cannot pass through the porins in the outer membrane to equilibrate with the environment. So these molecules contribute six particles for each molecule of MDO to the osmotic pressure, but only do so when the ionic strength is low.

How Organisms Solve Their Osmotic Problems

In this section we will outline the cell's problem and its solution, but the details are the topic of most of the rest of the book. Success in pumping small molecular weight constituents into the cell means that osmotic pressure is created that must be resisted or it would cause catastrophic swelling, especially when the cell is surrounded by nearly pure water. The prokaryote approach and that of plants is to construct an enclosing stress-resistant structure outside the cell membrane. Contrast this with the cytoskeletal structure inside many eukaryotes. These bear some stress but only to the degree that they are multiply attached to the cytoplasmic membrane. For free-living protozoa any internal scaffolding is generally supplemented by the formation and extrusion of water vacuoles to solve the problem in a kinetic (boat-bailing) way (see Fig. 5.1).

Although a strong wall is a simple solution to the osmotic problem, there is a price to pay for construction of an exoskeleton, as was noted in the previous chapter. The external location of the stress-bearing structure in prokaryotes means that the precursors have to be pre-charged with the energy to pay for the formation of the linkages needed to link the structure covalently, and in many cases as new growth occurs, old wall is discarded instead of being reutilized because of a lack of a better way to enlarge the sacculus.

The Measurement of Turgor Pressure

There are no good general ways to measure turgor pressure in such small cells as bacteria. For larger cells, pressure measurements can be made by inserting pressure-transducing probes into them. However, there are four methods that have been used for bacteria. I shall discuss three briefly. The fourth I shall present in more detail and discuss its, as yet, unrealized potential.

Osmotic Pressure from Cell Composition

The first method is simply to measure the concentration of the cell's constituents and its volume, and then from these compute the osmotic pressure. Let us see how far we can take such a calculation. A list of cellular components of typical Gram-negative bacteria and their concentrations in terms of the molecules per cell from the data taken from a number of experimental studies has been assembled

Prokaryote Eukaryote

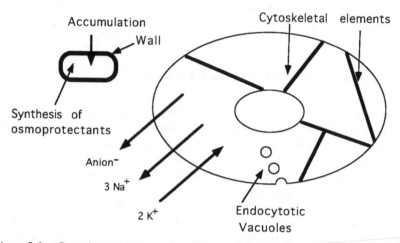

Figure 5.1. Osmotic strategies of the bacteria and animal cells. The prokaryote (shown on the left) resists osmotic pressure by building a strong wall. In a high osmotic pressure environment, it increases its internal osmotic pressure by accumulating (concentrating) solutes from the environment, seemingly to maintain a constant differential osmotic pressure and therefore a constant hydrostatic (turgor) pressure. Many cells also manufacture osmoprotectants; these are substances that increase the osmotic pressure without interfering with the use of cellular processes. The animal cell (shown on the right to approximate scale), in many cases, grows in an environment isotonic to its cytoplasm (i.e., it is bathed in a regulated internal milieu). Free-living cells (e.g., protozoa) resist the osmotic pressure differentials by forming and excreting water vacuoles and pumping out salt, which then leads to exodus of water. All eukaryotic cell have a cytoskeleton to maintain and modulate their shape and to resist to some degree osmotic stress. Also shown in the diagram is the electrogenic sodium/potassium exchange that animal cells use to energize the membrane by storing energy as in a charged electric capacitor, concentrating K^+ in the cell and extruding Na^+.

by Neidhardt *et al.* (1990, p. 4). Let us make a reasonable assumption about average molecular weight of the soluble components (200 daltons) and the inorganic ions (100 daltons). From this it follows there are about 24,100,000 soluble molecules (building blocks, metabolites, cofactors); 18,100,000 ions (predominantly potassium); 1,390,000 proteins not bound in ribosomes; 205,000 molecules of tRNA; 18,700 ribosomes with all their intrinsic proteins (but not the "factors" and the mRNA molecules); and 2.13 molecules of DNA per cell (see Fig. 5.2). Taking the volume of the cell water to be 6.7×10^{-13} ml, conversion of units shows the osmotic pressure is about 108 mosmolal and almost entirely (>90%) due to the small molecules. The term mosmolal is similar to mM; it refers to one thousandth the number of osmotically equivalent moles of an ideal substance

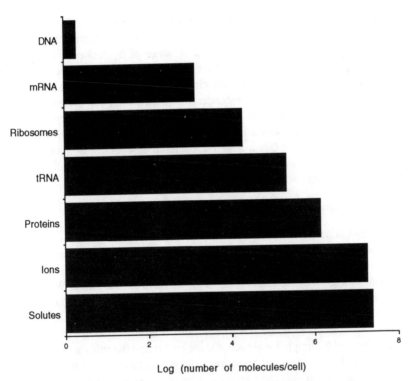

Log (number of molecules/cell)

Figure 5.2. Contributors to the osmotic pressure. Data from the literature (Neidhardt *et al.* 1990) were reexpressed in terms of the logarithm of the number of molecules per cell. It can be seen that solutes and ions are the main contributor to the osmotic pressure of *Escherichia coli* growing in a minimal glucose medium. It is noted in the text that these observed concentrations do not correlate with the osmotic pressure of the cell and, in fact, they predict a negative turgor pressure. A number of factors probably account for this discrepancy. A factor is the partition of water into domains, which in essence makes some regions have a much higher osmotic pressure than others. Additionally, water in some regions is bound and not free to serve as solvent. On the other hand the flexibility of some of the macromolecules would make each contribute as if it were more than one particle.

would give when dissolved in 1,000 g of the solvent. In the typical glucose minimal culture medium used to grow the *E. coli* analyzed to obtain the above data, the osmotic pressure is about 300 mosmolal. Surprisingly, from these facts the cell would appear to have a turgor pressure of −192 mosmolal. This would convert to measures in pressure units of *minus* 484 kPa or *minus* 4.84 atmospheres. Something is radically wrong because theory and the other kinds of measurements to be cited below all give positive numbers. As small ions are the big contributor, we could do much better with a more detailed analysis of the composition if the soluble part of the cell were more accurately and specifically analyzed. The major part of this inconsistency, however, is explained by noting that much of the water of the cell is not solvent, but is water of hydration of the cellular macromolecules. Such water is bound and does not contribute to the concentration of water in the mobile, osmotically relevant phase. Additionally there are domains where the water has a different concentration of osmolytes (Wiggins 1990). Consequently, we end up with a very vague impression of the turgor pressure and a distrust for calculations from chemical composition. Although this approach is very approximate for Gram-negative bacteria, it works reasonably well for Gram-positive organisms that have a much higher content of small molecular weight components and a higher internal osmotic pressure.

Osmotic Pressure by Water Content after Equilibration

The second method is the classic one of Mitchell and Moyle (1956). They prepared pastes of cells on tared vessels, weighed, dried, and reweighed them, and then placed them in individual closed chambers containing sulfuric acid solutions of known concentrations and allowed them to equilibrate. Water distills from the sulfuric acid solutions into the dried pastes, which eventually become hydrated to the equilibrium degree (Fig. 5.3). Because vapor pressure is another of the four colligative properties of solutions (the other two are freezing point depression and boiling point elevation), the pastes and the sulfuric acid solutions had to have the same osmotic pressures. The sulfuric acid solutions were present in large volumes so that their osmotic pressure would not change appreciably during equilibration. The amount of water taken up by the paste samples could be easily determined by rapid reweighing. When the sulfuric acid was too dilute, water would continue to distill into the paste to make them weigh more than they had originally (Fig. 5.4). If the acid was too concentrated, they would weigh less than original hydrated paste. In the series, some pastes were rehydrated to near their original wet weight; then the sulfuric acid solutions with which they had been equilibrated had to have an osmotic pressure approximately equal to that of the original bacterial cells. A more exact osmotic pressure could be obtained by interpolation. Because such experiments take a long time to achieve equilibrium, these experiments are flawed by the possibility, nay the probability, that the bacterial components had become degraded or aggregated and therefore the osmotic pressure had become altered. Correction for extracellular (interstitial) water in the original pastes also presents severe problems. Mitchell and Moyle's (1956) experiments established that the osmotic pressure was high (approximately 20 atm) in a Gram-positive organism, *Staphylococcus aureus,* and lower but still a very potent force in the Gram-negative *E. coli* (5 atm).

Osmotic Pressure by Plasmolysis

The third approach has been used in many guises. Basically, additions are made to the medium to increase the external osmotic pressure until plasmolysis is

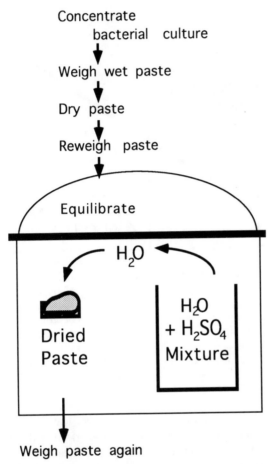

Concentrate
bacterial culture

Weigh wet paste

Dry paste

Reweigh paste

Equilibrate

H_2O

Dried
Paste

H_2O
$+ H_2SO_4$
Mixture

Weigh paste again

Figure 5.3. Osmotic pressure measurements by the method of Mitchell and Moyle (1956). A paste of cell is weighed in a tared vessel, dried, and reweighed. It is then equilibrated with a large volume of a sulfuric acid solution of known concentration. At equilibrium the osmotic pressure of the cell paste must equal that of the sulfuric acid solution. This is a thermodynamically exact relationship. This is done with a range of concentrations of the acid in different closed vessels with different water/acid mixtures. Then, from the known properties of the solutions, the original osmotic pressure can be calculated as indicated in Fig. 5.4. The problem with this method is that changes in the cell constituents are likely to occur during the drying and equilibrating process.

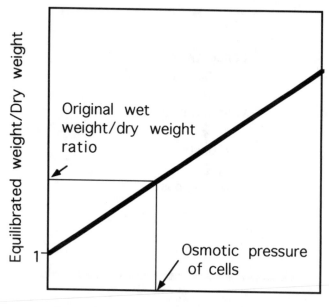

Figure 5.4. Weight of pastes equilibrated with solutions of different osmotic pressures. Experiments like that shown in Fig. 5.3 are set up in multiplicate, but with a range of precisely known sulfuric acid concentrations of the solutions, and therefore various known osmotic pressures. Then the ratio of the wet weight to the dry weight is plotted against the osmotic pressure. The osmotic pressure of the original paste of cells can then be calculated.

observed. Plasmolysis is the process in which water leaves the cell (because of osmotic challenge or by water deficiency due to drying) so that the cell must shrink, and because the wall itself cannot be shrunk beyond a certain point, the cytoplasmic membrane pulls away from the cell wall. Turbidity, light scattering, phase, and electron microscopy are some of the many techniques that can be utilized to follow the process of plasmolysis. Perhaps the most elegant method is that of Stock *et al.* (1977). They followed the space accessible to tritium-labeled water, to ^{14}C-labeled sucrose, and to ^{14}C-labeled inulin (inulin is a fructose polymer of 5,000 daltons molecular weight). Very concentrated suspensions of cells were diluted precisely with known volumes of the radioactive solutions. After equilibration and centrifugation, the radioactivity of a measured aliquot of the supernatant fluid was assayed to see how much it was diluted by the fluids in the paste. Because water can equilibrate across the cytoplasmic membrane, sucrose can equilibrate across the outer membrane into the periplasmic space but not further, and the high molecular weight inulin cannot penetrate the outer membrane and is therefore restricted to the outside of the cell, the dilution factor of the different radioactive probes is different (see Fig. 5.5). From this kind of data the volume of each space excluded from the tracer substance can be calculated. Finally, from the cell number or biomass, the aqueous volume per cell or per unit biomass of the cell of these cell compartments, called "spaces," can be calculated.

These spaces, when measured under a range of conditions, can be used to calculate the osmotic pressure; the actual experiments are to add nonradioactive sucrose, or other nonpermeant solute, in a range of concentrations to find out what external osmolality is just sufficient to cause a decrease in the volume enclosed by the membrane through which sucrose cannot penetrate. In the experiments of Stock *et al*. (1977) the other two radioactive substances serve as controls for interstitial volume and for the volume of the periplasmic space.

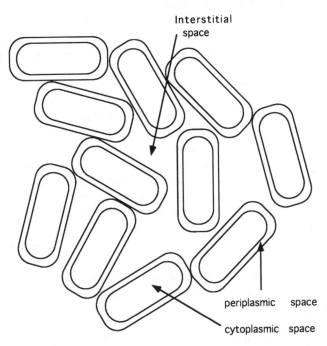

Figure 5.5. Fluid compartments in a paste of Gram-negative cells. The three compartments are: the fluid outside the cell, (in a packed suspension of cells thus called the interstitial space); the fluid between the outer membrane and the cytoplasmic membrane, called the periplasmic space; and the volume inside, called the cytoplasmic space. These spaces can be measured with radioactive probes that enter the three compartments. Typically the probes include: tritiated water, which can equilibrate with the entire aqueous space; radioactive sucrose, which can penetrate the outer membrane, but not the inner one; and inulin, a large nonutilizable polysaccharide (Mol.Wt. ≈ 5,000) that cannot penetrate at all. Between the three the cytoplasmic and periplasmic spaces (volumes), as well as the less interesting interstitial space, can be calculated.

Although the approach is valid no matter how the volume enclosed by the cytoplasmic membrane is measured, there are technical difficulties. For most procedures very dense suspensions are required. For example in the experiments of Stock *et al.* (1977) a sufficiently dense suspension is needed so that a large enough fraction of the volume is occupied with cells, greater than 10%. To put this number in perspective, in a typical growing bacterial culture of 10^8 cells per ml, only 0.01% of the suspension is occupied by cells. Consequently, important changes occur during the 1,000-fold concentration due to centrifugation, due to starvation of both nutrients and oxygen, and due to possible toxicity from cell excretion products.

Almost all experiments in the literature with this third approach are subject to the severe criticism that the cell's response mechanisms to osmotic challenge may have caused an increase in the internal osmotic pressure before measurements could be made or completed. All bacteria seem to be capable of some adaptation to different environments, but although the mechanisms are understood to a degree (see Csonka 1989; Csonka and Hanson, 1991), the speeds of response are virtually unknown.

To avoid these complications, I carried out some experiments of this third class, but without either of the liabilities mentioned. I did this by using cultures while they were in active balanced growth and following their turbidity in a precision spectrophotometer (Koch 1984c). I designed the experiment so that there was no centrifugation or concentration step. Instead, the growing and aerated culture was pumped, within a few seconds, from the water bath into a stopflow apparatus. The cuvette was inside a thermostatted spectrophotometer. The turbidity changes were recorded on a chart recorder and showed that the changes due to osmosis were completed in a few seconds, but that recovery took several minutes to become appreciable. Even these experiments are not fully unambiguous because of two additional complicating factors due to raising the osmotic pressure outside the cells: one is the changes in the periplasmic space (see below) and the other is the shrinkage of the wall because it is no longer under stress in the presence of an osmotic supplement (also see below). Shrinkage of the wall, as distinct from the retraction of the cytoplasmic membrane from the wall in frank plasmolysis, would have different effects on the measured turbidity. There is an equivalent process to the shrinking: as water leaves the cytoplasm the three layers of the Gram-negative wall may wrinkle, but not enlarge the periplasmic space. However, certainly the stopflow technique allows measurements to be made before the effects of biological accommodation can be exerted.

Measurement of Turgor Pressure by Collapse of Gas Vesicles

The fourth method to be considered consists of refinements of methods originally devised by Walsby (1971, 1994) to study the collapse of gas vesicles (also called gas vacuoles) within bacteria. Certain aquatic prokaryotes (both blue-green cyanobacteria and heterotrophic bacteria) have gas-filled vesicles. The vesicles are composed primarily of one protein that self-assembles into hollow cylindrical objects with conical ends. The protein sequence and the structure is such that the inner face of the vesicles is so extremely hydrophobic that water tends to leave and the space becomes filled with atmospheric gases of the same composition as that of the gas phase with which the cells have been equilibrated (Walsby 1980). (*Note*: Most gases are actually hydrophobic.)

The vesicles, no doubt, permit the bacteria to become buoyant. This permits the cells to float near to the surface of a lake or ocean. While especially advantageous for a photosynthetic organism, this feature is also present in some nonphotosynthetic heterotrophic organisms as well, two of which our group has chosen to study. With organisms containing such vesicles, by using Walsby's method (1980), it is possible to measure the turgor pressure. The technique is as follows. Pressure is gradually applied from the outside while measuring the light scattered by the cultures. Because the vesicles are rigid structures, they retain their size until the applied pressure is such that they collapse, analogously to the sudden crushing of a pop or beer can. When they collapse they scatter much less light. The experimenter raises the pressure until all of the vesicles have collapsed and then estimates the pressure at which half of the vesicles would have collapsed. The pressure actually applied to the gas vesicles throughout the measurement is the sum of the turgor pressure developed by the cell and the variable hydrostatic pressure applied by the experimenter (see Fig. 5.6). Collapse is resisted by the mechanical strength of the vesicles in combination with the hydrostatic pressure due to the gases within the vesicle. If we define C as the mean pressure needed to collapse vesicles inside the living cell, C_a' as the mean collapse pressure of the cell-free vesicles, P_t as the turgor pressure, and P_v as the partial pressure of gases within the vesicles, then these quantities are related by:

$$C = C_a' + P_t - P_v.$$

This expression is correct if one uses gauges that record absolute pressures. However, most pressure instruments are calibrated to read the pressure relative to atmospheric pressure. In addition, P_v cannot be measured routinely, but it can be expected to be the same as the total pressure of the gases in the environment of the organism. In our experiments, organisms are grown with gentle aeration with normal air; thus, P_v can be taken as the atmospheric pressure. These two circumstances afford a simplification because we can replace $C_a' - P_v$ with C_a, the mean total pressure above atmospheric pressure that causes collapse of vesicles when not in the cell. This substitution yields:

$$C = C_a + P_t \qquad \text{\textit{Collapse pressure}}$$

Once C has been measured, if C_a, the collapse pressure of free vesicles were known, then the turgor pressure could be directly estimated by subtraction.

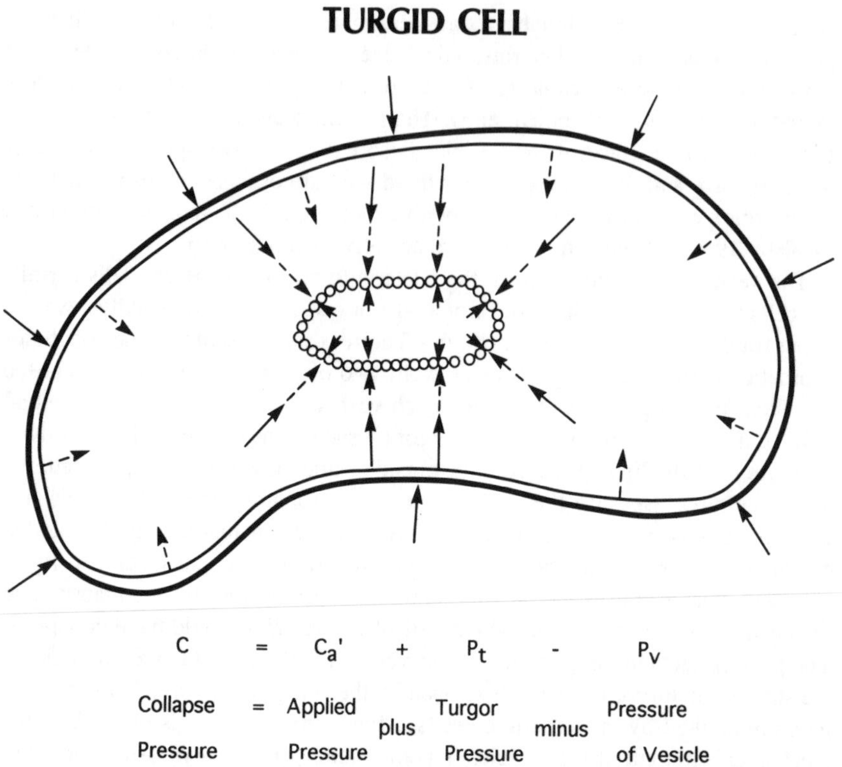

TURGID CELL

$$C \quad = \quad C_a' \quad + \quad P_t \quad - \quad P_v$$

Collapse	=	Applied		Turgor		Pressure
			plus		minus	
Pressure		Pressure		Pressure		of Vesicle

Figure 5.6. Turgor pressure and pressure causing the collapse of gas vesicles in bacteria while growing in their normal medium. A normal growing cell containing a gas vesicle (vacuole) is shown, not drawn to scale. This entity contains ambient gases at the ambient partial pressure. This pressure is opposed by the cell's turgor pressure and by any pressure applied by the experimenter. Experimentally the pressure is raised until the vesicles collapse and the light-scattering properties change. The pressure that causes half of the vacuoles to collapse is the applied pressure P_r. Besides knowing the value of P_t, P_v is known as it is equal to the ambient gas pressure under the conditions where the culture was grown. The mean pressure that collapses the average vesicle P_c in an absence of turgor pressure is determined by the experiment described in Fig. 5.7. Consequently the turgor pressure P_t, the quantity of interest, can be calculated.

However, it is difficult to isolate vesicles so that their collapse pressure can be estimated independently (they collapse even during gentle centrifugation). But it is easy to suspend the cells in a solution of high enough osmotic pressure so that the cell becomes plasmolyzed. This means that the turgor pressure is zero (Fig. 5.7). (*Note*: The turgor pressure cannot become negative because if it were to momentarily become negative, water would redistribute until there was no pressure differential across the flexible cytoplasmic membrane.) Consequently,

PLASMOLYSED CELL

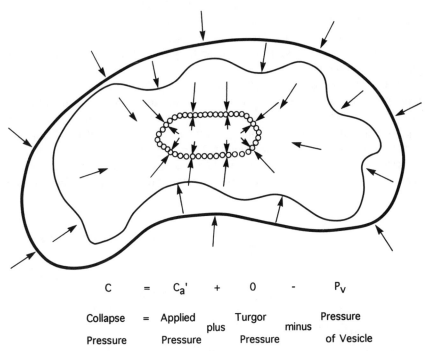

$$C \quad = \quad C_a{}' \quad + \quad 0 \quad - \quad P_v$$

Collapse Pressure	=	Applied Pressure	plus	Turgor Pressure	minus	Pressure of Vesicle

Figure 5.7. Effect of osmolyte (usually sucrose) on turgor pressure and the collapse of gas vesicles. When a cell is plasmolyzed with more than an adequate amount of an osmolyte, water leaves the cell, the wall may shrink a little, but the volume enclosed by the cytoplasmic membrane can shrink a great deal because the membrane is flexible. After movement of the membrane and water, it is assumed that the turgor pressure is zero. With $P_t = 0$ the equation given in Fig. 5.6 simplifies and the vesicle's critical collapse pressure is obtained by calculating $P_a - P_v$. This quantity is the applied pressure above atmospheric pressure and is directly measured by the pressure gauges and transducers. With P determined in this way, the turgor pressure in any set of experiments can be determined by the experiment shown in Fig. 5.6.

under these conditions the applied pressure giving collapse measures the mechanical strength of the vesicles themselves.

In carrying out such measurements it is necessary to be quick. Speed is essential for a number of reasons, but there are two overriding ones. As the pressure is increased, the partial pressure of gases within the vesicles increases and the presumption of the experiments is that it stays fixed at the ambient barometric pressure under which the culture was grown. But the more important problem is that the turgor pressure, although obliterated by the initial osmotic challenge, may redevelop as one or more of the cell's regulatory mechanisms comes into play and accumulates and/or synthesizes internal osmolytes. Evidently, to measure turgor pressure accurately such recovery must be avoided. There are only two valid approaches to overcome such changes in the internal osmotic pressure. One is to genetically delete *all* the regulatory mechanisms; the other is to speed up the measurements. It seemed to me and my coworkers that it was more to the point to do the latter.

On this basis we have assembled an apparatus (though several intermediate developmental stages) that now has the properties that allow us to measure the turgor pressure throughout physiological experiments without the criticism or objection that the internal osmotic pressure P_v has become altered before or during the experimental measurement (see Fig. 5.8). Our current apparatus has quick-connect fittings to attach the pressure source to the measurement cuvette. The sample can be mixed with various osmolytes or drugs, placed in the cuvette,

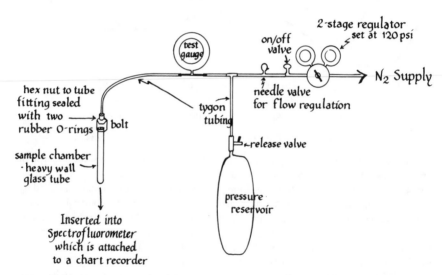

Figure 5.8. Diagram of apparatus used to measure turgor pressure. This schematic for the pressurized part of the apparatus contains a number of elements, mostly of an obvious nature. The pressure reservoir, or ballast, has the role of causing a gradual pressure ramp to be applied to the biological sample. The reservoir communicates with the rest of the system. When the on/off valve is opened, gas flow is limited by the needle valve. The pressure can rise only as it fills the reservoir. Because of the regulated pressure available from the cylinder P_{cyl}, the pressure on the sample starts rising according to the expression $P_a = P_{cyl}(1 - e^{-nt})$, where n is the rate constant corresponding to the setting of the needle valve. This expression is essentially linear in the initial part of the run before a pressure-protecting valve starts to act, limiting further increase in pressure. The result is that the pressure rises in a linear fashion and then plateaus at a constant value almost like the Blackman curve discussed in Chapter 10.

initial light-scattering measurements made, the pressure ramp started, and the final measurements recorded in less than 1 min. The physical arrangement of gas tanks, quick release valve, needle valve, pressure ballast, and relief valves gives a pressure increase that rises linearly to a specified value and then plateaus. The analytical part of the apparatus consists of a helium/neon laser whose beam is expanded slightly to illuminate a portion of a culture in a square cuvette. A solid state detector views the cuvette at right angles, and baffles are present so that the detector does not "see" the walls of the cuvette where the primary beam enters and leaves the cuvette. The output of the solid state detector and the pressure gauge are fed to a "smart voltmeter" and through an A/D converter to one of several available computers so that the collapse of vesicles versus the applied pressure can be graphically and/or numerically analyzed.

Gas vesicles have been found in several major groups of bacteria, including halophiles, cyanobacteria, and methanogens. In our initial experiments *Ancylobacter aquaticus* (also known as *Microcyclus aquaticus*) was chosen simply because it forms a large number of gas vesicles and is a Gram-negative heterotroph. We hoped that our results would apply to other Gram-negative organisms, such as the well-studied *E. coli*. Recent cloning of the gas vesicle protein gene together with ongoing attempts to induce gas vesicle assembly in *E. coli* may eventually allow direct measurements of turgor in *E. coli* by using these methods (Tandeau de Marsac *et al.* 1985). But for now, results for only two aquatic oligotrophic bacteria are available.

A second variation on the instrumentation is available. It uses the component parts of the first apparatus, but the analytical part is a microscope. With it one can observe the collapse of the gas vesicles within the cell. It was used to measure the variability of turgor pressure throughout the cell cycle.

Interpretation of Vesicle Collapse Curves in Various Concentrations of Sucrose

A. aquaticus grows slowly (doubling time, 4.1 h at 30 °C) in a very dilute medium whose main constituents are minerals and vitamins. Compared to *E. coli* growing in the usual minimal medium, little biochemical work is devoted to pumping substances to maintain an osmotic differential with the medium. These cells on the other hand have to do more work to accumulate needed potassium. In the experiments shown in Fig. 5.9, the samples from a growing culture of *A. aquaticus* were diluted with solutions of sucrose prepared in the growth medium so that the final osmolality of sucrose in each case was precisely known. Collapse curves were obtained immediately after addition of the sucrose solution. Diluents were brought to the growth temperature (30 °C) before use. The 50% collapse pressure (i.e., the pressure that causes the light-scattering signal to decrease halfway from its initial to its terminal value at high pressure) determined from these curves is plotted against the final osmolality in the sample

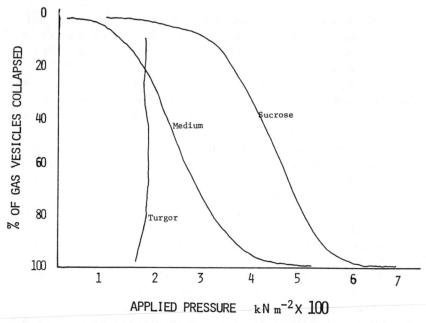

Figure 5.9. Collapse pressure curves under the conditions of Figs. 5.6 and 5.7. In the presence of sucrose the curve is displaced to higher applied pressures because in that situation there is no turgor pressure to add to the applied pressure in compressing the vacuoles. The lateral shift between the two curves corresponds to the turgor pressure indicated by the approximately vertical line.

in Fig. 5.10. It can be seen that as the osmolality of sucrose increases, C_a increases in a curvilinear relationship. The predictions of a variety of models are indicated by lines A through E.

The experimental results were reproducible and quite different from the predictions of several of the conceivable models. The simplest model is where the bacterial cell wall is rigid so that the cells have a constant volume and the collapse pressure is the same for all cells. Constant cellular volume implies that the internal osmotic pressure does not change as the external osmotic pressure changes because there would be no water flux in such a rigid container. For this case a biphasic curve (Curve A) is expected. At low external osmotic pressures, the turgor pressure decreases and C_a increases by the same amount that the external osmolality is increased; i.e., 2,520.5 kPa for every 1 osmolal increase in external osmolality. If, however, the external osmotic pressure were made high enough with sucrose to completely remove the turgor pressure, no further increase in the osmotic pressure of the medium would affect the observed 50% collapse pressure and C would equal C_a. Consequently, for the case of a population of identical bacterial cells of fixed volume, all with identical values of C and P_t,

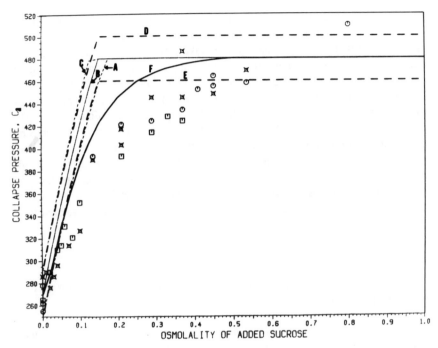

Figure 5.10. Changes in turgor pressure with the addition of sucrose to the suspending medium. Actual data from a number of replicate runs is shown for *Ancylobacter aquaticus*. Also shown are various theoretical curves. The important conclusions to be drawn are, first, that the cell wall is normally stretched and relaxes as sucrose is added, long before plasmolysis takes place and, second, that no model is adequate. This figure is taken with permission from Pinette and Koch (1988).

C_a would rise linearly with the increase in external osmotic pressure until the turgor pressure was reduced to zero, after which C_a would remain constant at C. Clearly, this prediction is very different from the actual experimental results.

The observed initial slope (590 kPa/mosmolal) is 23% of the predicted slope if the cell wall viewed as a container were rigid (2,520.5 kPa/mosmolal). This suggests that the cell wall is elastic and that as the external pressure is raised, water moves outward, the cell contracts, and the cytoplasm is concentrated. However, shrinkage explains only a portion of the difference between curve A and the experimental data points. Walsby (1971, 1994) assumed that the volumetric elastic modulus is constant (i.e., the change in the volume is proportional to the change in turgor pressure). Curves based on this assumption are marked B and D. We will return to them after considering a different model.

The curves marked C and E are the predictions of what has been designated the Hooke's law model (Koch and Pinette 1987). It assumes that the sacculus expands in each dimension with a constant value of Young's modulus. (These terms are discussed in Chapter 6.) This model proposes that the change in cell volume is a consequence of the effectively elastic behavior of the sacculus. It also assumes that changes in cell volume take place by the movement of water only. Consequently, the linear dimension of the cell wall changes with the net pressure acting on it in proportion to $(1 + kP_t)$, where k is the reciprocal of Young's modulus and P_t is the turgor pressure at a specified concentration of external osmolyte. The volume of the cell changes in proportion to $(1 + kP_t)$ raised to the third power. Define π_p as the osmotic pressure in the internal cytoplasm at just the point of plasmolysis where enough osmolyte has been added to the medium so that π_e equals π_p and hence $P_t = 0$. At this point the wall will not be stretched because it has no tension. During growth when the external osmotic pressure is essentially zero, the cell would be stretched and its contents diluted, decreasing the osmotic pressure. Consequently, π_i will be related to π_p and P_t by

$$\pi_i = \pi_p/(1 + kP_t)^3.$$

The relationship of internal and external osmotic pressure to the turgor pressure is given more generally by:

$$P_i = \pi_p/(1 + kP_t)^3 - \pi_e; \quad \pi_i > \pi_e. \qquad Hooke's\ law\ model$$

Because P_t is on both sides of the equation, it is actually a quartic in P_t which applies at external osmotic pressures below the point of incipient plasmolysis. Assuming that plasmolysis then takes place quickly at any external osmotic pressure above this point, the turgor pressure becomes:

$$P = 0; \qquad \pi_i = \pi_e.$$

Using P from the appropriate one of these two equations, the predicted values of C for given π_e can be calculated from values of P_t, k, and C_a, where $C = C_a + P_t$.

To analyze the experimental data, Suzanne Pinette and I required a means of estimating P_t, C_a, and k. To do this, the data was fitted by a minimization procedure. The sums of squares of the deviations between the observed data and the predicted data for assumed values of these three parameters were calculated. By systematically varying one parameter at a time, the best fitting value was obtained. This minimization is shown by fitted curves D and E in Fig. 5.10.

The program first solves for the value of π_p from the Hooke's law relationship for the case where π_e is zero:

$$\pi_p = P_t(1 + kP_t)^3; \quad \pi_i = \pi_e.$$

This value of π_t is then inserted into the *Hooke's law model* equation. The resultant quartic equation in P_t is then solved by an iterative algorithm. This process is repeated successively for increasing values of π_e. As soon as π_e is greater than π_p $(=\pi_i)$, the program sets P_t to zero.

The program was also used to fit Walsby's model using the *Hooke's law model* equation but without the cubic exponent. In this case, the new form of equation is a quadratic equation which was solved directly.

For future progress, once the mechanical properties of the Gram-negative wall are known in more detail; i.e., its shrinkage and wrinkling capabilities, the ability to adapt the available computer program to other relationships between turgor pressure and volume expansion of the sacculus should become important. The dependence of the volume contained by the sacculus on turgor pressure may be more complex than in any of the models considered so far. Let me outline some improvements and advancements that have yet to be carried out, but should be developed when newer, more precise data are obtained with a wider range of conditions and with a range of different organisms. This anticipated next model will assume that the wall expands with a Young's modulus that is the same in both directions in the wall fabric and that the normal stresses resultants are as indicated in Chapter 6; i.e., the stress is twice as great in the hoop direction as in the axial one.

Curves B and C are fitted to the assumption that the turgor in the growing cell increases the volume to finally become 1.75 times the relaxed turgor-free volume. This factor for expansion is consistent with our microscopic observations. Thus, when either model is fitted to be consistent with a degree of elasticity found in our studies of *E. coli,* the fit is poor. Curves D and E are for the same two models, but for fitted and much larger values of k. Curve E was obtained by minimizing the sum of the squared deviation of the *Hooke's law model* from the experimental data. The factor obtained corresponds to the case of a turgid cell with a volume 7.3-fold greater than that in its relaxed configuration. This expansion factor is much too large to be consistent for our microscopic observations.

The minimization procedure failed when applied to the Walsby model. This was because the sums of squared deviations became progressively larger as larger values of k were tried, and no minimum was found. The line shown as curve D had nearly twice the sum of squares as did curve E, although it corresponds to a similar degree of volume expansion of the growing cell (7.3 times that of the turgor-free cell). The value of k that gives the same sums of squared deviations as does curve E corresponds to the growing cell being expanded thousands of times above the volume possessed when the turgor has been dissipated. On this basis we feel that the Walsby model is inapplicable to the mechanical properties of the sacculus.

There are other possibilities to explain a gradual increase in C_a with increasing external osmotic pressure instead of an abrupt plateau when P_t equals zero. One is that the turgor pressure varies very much from cell to cell even when the sample has been drawn from a culture in balanced growth. To test this the

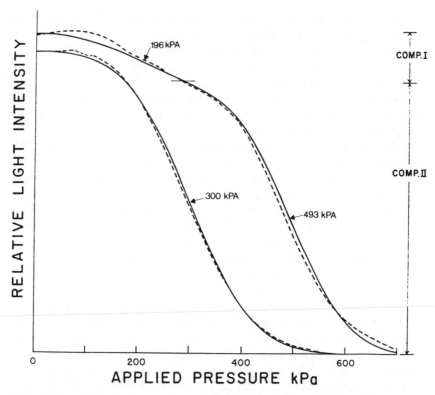

Figure 5.11.   Collapse curves before and after 20-min treatment with ampicillin. Before treatment the curve indicates that the sensitivity of the vesicles in the cells is distributed unimodally with a mean collapse pressure of 300 kPa. After treatment the sensitivity to collapse is bimodal with a large fraction corresponding to vesicles not under a turgor pressure and a smaller fraction under a heightened turgor pressure corresponding to a lower collapse pressure of 196 kPa.

computer program was modified to minimize the squared deviation between a model considering variability among cells and the observed data. In the fitting process it was assumed that Young's modulus is the same as in curve C, but that the turgor pressure is highly variable and is normally distributed from cell to cell in the population.

To take the possibility of variability in turgor pressure from cell to cell into account, the computer program was modified to compute the P_t averaged over 11 turgor classes, such that the central class corresponded to the mean turgor pressure and the others corresponded to the wings of the normal distribution. The two extreme classes correspond to cells that have turgor pressure two times the standard deviation from the mean or greater. In the averaging process, the classes were weighted according to the frequencies of the normal distribution. The best-fitting curve required that the standard deviation of the turgor pressure from cell to cell within the growing population be 105 kPa. Even this best fit does not approach the experimental data very closely, and for this reason it is not shown.

Our experimental results with *A. aquaticus* are quite similar to those obtained by Walsby with the blue-green alga, *Anabaena flos-aquae*. Both organisms have elastic cell walls, so both experience a smaller reduction in turgor pressure than expected based upon the amount that the external osmotic pressure is increased. This is because both quickly shrink (wrinkle) and concentrate the cell contents when placed in a medium of high osmolality. The shrinkage and/or wrinkling is much larger than it is for the cells of higher plants. Walsby's interpretation of quite similar data from *A. flos-aquae* assumes, as Suzanne Pinette and I have, that the cell shrinks as the external osmotic pressure increases. This happens because the stress in the wall changes in proportion to the induced change in turgor pressure. However, Walsby's model states that the change in cell volume is directly proportional to a change in turgor pressure.

Effect of Antibiotics on Turgor Pressure

A major reason for developing a method to measure the turgor pressure of bacteria was to study the effect of antibiotics of the penicillin class and in particular, ampicillin. Before experiments were done, the expected results were almost trivial. Ampicillin would prevent wall enlargement, but other cellular processes should continue, such as uptake of solutes from the medium and macromolecular syntheses. This unbalancing of cell processes should lead to progressively increasing turgor pressure. At some point the cell wall would rupture and spew out the cell's contents and the turgor pressure would abruptly become nil. As can be seen from Fig. 5.11, that is not at all what happened, at least not to all of the cells. Rather the collapse curves 20 min after treatment changed from monophasic to biphasic. From the pressure at the midpoint of the two components, it could be concluded that some bacteria had done what was expected and their vesicles now had a collapse pressure similar to that found in the presence of large concentrations of sucrose and equal to C_a. Unexpectedly, the collapse curve of the remaining cells had a stable turgor pressure but one higher than the untreated cells. This had to mean that in the face of a challenge preventing enlargement of stress-bearing wall those cells had implemented a strategy of stopping further increase in turgor pressure. To prevent a blowout, the cells must have blocked many active transport systems or opened leaks to let ions and solutes return to the medium. Probably they would also have to shut down synthesis of macromolecular species of all kinds. Such changes would seem to be as global in extent as those involved in the sporulation of Gram-positive bacteria.

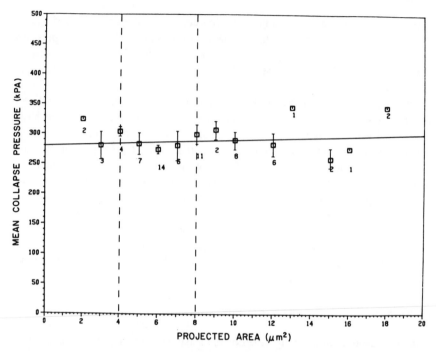

Figure 5.12. The mean collapse pressure of individual cells of *Microcyclus* (*Ancylobacter*) *aquaticus* measured as a function of projected area. It can be seen that there is essentially no dependency on cell size. This suggests, but does not prove, that the turgor pressure inside the cell does not change during the cell cycle.

Single Cell Studies

The pressure-regulating system was also adapted so that cells fresh from a growing culture could be examined under the microscope and the pressure collapsing the vesicles in individual cells determined. The 50% collapse pressures were then correlated with cell size. It was found that there was no significant change in pressure throughout the size range that covers the entire cell cycle (Fig. 5.12).

6

The Surface Stress Theory: Nonvitalism in Action

Cleavage of Wall to Allow Growth
Development of Cell Shape
The Aberrant Development of Bulges
How Can a Cell's Enzymes "Know" Where To Insert New and To Cleave Old
  Wall?

KEY IDEAS

Life accords with engineering principles.
Turgor pressure causes stress.
Surface tension is the amount of work needed to make a unit of area.
Isobaric processes mediate bacterial growth, in some cases.
Stable zonal growth is impossible for cylindrical elongation.
Stress lowers the energy of activation of autolytic enzymes.
The "make-before-break" strategies maintain the cell's integrity.
Tensile and compressive stresses in the wall determine growth patterns.
Enzymes respond only to the physicochemical nature of their immediate environment.
The growth of rod-shaped bacteria requires special strategies.

Beginnings of the Surface Stress Theory

D'Arcy Wentworth Thompson (1942) spent the two world wars writing and then rewriting *On Growth and Form*. His strictly reductionist thesis is that biology reflects engineering principles. He details two types of tenets. First, biological structures are constructed, whether by a "creator" or by Darwinian evolution, with the same design considerations that a human engineer or architect would employ. Secondly, biological systems use engineering principles to achieve biological ends.

I have applied this second concept to the case of bacterial growth. I have stubbornly taken the position that the bacterial cell has no stress-bearing contractile internal cytoskeleton to serve as guy wires or to do mechanical work, nor does it construct a temporary external scaffold that could support stress while the wall grows in a manner similar to the way in which some viruses are assembled (Casjens and King 1975). I have found no evidence for mechano-proteins able to generate stresses and organized in a way to lead to special shapes or to power cell movements; certainly there is no trace of those special mechano-proteins and mechano-enzymes that are present in higher organisms. These plus other sophisticated proteins enable the eukaryotic cell to build and modify its own cell shape in many varied and precise ways. Bacteria cannot do this.

The logical corollary to these negative statements is that the prokaryotic cell must rely on its turgor pressure in order to grow and enlarge. From my theoretical studies in the last dozen years, I have come to believe that this type of strategy is capable of sufficient variation to produce the variety of bacterial shapes. The thesis to be elaborated is that the bacterium causes growth in much the same way as a glassblower blows an object of desired size and shape by controlling temperature and pressure. This pressure expansion is a simple idea, and the assumption of a biophysics-based morphology is less mystical than endowing prokaryotes with characteristic and typical properties of eukaryotes. The execution of this prokaryotic strategy appears to be complicated and variations in different species do occur. In outline we understand quite well the growth of the Gram-positive rod. The biophysical mechanism for the case of the Gram-negative rod appears to be more sophisticated, involving more advanced biophysics and enzymology. Further, special modifications of the most basic strategies must apply for cells that are neither cocci nor short rods—fusiforms, budding bacteria, prosthecate bacteria, branching bacteria, filamentous organisms, etc.

In this chapter the elements from physics and engineering will be presented, but their application to different types of bacteria will have to wait until later. First let us consider the energy required to create a new unit of a fluid membrane surface, and later let us consider the stresses to form a solid. I will present the subject this way because the energetic argument requires only algebra. After these two aspects have been dealt with, we will add the constraints imposed by the growth process.

The Energetics of Pressure and Surface Tension of Fluid Membranes

Bacterial walls serve the vital role of containing the hydrostatic pressure within the cell. They are enlarged, in most circumstances, without rupturing or bulging. During growth new material is inserted into the wall. If this insertion were done in a random, nonsystematic way, the consequences could be the same as those resulting from blowing up a soap bubble in spite of the fact that the nongrowing wall is an elastic network of crosslinked peptidoglycan and not a fluid. Even if a part of the wall were to behave as a regular solid, which at some point must be ruptured and converted into fragments linked to newly inserted wall, the soap bubble analogy could be valid. Application of the rules determining the shape of a fluid membrane is crucial to the surface stress theory and this theory will be reconsidered below after the stresses and strains in the wall have been presented.

Consider the energetics of the fluid membrane of two idealized geometric closed surfaces analogous to those relevant for bacterial morphogenesis: the spherical and the cylindrical shell (see Fig. 6.1). Define the pressure as P (which has units of force per unit area). If ΔV is the volume change, then the work of expansion is equal to the product $P\Delta V$. Define the amount of work needed to increase the surface area by a unit amount as the surface tension T (which has units of work per unit area or force per unit distance). If ΔA is the increase in surface area, then the total work is the product $T\Delta A$. At equilibrium when the pressure-volume work done in expansion just equals the work needed to expand the surface, it follows that:

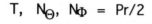

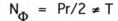

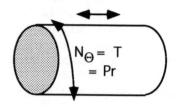

Figure 6.1. The stresses in solid and fluid spherical and cylindrical shells. The sphere and cylinder are the most basic elements of prokaryote morphology. They are shown together with the normal stress resultants developed in the thin walls of idealized nongrowing organisms of these shapes. If instead, the cell is growing in a mode that mimics the behavior of soap bubbles, then the relationship of the surface tension T to the pressure is given by $P = 2T/r$ and $P = T/r$ for the sphere and cylinder geometry, respectively. For the fluid spherical shell, T is equal to both N_Θ and N_Φ of the same shaped shell made of a solid. However for the cylinder, N_Φ is one half of T or N_Θ.

$$P\Delta V = T\Delta A. \qquad \textit{Energy conservation}$$

This is the basic equation for fluid membranes and will be repeatedly used in what follows and related to wall enlargement during bacterial growth.

Consider the expansion of a spherical shell under pressure. If the radius increases by Δr, then substitution in the above equation yields:

$$P(4/3)\pi((r+\Delta r)^3) = T4\pi((r+\Delta r)^2 - r^2).$$

On expanding, retaining only terms with Δr to the first power, and then canceling factors, we obtain:

$$P = 2T/r. \qquad \textit{Spherical equilibrium}$$

The significance of this equation is that a spherical shell formed of a fluid membrane will achieve a contained volume such that the resultant pressure depends on the physical properties of the membrane that fix the surface tension T and on the radius (of curvature) r.

Now consider a capped cylinder. The shape of the rigid caps is irrelevant. If the length of a cylinder increases by ΔL, then $P\Delta V = T\Delta A$ becomes:

$$P\pi r^2((L+\Delta L)-L) = T2\pi r((L+\Delta L)-L),$$

or:

$$P = 2T/r. \qquad \textit{Axial equilibrium}$$

This is the same as for the spherical case.

But there is another way that a cylinder can expand. For the increase in radius of the cylinder, we can write:

$$P\pi((r+\Delta r)^2 - r^2)L = T2\pi((r+\Delta r)L.$$

From this it can be similarly shown that:

$$P = T/r. \qquad\qquad \textit{Hoop equilibrium}$$

These three equations will be needed for very specific cases. Of course, if the caps are truly rigid, the last type of expansion cannot take place and the system must remain cylindrical.

A more general form was derived without any sophisticated mathematics by Thomas Young in 1804 and derived with slightly more elegant treatment than given above by Pierre LaPlace in 1806 (see Hiemenz 1986). They were able to show that:

$$P = T(1/r_1 + 1/r_2). \qquad \textit{Young-LaPlace equation}$$

Here r_1 and r_2 are the principal radii of curvature. One is the maximum and the other the minimum radius and it can be shown that they are to be measured perpendicularly to each other. The equation applies for any part of a membrane surface made with material having a surface tension T and supporting a pressure difference P. This is the general formula for determining the shape of soap bubbles and other fluid membrane surfaces at equilibrium. When the two radii of curvature are equal to each other, this expression reduces to $P = 2T/r$ and, as above, is the equilibrium condition of a spherical shell in which the radius tends to become neither larger nor smaller. Similarly, the general equation reduces to that for a cylinder by taking one radius of curvature to be infinite; then the *Young-LaPlace equation* becomes $P = T/r$. This is the expression for the static shape such that the radius tends to become neither larger nor smaller and the membrane remains cylindrical, but elongation of a cylinder can occur. (The expression for the case of dilation of a cylinder is not covered by the general formulation.)

Implications for Bacterial Growth

Now consider a bacterium that is rod shaped with hemispherical poles; this is the idealized shape often assumed for both *E. coli* and *B. subtilis*. The pressure is the same everywhere inside such a closed container because it has no completed rigid crosswalls separating it into compartments. For the simplest situation that can be imagined, the biochemistry is the same in all parts of the cell envelope and then the same value of T applies to all areas. In the case where the wall turns over at random, the wall will be effectively a fluid membrane, like the soap bubble, even when not enlarging. The analogy to the *Young-LaPlace equation* then requires, in the absence of other constraints, that the cell will have a spherical shape at equilibrium. No matter what the cell's initial shape was, if its walls became uniformly fluid, a growing cell or a cell turning over its wall material would assume a spherical shape. Even if rigid at any instant, the process of growth and metabolism would give an effective fluidity to the wall and enlargement would lead to a surface satisfying the *Young-LaPlace equation,* which would be spherical unless there are other constraints.

One possible constraint is that the poles are rigid, like the soap bubble pipe itself. In that case, turnover of the originally cylindrical region of wall would lead to an equilibrium shape that would satisfy the *Young-LaPlace equation.* It could remain as a cylindrical shell, if P remains exactly equal to T/r.

If growth of cytoplasm is occurring, if wall enlargement takes place only in the cylinder part, if the insertion of new wall occurs at random so that the *Young-LaPlace equation* holds, and if P remains equal to T/r, then the cylindrical region will elongate. This will happen whether or not turnover occurs.

But the chemistry of the poles is very similar, perhaps identical, to that of the side walls. Because it is not possible to satisfy $P = 2T/r$ and $P = T/r$ at the same time with the same values of the parameters, the rod-shaped organism must do something special to maintain its nonspherical shape. There appears to be only two resolutions of this seeming dilemma. One depends on the bacterium's ability to alternate conditions so that first there is growth of one part and then of the other; thus, making the two equations relevant at different times. The other strategy depends on the cell being able to vary T in time and space in a special way.

The first strategy is followed by the Gram-positive rod in which the cylinder part is elongating while suspended from the old poles that are essentially rigid and almost metabolically inert. As a result, the side walls can offer firm support during the laying down of the septum.

In fact, the three wall regions of the growing bacterial rod, whether they are Gram-positive or not, are metabolically very distinct. These are: (1) established poles, which turn over extremely slowly; (ii) the side wall, which grows by diffuse addition or insertion; and (iii) septa or constrictions, which are formed by rapid diffuse addition of new wall for a limited time. Clearly, there is no paradox if one part of the cell wall is a rigid solid while another can enlarge in response to increased synthesis of cytoplasm.

For pole formation in Gram-positive organisms, a crosswall starts to form from the outside; as long as it is incomplete and unsplit, then both sides of the septum are exposed to the same cytoplasm. Consequently, the analogy to the *Young-LaPlace equation* (or the equivalent rules for elastic material, see below) predicts that because there is no pressure difference across the crosswall, both radii of curvature will be infinite and a planar septum can form. Moreover, there are no radial stresses within the plane of the septum because the diameter of the outer wall remains constant. When the septum starts to split, an inside-to-outside pressure difference; i.e., the turgor pressure, will be developed across the split region of the wall. To the degree that the wall is extensible this will cause bulging, but not enlargement, because the completed wall is a metabolically inert solid. To the extent that the wall is elastic, the stresses unleashed by the splitting of the septum will cause stretching. Precise conditions and the ability of the particular kind of peptidoglycan to stretch determine the pole shape (see Chapters 8, 9, and 10).

The other possible solution is the Variable-T strategy, which appears to be characteristic of Gram-negative bacteria and depends in a different way on the timing of localized synthetic function. This strategy involves biochemical alterations so that different values of T apply to different parts of the cell. In particular, T must become smaller where and when constrictions are to occur, leading to division and pole formation (Koch and Burdett 1984). The Variable-T model postulates that the organism is able to adjust T to be smallest in the zone where division is to occur, to be larger in the side wall region, and to be larger still (or irrelevant) in the established pole region. The mathematics of membrane growth show that as little as a 2-fold change in T is enough to cause the cylinder region to grow inward and to lead to cell division (see Chapter 10). It appears that the Gram-negative rod-shaped organism must be able to vary the energetics of wall formation. Note that it is not sufficient to postulate that the cell endows different regions of its anatomy with different rates of wall enlargement, because that would not lead to rod elongation or cell division.

A recurrent theme in this book is that these two approaches are quite distinct. Gram-positive rod-shaped bacteria use the strategy of supporting new wall with old solid wall, and Gram-negative rod-shaped bacteria use the Variable-T strategy.

Consequence of the Relationships of Pressure to Radius

There are a number of important points that can be made from the simple equations $P = 2T/r$ for spherical shells and $P = T/r$ for cylindrical shells relevant to a variety of familiar circumstances in addition to their application to bacterial growth. The most important is that if T is fixed, the pressure inside a spherical or cylindrical shell is greater the smaller the radius (the length of any cylindrical region is immaterial). A number of physical phenomena derive from the inverse dependence on the radius. For example, this is the reason that pure fluids are difficult to boil and may become superheated before they explosively boil. This occurs because the temperature must become very high to generate a very high vapor pressure—any bubbles initially will have very small radii; consequently, a high pressure is required to get the bubbles to form. The equations explain the role of rough surfaces and boiling chips in the initiation of bubbles in boiling.

These relationships are also the reason that a high pressure may be needed to start an aneurism of a blood vessel, but once the bulge is formed, a lower pressure than normal will maintain it.

Consequence of Growth under Constant Pressure

The cell maintains its turgor pressure by linking wall growth to the increase in cytoplasm, but in addition, the cell has a variety of other control mechanisms to maintain turgor in response to environmental challenges. To simulate this constant pressure situation, let us imagine that we have connected a spherical soap bubble to a pressure regulator and set the pressure at some value (Fig. 6.2). If that value satisfies $P = 2T/r$, then a spherical bubble will be at equilibrium and will tend to get neither larger nor smaller. Now imagine that the pressure deviates for an instant to a slightly higher or lower value. Then gas will flow and the bubble will swell a little in the former case and contract in the latter. When the pressure returns to its original value, the system is no longer in equilibrium and the bubble will continue to expand in the first case and contract in the second. The momentary deviation has started a catastrophic process. In the former case, the bubble will accept gas under progressively lower and lower pressures as its radius increases and the difference between the pressure controlled by the regulator and equilibrium pressure of the bubble, $P = 2T/r$, becomes greater. Thus, the flow of gas into the bubble occurs at ever-increasing rates. In the other case, gas would be forced back into the regulator valve and the bubble would decrease in volume faster and faster. The conclusion is that a constant pressure system is unstable or metastable, and any excursion may lead to irreversible cataclysmic changes. Consequently, bacteria have needed to devise ways to keep themselves from either catastrophe. Bacteria may possibly use the catastrophic shrinking process to effect cell division.

The Soap Bubble Analogy to Bacterial Growth

As a child's soap bubble becomes larger, molecules of water of the film are forced to come to the surface from the bulk phase and hydrogen bonds have to be broken. The energy to do this is derived by the pressure-volume work as described above; that work, in turn, was derived from forcing air into the bubble. Likewise, as a bacterial cell accumulates solutes from the medium using metabolic energy, the osmotic pressure inside the cell is transiently raised until compensated by enlargement of the cell's volume. The expansion of the wall involves expenditure of pressure-volume work, affording new internal volume, which is largely occupied by water that passively accumulates following the actual uptake of osmolytes. The consequence is that, as the turgor pressure increases, the volume becomes larger and then the turgor returns to the previous value. Thus, the pressure-volume work goes into forcing the cleavage of bonds in the murein that are protected and allow safe enlargement. The pressure-volume work is mediated

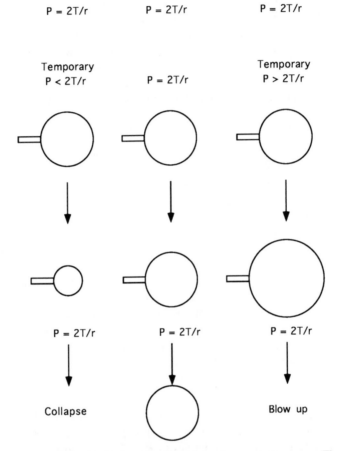

$P = 2T/r$ $P = 2T/r$ $P = 2T/r$

Temporary   Temporary
$P < 2T/r$ $P = 2T/r$ $P > 2T/r$

$P = 2T/r$ $P = 2T/r$ $P = 2T/r$

Collapse   Blow up

Figure 6.2. The metastable relationship of bubbles under constant pressure. The diagram imagines an experimental setup in which a pressure regulator supplies a gas at a pressure that is adjustable, but usually is maintained at a preset value P. Initially a bubble is blown to a radius r that satisfies $P = T/2r$, where T is the surface tension. The system, when considered to include the gas tank and the regulator, is unstable or metastable. If the pressure increases or decreases momentarily, then gas will enter or leave the bubble, increasing or decreasing the radius slightly. When the fluctuation ends and the pressure returns to its previous value, the system is no longer in equilibrium and the expansion or contraction will continue and lead either to a bursting of the bubble (or one of infinite size) or to the shrinking and disappearance of the bubble entirely.

by the pressure-produced stresses that lead to strains and the expansion of the bonds to the point that they facilitate enzymatic cleavage.

There is an apparent paradox, which will be explored extensively in the remainder of the book. The paradox is that when considered as a solid the stresses in a cylindrical shell are greater in the hoop than in the axial direction, yet the fluid membrane adjusts itself at constant pressure as if it were pulled more in the axial direction, as shown above. This is resolved for the case of a growing wall by noting that new wall is inserted independently of stresses, but the stress affects the later process of the cleaving of elastic stress-bearing wall.

Zonal Growth of Poles and Cylinders

In the zonal growth pattern, wall growth occurs only in a narrow region of the wall and the remainder is a solid, subject to elastic deformation due to the stresses that develop. The new material inserted into the zone is effectively plastic, as inserted, and effectively subject to the fluid membrane constraint $P\Delta V = T\Delta A$. For these regions we can use the $P\Delta V = T\Delta A$ relationship given the dimensions of the neighboring rigid part to calculate the shape of the newly formed zone. To do this we can imagine the new wall to be made of two identical sections of a bi-cone as shown in Fig. 6.3. Patterned after the treatment of the sphere and the cylinder, expressions for the dimensions and slopes of the new wall can be written down and substituted into $P\Delta V = T\Delta A$. Choosing the axial distance of the cylinder to be z, the amount of new wall surface of the bi-cone is $\Delta A = 4\pi r(dr^2 + dz^2)^{0.5}$ or $4\pi dz(1+S^2)^{0.5}$, where S is the tangent of the angle that the new zone makes with the cell axis. The new volume is $\Delta V = 2\pi r^2\,dz$. Then algebra allows the slope to be expressed as:

$$S = \pm[(Pr/2T)^2 - 1]^{1/2}. \qquad \textit{Zonal growth}$$

Here r, P, and T are as defined above and in Koch et al. (1981b). This equation is consistent with the formula for cylindrical elongation, because when the slope is equal to zero, the equation simplifies to $P = 2T/r$ as above. If the cell has ways to precisely enforce $P = 2T/r$, then stable cylindrical growth will occur, because S could not be other than zero.

The Zonal growth equation was taught to a computer (Koch et al. 1981b), which carried out the calculation over and over of computing the slope and then the new radius for the next growth increment. This was used to model the resultant shape of cells formed by zonal growth for a number of cases. The key result from these simulations (Koch 1982c) was that stable cylindrical growth cannot take place zonally. If there is even a single temporary deviation from $P = 2T/r$ (see Fig. 6.3) cylindrical growth will no longer occur. Such a result is not unexpected as it follows exactly the logic used to show that soap bubbles under constant pressure are unstable, as illustrated in Fig. 6.2.

The finding that zonal growth is therefore an extremely dangerous strategy

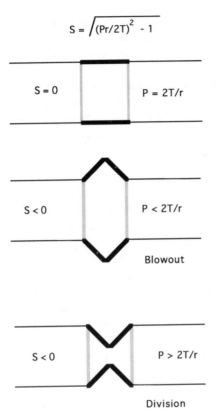

$$S = \sqrt{(Pr/2T)^2 - 1}$$

$S = 0$ $P = 2T/r$

$S < 0$ $P < 2T/r$

Blowout

$S < 0$ $P > 2T/r$

Division

Figure 6.3. The metastable nature of zonal cylindrical growth. The slope of growth of new wall to be inserted in a narrow region of a cylinder of radius r is given by the equation shown in the figure. The slope is zero if $P = 2T/r$. If $P > 2T/r$ or $P < 2T/r$ then the new zone will bulge out or in. As in Fig. 6.2 the system is metastable because if the pressure returns to its original value after a momentary perturbation, growth will continue to bulge further outward or continue to constrict inward. Only if there never is a fluctuation in pressure, radius, or the value of surface tension will cylindrical growth continue.

for a bacterium to chose and can only occur if P and T are rigidly constant has been important to our understanding of bacterial growth. The conclusion that zonal growth could not be expected to function during cylindrical elongation (Koch et al. 1982) led me to reexamine the literature and to uncover errors of interpretation in earlier experimental work (see Koch 1983), leading to the conclusion that cylindrical elongation occurred by diffuse growth. Subsequent experiments showed that rod-shaped bacteria did not grow zonally (Burman *et al.* 1983; Woldringh *et al.* 1987), and, in fact, growth occurs diffusely. [However, this metastability may be an advantage during the formation of crosswalls and new poles in the cell division of rods or such processes may be initiated by momentary deviation, giving brief inward growth and resulting in further inward growth that, in turn, leads to cell division in *E. coli* (Koch 1990g) (see Chapter 10).]

A related equation was important because it applied to the growing poles of streptococcus, *Enterococcus hirae,* then called *Streptococcus faecium* (Koch *et al.* 1981b). For this purpose a minor modification of the above treatment was all that was necessary to take into account the contribution of the wall from the splitting septum (see Chapter 8).

Diffuse Growth of Bacterial Wall

The alternative strategy to zonal cylindrical elongation is random insertion of new peptidoglycan all over the nonpolar wall surface; i.e., diffuse enlargement over a large region. In zonal growth, once growth has taken place the wall becomes rigid and the growth zone functions elsewhere, but in diffuse growth there are continuing changes in the shape of the wall. Mathematical analysis is now a much more difficult problem. The formal problem is to reexpress and then integrate the basic *Energy conservation* equation $P\Delta V = T\Delta A$ subject to the further requirement that the cell always has a shape that minimizes the energy. This type of problem is in the domain of the calculus of variation; the derivation for the case of cylindrical symmetry is given in Koch *et al.* (1982) and Koch (1984d). One form of the result is:

$$S = \pm[(2Tr/\{C_i + Pr^2\})^2 - 1]^{1/2}. \qquad Diffuse\ growth$$

Here C_i is a constant to be determined by the slope of the adjacent surface. This equation is quite different from that for *Zonal growth* both in form and because there is an additional parameter.

This equation describes the slope of a shell surface having cylindrical symmetry from the slope S_{i-1} that the surface had as it passed through the point r_{i-1}. Given these two values and values of T and P, the value of C_i can be calculated. Then from C_i, a new S_i and r_i can be computed for the adjacent axial interval, and then iteratively from these values a new constant for the $i+1$ interval C_{i+1} can be generated. This process repeated over and over again yields the profile of the entire cylindrical surface. Such simulations have been used for a variety of purposes in my studies during the last decade. During this time better computers and better numerical methods have progressively become available. Consequently, I have written various equivalent programs to calculate the slope to generate potential cell shapes. With any of the programs, the shape of cylindrical objects that would form can be projected given T, P, the slope, and a radius at any single point. The reason that this was worth the effort and computer time was that I could test assumptions about possible control mechanisms of wall formation. In addition to generating potential cell profiles, the simulations have been important because they allowed the calculation of the stability of various shapes of cells that might be produced by diffuse growth.

From the aspect of numerical calculation, these computations present many difficulties. These can be overcome by choosing suitable formulations of the solution; some of these involved high precision arithmetic, and utilization of subroutines that use expansion of square roots, etc., to lead to simplified expressions. Applications will be presented in later chapters. I have tried various algebraic forms. The best and most robust and its derivation are given in Koch and Burdett (1984) and repeated in Chapter 10. A more efficient but less useful form for our purposes is given in Koch (1983).

The major problem with the use of this approach is that it does not provide an analytical solution to the various problems we should like to study. Instead, it only gives one a license to "hunt and shoot." That is, we can start with guesses of the slope and the radius of the cylindrically symmetrical membrane at some axial distance and then proceed to project the shape of the wall on one or both sides. If our projection fulfills the boundary conditions at both ends, then we have found the desired solution; if not, we must start with another guess for the initial slope and try again.

In spite of these computational problems, the basic problem of the stability of cylindrical elongation could be tested. Fig. 6.4 shows the consequence of elongation when an occasional fluctuation of turgor pressure has occurred. This is a diagrammatic presentation; a precise simulation is given in Chapter 10.

The Surface Stresses and Strains of Elastic Wall

The paradigm of the soap bubble and a fluid membrane is appropriate for a growing wall, but not for a nongrowing, non-turning over, static wall. The

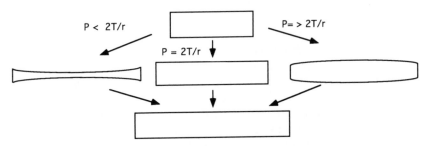

Figure 6.4. The stable nature of diffuse plastic cylindrical growth. A cylinder is shown at the top that either grows as a cylinder to a length shown in the center or bulges to an increased maximal diameter (right side) or bows inward to a decreased diameter (left side) dependent on the momentary pressure fluctuations. Diagrammatically three possibilities for the resultant shape are shown. Computer simulation shows that further growth from any of the three shapes, once conditions are returned to $P = T/r$, leads to further elongation with a return to a cylindrical form with the original radius.

bacterium is a closed container under pressure, no different in that regard from any other static pressure vessel (such as a steam boiler). Thus, the presence of a hydrostatic pressure means that there is an outward force per unit area independent of the shape of the container. This force is always normal to the surface. It tends to push out the wall until the wall is stretched doing work and produces counteracting stresses within the wall material. (The wall may stretch only infinitesimally if it is made of a very rigid substance in order to produce the balancing stresses.) The forces due to pressure are always normal to the surface; however, the stresses produced are largely in other directions. For thick, rigid-walled rectangular vessels, pressure on opposite faces would be balanced by the stresses developed in the side wall faces (Fig. 6.5A). If the wall were elastic enough on two opposite sides to bulge then some of the stress in the plane wall would be expressed in these faces (see Fig. 6.5B). If the wall were elastic, it could not resist stresses in the normal direction and would hence bellow out until the stresses in the plane of the wall had a normal component sufficient to balance the forces due to the pressure (Fig. 6.5C). Thus, at equilibrium the components of the stresses in the plane of the wall at all points must just balance the force due to the internal pressure.

The stresses, shears, and bending and twisting moments from point to point inside the wall of a pressurized vessel can be very complicated functions and depend on mechanical properties of the wall, the geometry of the vessel, and the pressure difference between the inside and the outside. We can avoid most of these difficulties because certain properties relevant to practical problems in mechanical engineering and architecture are irrelevant for our microbiological purposes. For example, the weight of the structure or its contents and how the vessel is supported are very important to the engineer and architect, but not for our needs. Many other complicating factors can be eliminated because we have restricted interest to thin elastic wall structures that cannot resist a net normal force. A further simplification results because only objects that are shells of revolution are to be considered. In Fig. 6.6 a small segment of a shell of revolution is shown cut by meridians and by circles normal to the axis of the cell. In your mind's eye, imagine that forces within the rest of the wall of the vessel are pulling at all four imaginary planes cut through the shell defining this wall segment. The stresses have the mathematical properties of a pressure in that they are forces per unit cross-sectional area. If we divide them by the thickness of the segment, the quantity so produced is called the "stress resultant," and if they are normal (perpendicular) to the plane cutting the surface, they are called "normal stress resultants," for which we will use the conventional symbol N with an appropriate subscript. In Fig. 6.6, the fragment of a cylindrical shell has been chosen in such a way that a pair of faces has been cut with meridians and the other pair by coaxial parallel circles. The wall segment perpendicular to the cell axis (axis of rotation) covers an angular span $d\Theta$, and N_Θ is called the hoop stress, or more properly, the normal hoop stress resultant. *Theta* will always be

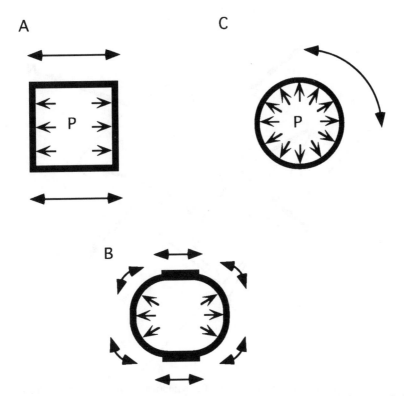

Figure 6.5. Stresses in rigid containers of various cross-sections under pressure. Only the hoop stresses normal to the cross-section of three pressurized vessels are shown. A. Square cross-section. Although pressure acts on all surfaces, only those acting on two opposite walls are shown (by small, short arrows) inside the walls. The pressure operating over an area produces the forces that are the stresses in the other two side walls, in the case where the walls are rigid and do not stretch due to the pressure; i.e., they are made of a solid with a very high value of Young's modulus. B. Bulged faces. Imagine that two parallel faces are made thicker (or of a stronger material). Now a portion of the stress is expressed in the plane of the curved region. C. Circular cross-section. For a circular cross-section the pressure produces hoop stress in the plane of the cross-section. The hoop stress is the same at any point in the wall.

used to designate the angle in the hoop direction. Likewise, the axial or meridional normal stress resultant is measured in a plane parallel to the cell axis and is designated by N_Φ. *Phi* will always be used to designate the angle in the axial direction. If the surface is a cylinder of length L, radius r, and thickness τ, the total force tending to rip it axially into two half-cylinder segments is $N_\Theta 2L\tau$ and that tending to rip the cylinder longitudinally is $N_\Phi 2\pi r\tau$.

Our quest is to relate these stresses to other relevant properties: the dimensions and the pressure difference from inside to outside. The forces differ by an

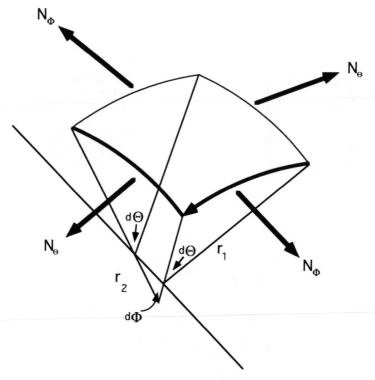

Figure 6.6. Stresses in a small section of a cylindrically symmetric shell. The normal stress resultants are shown by the symbol N. The hoop angle Θ around the axis is designated by Θ and the meridional angle is measured from the axis and is designated as Φ.

infinitesimal amount on the opposite two faces of the segment of Fig. 6.6. Partial differential terms should actually have been added to one side of each pair. From these it is necessary to compute the component of this difference; note that there would be none if the section were perfectly flat (as in Fig. 6.5A). But for curved surfaces there are components in the surface in both the Θ and the Φ directions. The sum of these at equilibrium must equal and balance the force due to the pressure applied over the inside area of the little hypothetical segment. I will not go through the trigonometry of the analysis, which is presented in Flügge (1973), in Roark and Young (1975), and in Timoshenko and Goodier (1970). The resulting equation is:

$$P = N_\Theta/r_1 + N_\Phi/r_2. \qquad \textit{Pressure vessel stresses}$$

In this equation r_1 and r_2 are the two principal radii of curvature, as before. This expression is reminiscent of the *Young-LaPlace equation*, where both N's were replaced by T, but this equation has a very different meaning. One difference is that each principal radius of curvature is associated with the different (hoop and axial) stresses in the *Pressure vessel stresses* equation, but the same value of T applies in the *Young-LaPlace equation* no matter which radius of curvature is being considered. Note that N_Θ, N_Φ, and T have the same units; i.e., either work per unit area and unit thickness or force per unit length and unit thickness. However, although N_Θ and N_Φ also have the units of T, they are not to be thought of as the work to create a new unit of area, but they are tensile forces per unit normal length tending to rip the structure apart. If their magnitude is negative, then the resultant force is compressive. Both concepts are foreign to fluid membranes. The most important difference is that the *Young-LaPlace equation* applies to fluid membranes and the *Pressure vessel stresses* equation applies to a solid membrane.

Strains in Solid Materials

The consequences of stress is the deformation of the object. To calculate the dimensional changes it is necessary to review concepts going back hundreds of years: Hooke's law and the Poisson coefficient. Imagine a rod with a unit cross-sectional area to which an incremental force ΔF is applied in its axial direction that would tend to stress and stretch it. Newton's first law assures us that the bar will develop a counter force just equal to the applied one. If the bar is made of material obeying Hooke's law (an elastic material) and E is the Young's or elastic modulus, then a length of the bar L will get longer by an amount given by $\Delta L/L = F/E$, and we say that the force F (it really has the units of force per unit area and is therefore a pressure) causes a stress equal to the applied force and produces a strain ΔL.

Evidently, if the bar elongated, it would get thinner and narrower. Poisson's coefficient v is the measure of this and is the ratio of the negative of the relative lateral strain to the relative longitudinal strain under the conditions of uniform and uniaxial longitudinal stress applied within the range where Hooke's law is obeyed.

If the density of the material is unaffected by a linear stretch in the x direction, then it must compress in the y and z directions so that:

$$(1 + \Delta L_x)(1 + \Delta L_y)(1 + \Delta L_z) = 1.$$

If it is distorted equally in the y and z directions, then $v = -\Delta L_y/\Delta L_x = -\Delta L/\Delta L_x$. Both ΔL_y and ΔL_z can be replaced by $-v\Delta L_x$. This leads to a quadratic equation in v. For small extensions of a rod, we can drop the terms containing the square of ΔL_x and obtain:

$$v = 0.5$$

and therefore the material of which the rod was made has a Poisson coefficient approaching 0.5. For the cases that interest us the strain is much larger; i.e., the stretch may double the length. The same quadratic relationship, if ΔL_x is 100%, gives $\nu = 0.293$, and if ΔL_x is 200% it gives $\nu = 0.211$. The value of ν approaches zero upon extensive stretching. Although when ΔL_x is small, an incompressible material, behaving as a constant-volume fluid, has a Poisson coefficient approaching 0.5, many materials, for example, steel, have values of ν of approximately 0.3. However, the conceivable range of materials can have values ranging from negative values to $+0.5$. Negative values hold particularly for space age, man-made materials that expand in total volume when stretched, and more positive values hold for materials that are compressed on stress, such as many woven fabrics.

Stresses in Spheres and Cylinders

If we knew one of the normal stress resultants we could calculate the other from the *Pressure vessel stresses* equation. The stress N_Φ is the easier of the two to calculate by analysis of the geometry given in Fig. 6.6. It is given in Flügge (1973) by:

$$P = 2N_\Phi/r.$$

Notice that, if N_Φ were to replace T, this is exactly the expression we had before for *Spherical equilibrium* or *Axial equilibrium*. For the case of a spherical shell, r_1 and r_2 are equal and it is easily seen by substitution into the *Pressure vessel stresses* equation that:

$$P = 2N_\Theta/r$$

or

$$N_\Phi = Pr/2 \quad \text{and} \quad N_\Theta = Pr/2. \quad \textit{Spherical shell stresses}$$

The two normal stress resultants are equal, as could have been concluded from symmetry. Similarly, for the case of a cylinder, it can readily be shown that:

$$P = 2N_\Phi/r \quad \text{and} \quad P = N_\Theta/r,$$

or

$$N_\Phi = P_r/2 \quad \text{and} \quad N_\Theta = P_r. \quad \textit{Cylindrical shell stresses}$$

Comparison of the equations for the hoop and axial stresses of the elastic cylinder shows that the normal hoop stress is twice as great as the meridional stress. This mechanical fact has profound implications for microbiology, as is discussed in later chapters. The practical consequence of this 2-fold difference is that cylindrical vessels are more at risk of being ripped lengthwise than in the hoop direction. This is a commonplace observation. Some examples: when cylindrical balloons are burst with a pin prick or when sausages are cooked until they split, the fractures tend to propagate along the axis. This is simply because once a brittle fracture starts, no matter what its initial direction it deviates to become parallel to the long axis because of the greater stress normal to that direction.

Although a full proof of relationships of normal resultant stresses has not been given here, it can be found in the mechanical engineering literature cited above; in addition a proof in microbiological context has been provided by Chatterjee *et al.* (1988). An easier way to show these relationships, however, is as follows. Consider a cylindrical shell of length L, radius r, and a thickness τ. First imagine a plane including the axis of the cylinder. The pressure times the area on this plane creates a force of $P2Lr$. The force is supported by the thickness and length of the cylinder wall; i.e., by an area of $2L\tau$. Therefore, the normal hoop stress resultant, N_Θ, is the quotient, $P2Lr/2L\tau$, times τ. This simplifies to Pr. Similarly, starting with a plane at right angles to the axis, the pressure times the cross-sectional area is $P\pi r^2$ and the area of the wall supporting stress is $2\pi r\tau$. Consequently, N_Φ is $Pr/2$ and thus is half as large as N_Θ.

Calculation of Stresses for More General Geometries

Engineers need to know the stresses that would be generated in a variety of shapes of vessels subject to internal pressure and to additional stresses. There are several shapes other than the cylindrical shell and spherical shell that are relevant to the engineer and to biologists. If the radii of curvature can be calculated, then the stresses can be calculated from the *Pressure vessel stresses* equation.

ELLIPSOIDS OF REVOLUTION

For the particular shapes of ellipsoids of revolution, the equations for the radii of curvature in terms of semimajor and -minor axes (a,b) can be found in textbooks and substituted into the *Pressure vessel stresses* equation, to yield:

$$N_\Phi = \frac{Pa^2}{2(a^2\sin^2\gamma + b^2\cos^2\gamma)^{0.5}} \quad \textit{Ellipsoid meridional stress}$$

and

$$N_0 = \frac{Pa^2(b^2 - (a^2-b^2)\sin^2\gamma)}{2b^2(a^2\sin^2\gamma + +b^2\cos^2\gamma)^{0.5}} \quad \textit{Ellipsoid hoop stress}$$

The derivations of these equations are presented in Flügge (1973, p. 22); these equations will be fundamental in considering the design features of the poles of variously shaped cells in later chapters where P is the turgor pressure, a is the maximum distance along the axis of cylindrical symmetry, b is the maximum distance in the radial direction. In these equations γ is the azimuth angle ($0°$ at the tip of the pole and $90°$ at its base). The ellipse of the axial section is given by $(a/z)^2 + (b/r)^2 = 1$, where r is the radius of the circular cross-section and z is the axial height of the completed pole.

STRESSES AND STRAINS IN SEPTA

If a thin circular disk of material has outward force applied at very many points along the periphery, then there are tensile stresses at all points on the two-dimensional sheet. This represents the circumstance of a completed, partially split septum. A circular annular ring is the model for an incomplete septum that is partially split (see Fig. 6.7). From equations in the literature (Timoshenko and Goodier 1970, pp. 70–71), we can easily calculate the stresses and strains

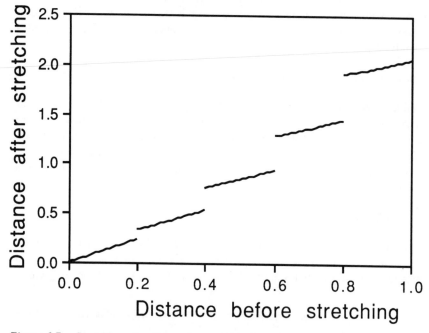

Figure 6.7. Stretching of a disk and an annular ring when pulled uniformly from the outside. See text. At the right-hand side of the figure is represented the stretching of an annulus whose innermost radius was 0.8 and outermost radius was 1.0 of the radius of the cell. The next line to the left represents the stretching of an annulus whose innermost radius was 0.6 and outermost radius was 0.8 of the cell. Correspondingly, the next line represents the stretching of an annulus whose innermost radius was 0.4 and outermost radius was 0.6. At the left-hand side a disk 0.2 of the radius of the cylinder is shown after being stretched by the remainder of the completed pole.

at any point. For a disk the stresses are constant at any distance from the center. I will omit the stress equations and give just the equation for the produced displacement u:

$$u = (r/E)\{(1 - v)(a^2 p_i - b^2 P_0)/(b^2 - a^2) \quad \textit{Annular ring displacement}$$
$$- (1 + v)(a^2 b^2 (P_i - P_0))/[(b^2 - a^2)r^2]\}.$$

To reiterate the significance of the symbols, u is the radial stretching at radius r of a point of interest in the original unstretched septal disk or annulus, P_i and P_0 are the inner and outer pressures of the cell, and now a and b are the inner and outer radii of the annulus ($a = 0$ for a solid disk). Note that the radial array of outward forces applied at the edge of the disk corresponds to a negative pressure. For a solid disk the equation reduces to:

$$u = (r/E)(1 - v)(-P_0). \qquad \textit{Disk displacement}$$

Thus, a solid disk is stretched directly in proportion to the radius and to $-P_0$, which is the uniform tension pulling outward. A feeling for this result can be obtained logically in a qualitative fashion as follows. The center of the disk is not displaced by the radially symmetrically applied forces. The farther from the center, the more the disk is enlarged, because there is more area of material stretched below it. The larger the value of the elastic modulus E, then the stiffer is the material and the disk stretches less. Similarly, the larger the value of the negative of P_0, the greater the stretch (see Fig. 6.7). If the Poisson coefficient v were zero, then as the disk expanded radially, the circumferential stretching would not influence the compression of the thickness of the disk. Therefore, just by thinking about it we could have almost written down the *Disk displacement* equation as given above.

In some cases as the bacterial septum is formed, it is simultaneously split. Therefore the septum is completed only shortly before the splitting separates the cell into two daughter cells. Consequently, during cell division the intact septum is an annular ring, being stressed outwardly. Consider a thin annular ring (see Fig. 6.7), under radial stress pulling from the outside. Each ring has enough material to have a width corresponding to 0.2 of the cellular ring diameter. The stress pattern will be much different depending on the Young's modulus of the material. If it is made of a material with a very large value of E, it will not expand appreciably, but will develop counterstresses. Then, because the displacement is very small, it would not enlarge appreciably and all the stresses would be taken care of in the ring itself. In Fig. 6.7 the parameters have been chosen so that there is a 10% stretch when the completed disk whose outermost radius was 0.8 of the cell has been split, so that the pressure transmitted from the split wall to give tension to the disk is -1.

The Make-Before-Break Strategy

Role of Stress in Autolysin Action

A key role in the growth process of bacteria is the selective enzymatic cleavage of wall by the bacterium's own autolysins. Although essential for normal growth, under some conditions these enzymes may cause the bacteria to self-destruct, hence the name. Higher organisms produce similar lysins, such as the lysozyme from egg white, tears, phagocytic cells, or the umbilical cord. These enzymes serve a protective, antibacterial function for the organisms that manufacture them.

A cardinal axiom of the surface stress theory (borrowed from the field of enzyme kinetics) is that enzymes can respond only to the circumstances in their immediate environment. But enzymes do the joining and splitting. Consequently, there is need for a way to relay information about global aspects of cell morphology to the submicroscopic level where enzymes function. The relay system is the stresses developed in the wall. The prokaryotic strategy depends on autolysins cleaving stress-bearing peptidoglycan so that enlargement is possible. But it also depends on the prevention of inappropriate hydrolyses of the same chemical fabric; hydrolyses that sometimes do kill the cell. Much more will be said about such "guarding" below, but in this section we must deal with the role of stresses on their substrate affecting the enzymology of the autolysins, because it will become evident that the responses of autolysins to stress is necessary to understand bacterial morphology.

This subject brings us into an arcane branch of enzymology. Microbial physiologists have usually considered that autolysins are specific hydrolases that combine with the sensitive bonds just as Michaelis and Menten eighty years ago envisaged the combination of sucrose with invertase. The enzyme binds, it is presumed, similarly to many other enzymes in a step that depends on the groupings present in the substrate that may have nothing to do with the covalent chemistry that is to take place. Those groups that favor binding, however, give the substrate its specificity. The enzymologist has concluded that for the invertase example (e.g., see Metzer 1977), a nucleophilic group on the enzyme reacts with the sucrose to become glycosylated, thereby releasing fructose, which now has the hydrogen atom that originally (nominally) was present on the nucleophile atom of the enzyme. Then the enzyme-glucosyl entity reacts with water. This second reaction is equivalent to the reverse of the first from a chemical viewpoint in that the oxygen of water is just another nucleophile. At the end, the nucleophile in the enzyme has regained a hydrogen atom at the expense of the water and the remaining hydroxyl group is now part of the liberated glucose.

From the thermodynamic point of view, there is quite a difference between the first and second steps because water is the solvent, and in a dilute solution, water is present at a high concentration of 55.5 M. This means that mass action

(LeChatelier's principle) drives the hydrolysis. Only in a concentrated glucose and fructose solution would hydrolysis not be spontaneous, and sucrose be formed from glucose and fructose (but then the sucrose could inhibit the reaction and possibly denature the proteins). Synthesis could be favored in several other ways. One would be to attach scaffolding groups (away from the binding site). These groups bind to the substrate and in turn must be connected to each other in such a way that they hold the glucose and fructose moieties close together, but in addition, they must hold the groups in the correct orientation. Proper orientation increases the probability of fructose serving instead of water in the reaction of the glucosylated enzyme. In some model systems, this type of propinquity and adjustment of the orientation of the groups has been shown to increase the rate of a chemical reaction 10^{10}-fold.

Now we can consider the case of hydrolysis where the substrate may be under tension, as in the case of the stress-bearing peptidoglycan of the living cell. Obviously, once the stress-bearing bond is cleaved, even if this process trades one nucleophile for another and forms an enzyme linkage to one portion of the substrate, then bond rotations will take place and the gap between the two elements will lengthen, and the original partners will not be able to get back together. Only very rarely would some extremely uncommon random thermal fluctuation bring them back together and allow resynthesis.

Stress and the Rate of Autolysis

Chemical reactions proceed through a transition state. The probability of the reactants having enough energy to form this excited state is all-determining for the rate of reaction. If k is the rate constant, $\Delta S\dagger$ is the entropy of activation, $E_a\dagger$ is the activation energy, R is the gas constant, and T is the absolute temperature, then:

$$k = C\, e^{(\Delta S\dagger/R)} \times e^{(-E_a\dagger/RT)}. \qquad \textit{Arrhenius equation}$$

In this equation, e is 2.718 . . . and C is a constant. Because of the negative sign in the exponential functions argument, the smaller the $E_a\dagger$, the faster the reaction will go.

Let us calculate the lowering of the energy barrier to hydrolysis due to the tension on the sensitive bonds in the peptidoglycan. The mathematics of this situation is not difficult (Koch *et al.* 1982; Koch 1983). I will repeat the quantitative aspects again here (with a numerical correction to the Koch (1983) paper. Let us idealize the situation and imagine that *E. coli* is a spherical cell with a radius of 0.8 μm and a turgor pressure of 4 atm (or 4.052×10^6 dyn/cm^2). Then the surface tension and both normal stress resultants are given by:

$$N_\Theta = N_\Phi = Pr/2 = 4.052 \times 10^6 \text{ dyn/cm}^2 \times 0.8 \text{ μm}/2 \times (10^{-4}\text{cm/μm})$$
$$N_\Theta = N_\Phi = 162 \text{ dyn/cm}.$$

Now, consider a region of wall in which: (i) the glycan chains are packed 0.45 nm apart (closest packing), (ii) the wall is only one layer thick, and (iii) all chains are infinitely long. Then because one cm contains 10^7 nm, there are 2.22×10^7 packed chains/cm. In the direction of each chain, every bond would have to support a stress of 162 dyn/cm/2.22×10^7 chains/cm = 72.9×10^{-7} dyn/chain. In order for the chain to split, some bond in the chain must stretch by about 0.05 nm. The work to stretch one mole of such bonds would be

$$\Delta E = 72.9 \times 10^{-7} \, \text{dyn/chain} \times 0.05 \, \text{nm} \times (1 \, \text{cm}/10^{-7} \, \text{nm})$$
$$\times \, (6.023 \times 10^{23} \, \text{molecules/mol})$$
$$\Delta E = 21.9 \times 10^{-9} \, \text{dyn-cm/mol} \times (1 \, \text{J}/10^7 \, \text{dyn-cm}) = \text{or } 2.19 \, \text{kJ/mol}.$$

These estimates are a significant fraction of the free energy of activation for the hydrolysis of stressed wall. The typical free energy for enzyme hydrolysis for polysaccharides is in the range of 40 to 80 kJ/mol. The salient number is the value of $e^{(-E_a/RT)}$; substituting 21.9 kJ/mol, 8.31431 J/K mol, and 298 K (for 25 °C) for E_a† and R and T, it is 2.4-fold. This is the minimal estimate of the increased rate of hydrolysis relative to that of the unstretched wall. The actual energy favoring hydrolysis must be greater for several reasons. The glycan chains are not very long and the gap between chains is probably three or four times larger in the stressed wall, because then the spacing in unstressed peptidoglycan is about 2 nm instead of 0.45 μm. The energetics of autolysis of the murein peptide bonds would be similar to those of carbohydrates because the free energies of hydrolysis are similar. The pull on existing peptide crossbridges should be larger because peptide chains in the same plane are farther apart because all the possible crossbridges are not formed and therefore do not share the stress. An estimate of a maximum effective energy of activation factor would be 22 kJ/mol, which corresponds to a 6,700-fold greater hydrolysis rate when the substrate is stressed than when it is unstressed.

Once the wall is breached and pressure falls to zero, the remaining bonds in the chain will not be under tension and the lytic rate for them will decrease. This decrease in reaction rate is the reason that it is very difficult to study the enzymology of mechanically stressed substrates. Therefore, it is not surprising that there are no published experimental studies of the influence of mechanical stress on the rate of enzymatic and nonenzymatic reactions. Evidently, stress should stimulate enzyme-catalyzed reactions as well as noncatalyzed ones. Although there are many studies showing that enzymes open up compounds containing strained rings faster than they hydrolyze substrates that are not internally strained, at present there is no direct evidence, and we must depend on the theory given here that linear stress lowers the Arrhenius energy of activation of autolysin-catalyzed hydrolyses. The effect of stress on degradative chemical reactions, however, is part of common experience. As an example consider a rubber band stretched around a pack of cards and left in a drawer. In years, the rubber disintegrates; it does so more rapidly than a control rubber band left loose in the same drawer.

Alternatives to the Surface Stress Theory

Up to this point, we have considered the chemical, biochemical, and physical basis of the surface stress theory. In this section alternates to the theory will be collected together. *A priori* these possibilities consist of: (i) mechano-proteins doing work, (ii) fibrous proteins acting as ties or guy wires, (iii) the wall having a very regular structure so that old wall can serve as the template for a very accurate one-to-one insertion in the form of hoops or spirals, and (iv) possession of a sophisticated control system so that the cell can sense its own dimensions and then regulate the growth (and directionality of that growth) in regions of the cell so that morphology is determined and maintained.

Growth with Contractile Protein

This method clearly functions in eukaryotic cells, but it cannot be accepted as a candidate until functional mechano-proteins and enzymes are actually demonstrated. So far none have been demonstrated (see Chapter 2).

Growth with Struts, Braces, and Guy Wires

If the wall shape is determined by nonelastic elements that are engineered in place before the wall is permitted to enlarge, then zonal growth could function because the forces leading to bulging during wall growth are prevented from acting. These elements would be equivalent to cables and guy wires, unable to contract or perform mechanical work. That does not solve the basic problem, however, because the wall will bulge or invaginate between such braces to a degree dependent on the pressure. Then when new braces are formed in between existing braces, they will be a little longer or shorter than the previous ones, and that will lead to inappropriate length, leading in turn to further deviations in the same direction. In the long run this will necessarily lead to the catastrophe of the cell's rounding up and becoming bigger and bigger, or inappropriately dividing.

Growth through Template Action

Arguments against the bacterial wall having a regular structure have been assembled in a review article published in 1988. Basically all evidence points to noncrystalline structure and against the wall consisting of regular, complete hoops or spirals of precise diameter that could serve a role of maintaining a precise diameter. In Chapter 10, however, the Three-for-one model of Höltje, which assumes templating on a local scale, will be discussed in detail.

Cellular Control of Localization of Regions of Secretion, Crosslinking, and Autolysis of Peptidoglycan Units

The still-mysterious part of the bacterial growth process is that wall grows in some, but not all, regions of the cell. That is a fact that might imply a further role of the cell's ability to localize wall growth. Although the surface stress theory tries to explain the phenomenon simply on the physical basis of an enlargement controlled by stress favoring autolytic activity, this alone is clearly not sufficient. To extend cellular control, there could be regulated control sites where the water-soluble precursors are attached to the membrane-bound isoprenoid carriers; where their movement across the membrane (supply side control) takes place; or where autolysins are permitted to act to enlarge the wall and convert new intact, but unstressed, wall into stress-bearing wall (demand side control). If in some way or ways cellular control mechanisms can sense growth requirements and restrict wall growth to certain regions or zones of the cell, this could suffice to allow measured and controlled growth. In fact, only a few types of controls would suffice to make cocci-, rod-, and fusiform-shaped cells. So to some degree this is a valid mechanism, but one about whose details we have no evidence. For the Gram-positive organisms there would be signals and controls to localize metabolic activities to cause: (i) rapid growth in narrow zones where a septum is invaginating, (ii) diffuse, very broad growth zones to function where the side walls of cylindrical cells are elongating, and (iii) minor (very slow) turnover processes to replace pole wall lost by degradation. For the Gram-negative organisms there would be: (i) a very active site where wall growth leads to a constriction that leads to cell division, (ii) a less active region where the side wall is elongating, and (iii) still less active regions of the completed poles that are essentially quiescent.

Enzymes for crosslinking, as represented by the known penicillin-binding proteins (PBPs), appear to be associated with all regions of the cell membrane of both Gram-positive and Gram-negative organisms (Buchanan 1988). There is only one known possible exception (Dolinger via personal communication from G. Shockman) to these enzymes being membrane bound. It is possible that some autolysin may be retained in the cytoplasmic membrane or possibly on the inner face of the outer membrane of Gram-negative bacteria (Höltje and Keck 1988).

The severe criticism of this class of model is that for cylindrical growth, the model demands that either the enzymes "know" how to elongate the side wall and not permit it to bulge (i.e., act meridionally and not in the hoop direction), or the enzymes are regulated in function via unknown cellular mechanisms that sense the cell's diameter and transmit the information to turn off growth that might lead to dilation or inappropriate constriction. Not only is there no evidence for such a sensory mechanism, simply turning off and on a growth mechanism probably would not in the long run prevent morphological malformations.

Surface Stress Theory and Growth Strategy

In this section only rod-shaped bacteria will be considered. The rod shape is the critical morphology for assessing models because this shape seems most difficult to produce and maintain. The theory is that the bacterial cell possess only an ability control the localization of enzymes in the membrane and the sites where they may be excreted. Because of the sophistication of the intermediary metabolism, the transport of peptidoglycan pentapeptide, the crosslinking and autolysis of the wall, wall growth occurs in the proper places and at a rate consonant with the needs of the bacterium. Part of the strategy is a mechanism to initiate cell division and follow through with a minimum of sophistication.

Stable Cylindrical Elongation

How can the cell arrange the physical and biological situation such that wall growth spontaneously generates cylindrical side walls during the elongation phase without constriction or bulging? The analysis presented earlier shows that the ability to enlarge the surface can happen only if four conditions are met. Three of them are: (1) when a rod-shaped cell grows by diffuse addition over the entire cylindrical surface, (ii) when the ends are relatively rigid and turnover of the poles is very restricted, and (iii) when the pressure is appropriate for the diameter of the cylinder. But beyond these three conditions is a yet more important one: (iv) the wall growth process must be one in which the mathematics for fluid membranes apply and not the mathematics for the stresses in a static solid.

Parts of the argument for the appropriateness of fluid membranes as a model wall have been presented in different sections above. Simple rupture and repair would not work because the hoop stress is stronger than the meridional stress and thus a cylindrical shell of a solid material would tend to rip longitudinally. If it was repaired by insertion of wall material, the object would get rounder and bigger indefinitely. If, however, the cylindrical shell was effectively fluid with the above set of constraints, then it could elongate through a series of equilibrium states as it grew bigger. These considerations point to the necessity that the covalent structure of sacculus grows so that it can be the equivalent of a soap bubble and not the equivalent of the stretching of a solid tube or pipe.

There appear to be two mechanisms that would permit stable cylindrical growth with the fluid membrane paradigm. The Gram-negative Variable-T mechanism is the easier to understand. Diffuse insertion at random into the single thickness of the stress-bearing part of the sacculus is the equivalent of inserting more fluid surface in between islands of more rigid material (see Chapter 10) and by varying T in different portions of the cell, pole formation as well as cell extension can occur. The Gram-positive rod is only a little more difficult to comprehend. Here the wall is added diffusely to the inner surface, crosslinked, and later, as a solid, it is subject to stresses and autolysin activity that create cracks in the outer portion of the elastic solid wall. The effect is that the wall, after it is crosslinked, is stressed to its limit in the axial direction before being stressed in the hoop direction. The result is that further out from the intact region, cracks form that will be helical in nature (see Chapter 9). Something related to this was first suggested by Previc (1970); his analogy was to the Chinese finger puzzle. I have augmented this idea by noting that outside of the intact region of the wall there would be bands of wall material supporting the inner regions of wall as helical wall wrapped around the intact regions. These reinforcing residues of the wall laid down at an earlier time would serve the role of balancing the hoop stress, as is common in modern engineering practice.

Zonal Growth

The simplest form of the surface stress theory is zonal growth. If there is a narrow cylindrical zone of growth that is fluid membrane, subject to the *Young-LaPlace equation,* and other regions of the cell that are not in the growth zone behave as rigid or elastic solids, then given other constraints, such as the presence of a septum, or no momentary deviations inward, growth in characteristic patterns is possible.

Synthesis Before Degradation

The cornerstone of the surface stress theory is the make-before-break strategy. The requirement is that new wall be added and crosslinked prior to autolysin action. Not only must the growing prokaryotic cell have the synthetic ability (for oligopeptidoglycan chain formation and crosslink formation) and the ability to hydrolyze old wall, it must have a way to regulate these processes to match each other and to respond to the cell's current success in producing cytoplasm.

The crosslinking enzymes (PBP 2 and PBP 3) permit nascent oligopeptidoglycan chains to be linked to other nascent chains or to the existing wall (the linkage to other nascent chains is predominant in the layer structure of the Gram-positive and to existing wall is predominant in the Gram-negative bacteria). The major constraint is that crosslinking can occur only in the immediate vicinity of the membrane, because this is where the donor oligopeptidoglycans are secreted and where the membrane-bound transpeptidases can act.

Cleavage of Wall to Allow Growth

But additions of murein by itself are only preparatory to making the cell larger. Autolysins need to have special properties to be able to cleave wall and still not destroy their cell of origin. Ordinary enzymes cannot act in response to conditions even a few tens of nanometers distant from their active sites; they respond to the local pH and ionic environment and to the chemical nature and conformation of their potential substrate. Obviously, a major problem for prokaryotes was to develop strategies to keep their autolysins in check and not to propagate cracks once a portion of the wall had been weakened, but instead to act in such a way that unstressed peptidoglycan is able to take up the stress. Finding out and proving what these solutions are is the major problem for the microbial physiologist today. In the following chapters, mechanisms—some proved, some obvious, and some speculative—will be presented.

Development of Cell Shape

The surface stress theory postulates that growth and enlargement lead to shapes that utilize the turgor pressure and the current shape and size. These factors create the stress pattern that autolytic enzymes can sense in the stresses of the bonds that are to be cleaved. Although continuing the make-before-break process allows the surface to enlarge, additional factors that evidently depend on the localization of enzymatic process must also be involved in determining the geometry of that expansion. Enlargement must take place in such a way that the cell elongates to extend the cylindrical part of rod-shaped organisms or extend the curve of a pole inward, while allowing neither bulges nor indiscriminate enlargement of the cell.

The Aberrant Development of Bulges

Enlargement of the freestanding walls of a nonspherical pressurized object frequently leads to bulges. A soap bubble can be blown into a cylinder shape, if suspended between two caps and if the pressure is appropriate. But if additional air is blown into the bubble and the rigid caps remain the same distance apart, the bubble bulges (and the pressure decreases!) That is exactly what happens during the development of bulges of a Gram-negative organism when treated with low levels of penicillin. In the latter case, presumably the pressure initially increases because wall synthesis is partially impeded, but growth of cytoplasm occurs normally. This transiently increases the turgor pressure and mandates wall expansion. That in turn changes the geometry and stress relations and leads to a concomitant decrease in turgor pressure. The decrease, the mathematics show, can lead during further growth to enlargement above the original diameter characteristic of balanced growth of cells.

Thus, the consequence of drug action that blocks wall formation can be a bulge where there should be a constriction. Eventually, the cell wall may burst and form the classical "rabbit ears" observed during protoplast formation in hypertonic media in the presence of penicillin or similar antibiotics.

How Can a Cell's Enzymes "Know" Where to Insert New and to Cleave Old Wall?

Although at present this is an unanswerable question, some suggestions can be made concerning the pole. To inhibit pole growth, perhaps there are smaller quantities of substrate localized at poles. Possibly the pole wall is refractory because of secondary chemical factors. Possibly physiological factors, such as the protonmotive force which affects the local acidity, are differently localized in different parts of the cell and this plays a role. One factor that must be involved certainly is a physical one. The stress on an average bond in the completed pole is half of the hoop stress in the side wall. That would be true if the stress were distributed in the same way throughout the thickness of the wall. However, it is distributed much more evenly throughout the thickness of the polar wall, at least in the case of Gram-positive cells. On the side wall, the stress is localized in the intermediate region of the thickness of the wall. Inside this layer, the newer wall has not been maximally stressed, whereas further outside it has been partially fragmented. The stresses within the single layer of the side walls of Gram-negative cells may not be uniform but modified by the effects of regions where growth or turnover is occurring.

But the primary presumption is that wall enlargement by insertion and cleavage is a passive process dependent on the availability of substrate and of enzymes forced to function by the turgor pressure.

7

Mechanical Aspects of Cell Division

KEY IDEAS

*Bacteria are small enough not to need mechanical methods for distribution of
   substances.*
Mechanical forces, under appropriate circumstances can bisect the cell.
*There are difficulties with all published models that attempting to explain how rod-
   shaped bacteria locate their midpoint in preparation for the next cell division.*

Many bacteria find their middle quite accurately, even though the distances involved are much greater than the dimensions of globular proteins that might somehow serve as rulers or measuring sticks.

A number of inadequate models for the control of cell division have been proposed based on aspects of molecular genetics, ultrastructure measurements, or biochemical kinetics.

The new "Central stress" model which depends on the partition of wall tension between the cytoplasmic membrane (CM) and the murein (M) layer in such a way that the CM at the center of the rod experiences a higher stress than near the poles and that this peak stress increases through the cell cycle.

The Central stress model is based on the fact that the M and CM layers interact physically with each other.

Because of the fluid nature of the lipid bilayer there is a flux of lipid from the established poles towards the cell center as the murein wall elongates.

The flux arises because at the pole the old M and OM layers are inert but the CM enlarges.

The flux phospholipids from the pole lowers the tension in the CM at the ends of the sidewalls and creates a peak tension in the exact center.

A special two-component sensory system is postulated to respond to tension in the CM and trigger a cell division event.

Cell Size and the Surface-to-Volume Ratio

The basic strategy of bacteria is to be small enough so that they can acquire needed materials from the external environment and also avoid the need for very sophisticated and expensive mechanisms that distribute materials internally. Because bacteria are small, they have a high surface area for transporting substances to service only a small cytoplasmic volume. Additionally, the cell's small volume permits diffusion to distribute metabolites from where they are made in the cell to where they are needed. Larger cells need a plumbing system like the endoplasmic reticulum, and multicellular organisms need hearts and blood vessels for internal distribution and need apparatuses, such as appendages for motility and for capturing prey. How small and how big a single-cell organism can be is discussed in much more detail in Koch (1995cd).

The Variation of the Surface Area-to-Volume ratio with Shape and its Size Dependence

A very thin rod-shaped organism that grows by elongation and binary fission is the only microbe that maintains an essentially constant surface area-to-volume ratio (S/V) as it grows and divides. In this extreme case we can neglect the ends; consequently, S/V is $2\pi rl/\pi r^2 l$ or $2/r$, where r is the radius and l the length. At the other extreme is the hypothetical case of a spherical organism that grows bigger and suddenly changes shape and divides into two spherical cells with half the volume. The S/V ratio is $4\pi r^2/(4/3)\pi r^3$ or $3/r$ at birth. Just before division the radius r has increased by $2^{1/3}$ or 1.26-fold; then S/V is lowered from the birth value by a factor of 1.26.

Clearly, much more important than choice of shape and changes of shape during the cell cycle is the fact that r appears in the denominator in these expressions for S/V; therefore, smaller objects have higher S/V ratios for any shape of the cell. The key point regarding the adaptability of the small size organisms is that with microbes transport can be simple and still be adequate. This simplicity leads to the generality that small organisms frequently can and do grow fast. No doubt as evolution has tuned organisms to their particular chosen niche, in many cases the optimum growth strategy is to grow as fast as possible. For many microbes under many circumstances, this is the important issue and therefore being as small as possible is usually a great virtue.

Only for a very thin rod is the S/V constant. Are the variations within the 26% range possible from long rods to spherical cocci important biologically? Just how important is shape? How much does the growth rate vary during the cell cycle because of systematic variations in the S/V ratio for variously shaped organisms? Consider a typical Gram-negative rod-shaped organism. For this calculation the approximation that these cells are cylindrical with hemispherical poles should be adequate. From a simple computer program, the dashed line in Fig. 7.1A was generated; it shows how the S/V ratio varies during the cell cycle if the newborn cell has a cylindrical region of length equal to the radius, i.e., the total length of the cell at birth is 1.5 times its radius. It can be seen that as this short rod-shaped newborn cell grows longer, S/V progressively decreases, but then rises as central constriction and new pole formation take place. For the case shown, constriction was assumed to start when the cell was at 80% of the doubling time. The S/V at this time was 1.24-fold smaller than at division or at birth. Also shown in Fig. 7.1A by the dotted line is the case of idealized coccal shape that at birth grows by creating new polar wall. Again, there is a variation of the S/V, but the range is 1.26, not much different than for other-shaped organisms. As a backdrop the solid line in Fig. 7.1A is for the case of the sphere that increase in radius during the cell cycle by 1.26-fold.

In Fig. 7.1B, growth curves are shown for the rod and the coccus for the case when growth is always limited by the cell surface. For the three curves it is assumed that the rate of transport across the cell surface depends directly on its area. It can be seen that the growth curve is not much affected by cell shape. The assumption of limitation by surface area overestimates expected deviations from exponentiality because the concentrations of membrane-bound pumps and their function would also be expected to change. The activity of the pumps should change in ways to damp out even these small effects.

I had hoped that the limitation by the surface area would cause a large effect that might lead to an understanding of the interesting experiments of Kepes and Kepes (1980) in which cultures were propagated in a specially developed dilution

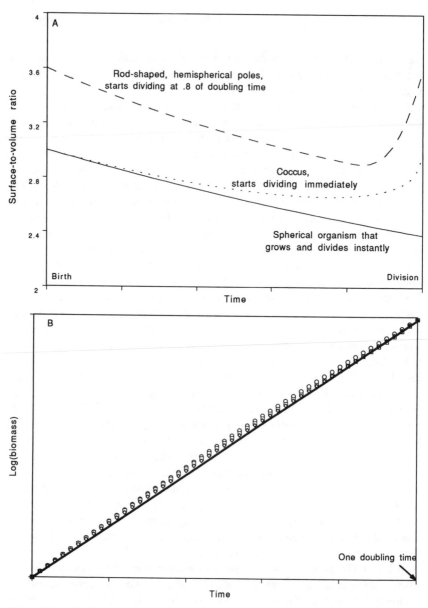

Figure 7.1. Surface-to-volume (S/V) ratio and growth limitation. Three morphological types were considered: a spherical cell that increases in radius until the volume doubles and then instantly divides into two equal-sized cells (thin line and circles); a coccus-like cell that at birth starts to form two new poles (dotted line and diamonds); and a rod-shaped bacterium that at birth is 1.5 times its radius in length and starts to form new poles at 0.8 of the cell cycle time (dashed line and squares). A shows the S/V ratio for the three hypothetical shapes of cells from birth to division. B shows the corresponding growth curves on the hypothesis that growth rate is always proportional to surface area. Pure exponential growth is shown by the thinner solid line.

machine that diluted the culture 2-fold once a nominal cell cycle. At the end of each doubling time, half the culture was removed and replaced with fresh medium. They used a medium containing a limiting amount of phosphate and found that after many cycles of this kind, the culture could become synchronized. When diluted into medium containing an excess of phosphate in batch culture it would undergo up to 12 synchronous divisions. I had thought that change in S/V would be of sufficient magnitude to have led to this synchronizing effects because cells in different times of the cell cycle would be more or less efficient in scavenging the phosphate, and gradually cells would have been driven into the optimum timing. Unfortunately for this idea, my calculations (Koch 1986a) showed that the changes in the aspect ratio (i.e., the ratio of length to width) were inadequate to account for the observed synchrony. Moreover, no other explanation for how this technique yields a synchronized culture has been forthcoming. The question is moot, however, because the technique is not sufficiently reproducible and its use has been discontinued. This is too bad, because the occasional achievement of synchrony by this and other techniques means that there are very interesting phenomena left to be uncovered and understood.

The Unexpected Precision of Cell Division in Some Bacteria

Rod-shaped bacteria such as *Escherichia coli* (Trueba 1981) and *Bacillus subtilis* (Burdett and Higgins 1978) divide very precisely at their midpoints to create nearly equal-sized daughters. The coefficient of variation (*CV*) of the ratio of daughter cells' lengths to those of their mothers in one observed case is only 4%. Inasmuch as the electron microscope data contain experimental variation and include a few aberrant cells, the actual mechanism determining the midpoint of the cells must be more accurate than this. The observed precision is a significant and a well-documented fact in the field of the physiology of Gram-negative and Gram-positive organisms. But why the cells bother to achieve this precision is a mystery because the calculation presented above and in Koch (1986b) seems to show that this would not significantly increase the S/V averaged over the whole growing culture. Struggling with this fact, Höltje and I (1995) came to suggest that division is precise because precision is simple and easy to achieve. Although the relevant experimental observations will be presented below, it is to be emphasized that just how the cells can achieve this precision or how they even approximately fix the place where the next division is to take place are, at present, only matters of speculation. Before going into the models we must describe two additional phenomena. In rare cases and more often in certain mutants, division takes place at one of the poles to pinching off DNA-less spherical minicells (Adler *et al.* 1967). Thus, in addition to the midpoint of the cell, the poles are potential sites for cell division to take place. Normally the polar sites are prevented from being used by the *minCD* cell division control system (de Boer *et al.* 1989).

The other point to be mentioned before we proceed is that some bacteria divide very unevenly; for example, many filamentous blue-green algae (bacteria) vary the length of cells in a systematic way, frequently in relationship to their distance from heterocyst cells, i.e., those cells that fix atmospheric nitrogen and supply amino groups to the entire filament.

Other Proposed Mechanisms for Fixing the Position of the Next Cell Division

Four models have been put forward in the literature to explain the positioning of the site of division. All, however, are incomplete—three fail to explain how cell division occurs exactly at the center, and the fourth model presumes that the origins and termini remain attached to the tips of the poles of the cell throughout the cell cycle, which may or may not be the case. After these are considered, a new proposal based on an entirely different principle will be discussed for this essential and unique process of prokaryote biology.

The Periseptal Annulus (PSA) Model

The model of MacAlister *et al.* (1983) (see Fig. 7.2) postulates that a pair of periseptal annuli are created adjacent to a previously established constriction site where active division is ongoing. These annuli are regions of tight attachment of the cytoplasmic membrane to the peptidoglycan/outer membrane layer. The annuli then move during elongation of the cell to sites that are ¼ and ¾ of the cell's length. After this positioning, they remain at the ¼ and ¾ proportions as the cell continues elongating. After the next cell division, both become the midpoint positions of the resultant daughter cells and then cause localized changes that fix the sites of the next cell division (Cook *et al.* 1987, 1989; Rothfield *et al.* 1991). The model is based on the properties of mutants and especially on the electron microscopic appearance of cells that have been plasmolyzed with sucrose. Plasmolysis spaces develop on osmotic challenge and their positioning suggested to Rothfield's group the idea that their margins defined the junction of those regions of the cytoplasmic membrane that have a higher affinity for the murein/outer membrane layer compared with other regions where the attachment is weaker. This boundary between the zones of adhesion and absence of adhesion was taken to be the location of the annuli, which then relocate during growth and division and finally mark the sites of future cell division.

The fundamental criticism of the PSA model as a mechanism for site selection is that it is not a complete model because it does not attempt to consider how these objects, nor the plasmolysis "bays" (spaces) that are presumed to mark them, can "know" when to stop moving. Only with the addition of other elements (perhaps the model presented below) to spatially define the sites could an augmented PSA model be a candidate model for cell division.

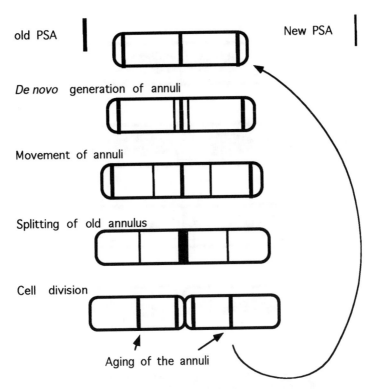

old PSA

New PSA

De novo generation of annuli

Movement of annuli

Splitting of old annulus

Cell division

Aging of the annuli

Figure 7.2. The periseptal annulus model. A structure exists that determines the division site. Not only must it split in two with the cell division event, but it must generate two new structures much earlier (the periseptal annuli, or PSAs). These must move to the center of their half of the cell and remain in this relative position as the cell grows until they are triggered to initiate division.

The Nucleoid Occlusion (NO) Model

The nucleoid occlusion (NO) model (Mulder and Woldringh 1989, 1993; Wold-ringh *et al.* 1991) (see Fig. 7.3) posits that the nucleoid produces a signal that inhibits cell division by blocking murein (peptidoglycan) synthesis. This model is a complex one with many additional elements, but it is an incomplete model because it does not specify how the nucleoids "know" when and how far to move from each other after the sisters have become separated. Additionally, because the model has been formulated to explain why cell division occurs between the nucleoids, it does not necessarily explain the experimental finding of the near equality of the length from pole to pole of newborn sister cells.

The Renovated Replicon Model

The renovated replicon model (RRM) (Koch *et al.* 1981a; Koch 1988b) (see Fig. 7.4) is in principle capable of the observed precision. The model is based

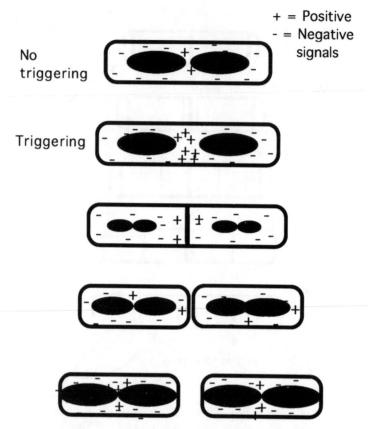

Figure 7.3. The nucleoid occlusion (NO) model. A positive signal is released upon completion of chromosome replication. It triggers constriction at a site where it can overcome the negative signal effects. The negative signal is generated as part of transcription/translation activity, but quickly turns over and therefore is weaker in regions remote from a nucleoid.

on the fact that the replicating bacterial chromosome develops a symmetrical "theta" structure that during DNA replication, if the origin and terminus regions were attached to the tips of the poles, could develop forces that would tend to center the replisome on the cytoplasmic membrane midway between them. For this model to function, the origins of the replicating chromosomes need to be physically attached to the pole tips for at least part of the cell cycle. There is evidence, however, that the origin region of the chromosome binds to the wall for only a short time after synthesis (Ogden *et al.* 1988). Without an explanation for the precise rebinding to the polar regions so that these loci at both ends of the cell can serve to center the replisome, this model, too, is incomplete.

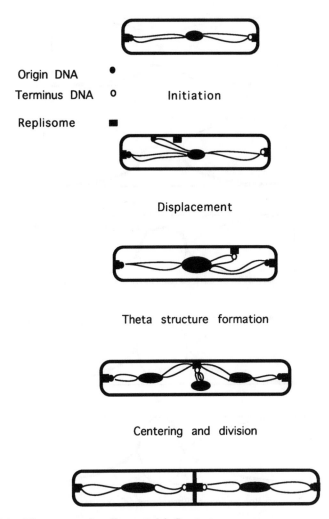

Origin DNA   ●

Terminus DNA   o   Initiation

Replisome   ■

Displacement

Theta structure formation

Centering and division

Figure 7.4. The renovated replicon model. See text.

The Activator-Depletion Substrate Model

The activator-depletion substrate model (Meinhardt 1982) (see Fig. 7.5) is a special variant of a more general biochemical theory that describes pattern formation by an autocatalytic system that produces long-range inhibition. Accordingly, metabolic processes of special classes functioning within a confined space, e.g., a container, like the cell itself, could generate spatial maxima and minima through enzymatic action. These processes have the property of creating spatial gradients where there were none before. In principle, they could define the midpoint of a cell, given a number of additional special circumstances such as appropriate

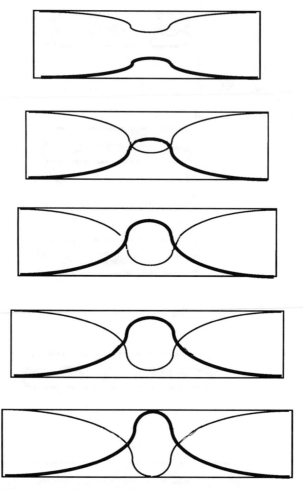

Figure 7.5. The activator-depletion substrate model. See text.

diffusion constants for the substrate and activator. This model would be very attractive if it also explained how the system regenerates itself at the end of each cell cycle by creating anew a uniform substrate concentration in the cell and by destroying essentially all the activator from the previous cycle. Also, because the kinetic system produces a broad optimum, the model, as it stands, may not in principle account for the observed evenness and precision of cell division.

Role of Mechanical Forces in Locating the Geometric Center of a System

Dividing the cell into two unequal parts may be a poor strategy because, if there is not an advantage to being either smaller or larger than the mean size-at-birth, then uneven division will allow slower growth to some and faster growth to others, but on the average slower growth to the total population (Fig. 7.6). The difference is not very great when surface area is limiting and therefore this argument is weak. A stronger hope, more than an argument, is that binary cell division is an essential feature of bacterial systems, even of the least developed ones, and therefore the basis of equipartition is likely to be a quite simple principle. Koch and Höltje (1995) therefore asked whether the basis could be the simple physics of mechanical forces. Consider a structure being subject to tension only from two points near its ends as indicated in Fig. 7.7. With stresses resulting from forces pulling against each other, tension will develop between the two points and will be uniform throughout the intermediate region, assuming that the structure is of uniform thickness and physical properties. The higher the forces, the greater will be the tension. With sufficiently high tension, the structure may fail and tear. Failure will happen at the point where there is an imperfection or defect in the structure (see Fig. 7.7B), according to the dictum: A chain is as strong as its weakest link. This is actually a form of the key principle of engineering mechanics presented several times above. It states that under stress, the first bond to break will be adjacent to a defect because there the stress is not evenly shared (Griffith 1920). It will be suggested shortly that in a living biological structure, high tension could lead to the triggering of some local response, such as initiating a cell division event. Triggering cell division might forestall the physical destruction of the cell because the tension is decreased through the wall growth response and the alteration of the cell structure.

With opposing forces applied only at two sites, however, this process would occur anywhere along the length of the structure and be dependent on structural abnormalities or the vicissitudes of bombardment of the cell by Brownian motion, etc. Mechanical forces can, however, serve to localize the center of a structure under special conditions. Fig. 7.7C and Fig. 7.8 show the results if the stresses are applied in segments along the length of the body. This can be best understood by analogy to a tug-of-war in which two teams pull on a length of rope in

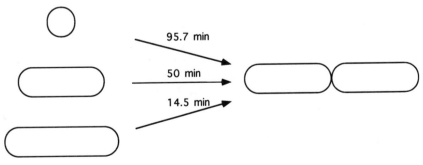

Figure 7.6. Lack of effect of equality of cell division on surface-limited growth.

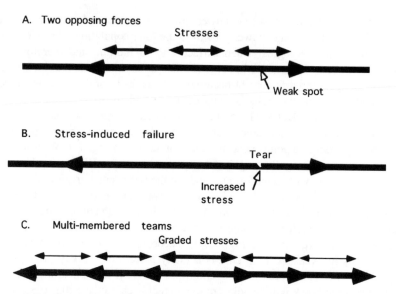

Figure 7.7. Development of stress in a stretched rope. A. If a physical body is stressed from two points, is of uniform cross-section, and is of a uniform material, then a uniform tension will be experienced throughout the region in between. The magnitude will depend on the forces applied and on the cross-sectional area of the body. B. If there is a weak spot, the stresses may exceed the strength of the rope and it may tear. C. If force is applied at a series of points within the object as in a tug-of-war, then the stresses are larger in the center of the object.

opposite directions (Fig. 7.7C). Although the stronger team wins, the tension in the rope reaches a maximum between the two teams. The high point of stress is at this point because a certain amount of tension develops in the rope between the most peripheral player and the adjacent partner. Between that person and the next there would be approximately twice as much tension because the forces of both players contribute and add to each other. If the teams have the same number of equally strong players and they are uniformly distributed along the rope, the stress maximum is in the rope's exact physical center and this is where it will break unless there is a very weak spot elsewhere. This concept is not new in biology. Hershey used the principal to shear DNA into half- and quarter-sized molecules in the days before restriction enzymes. He passed the DNA through a fine hypodermic needle. A shear stress developed in the tug-of-war as different regions of the linear molecule were stressed to differently, and mechanical breakage occurred almost exactly at the midpoints.

A Physical "Tug-of-War" Model of the Central Site for Cell Division

What could generate the equivalent of the tug-of-war? Koch and Höltje (1995) suggested a "central stress model" that could be a result of the pattern of stresses

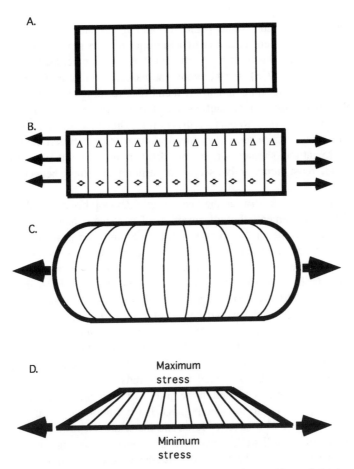

Figure 7.8. Development of stress in an elastic sheet. A shows a sheet of elastic material with parallel strip marks. Uniform stress on the ends of a body lead to uniform expansion, shown in B. It does not matter if expansion results from new insertion, shown by the Δ, or by elastic stress, shown by the ⇐ ⇒ symbols due to force applied at the ends shown with the arrows. C shows that force applied medially produces greater tension in the center because the regions on the side support part of the stress. D shows the same phenomenon of higher tension in the center for force applied from the two ends on one edge.

that would develop between the murein and the underlying cytoplasmic membrane. Differential stress might build up because the bacterial wall is a composite in which one layer interacts strongly with another layer. Let us neglect the outer membrane and consider only the other two layers, the murein sacculus (M) and the cytoplasmic membrane (CM). The interaction results in large part from the turgor pressure pressing the CM against the M layer. It also results from the membrane proteins that specifically associate with the murein. In particular, these are the membrane-bound penicillin-binding proteins (PBPs), which are involved in the covalent insertion of additional murein subunits into the pre-existing murein sacculus during growth but act as well in other aspects of murein metabolism (Spratt 1975). Furthermore, the nascent murein may remain covalently linked for quite some time to the membrane-residing hydrophobic undeca-prenylphosphate carrier that shuffles the murein precursors across the cytoplasmic membrane (see Rogers et al. 1980). Thus, the machinery for wall growth provides a major temporary physical connection between the two layers that tend to inhibit slippage of the two from point to point on the cylinder portion and, most importantly, serves to partition the stress between the two layers.

Differential stresses between the layers might develop because wall growth is actuated by enlargement, through insertion of new material into the layers of the cell envelope. The elastic properties of the CM and murein are quite different and it is likely that the timing of insertion is different as well. In the absence of insertion, increased turgor due to new cytoplasm formation would stretch the murein and CM, increasing stress in both. Then when insertion in one took place, tension would be transferred to the other. The cell must eventually enlarge the lagging member of the paired layers; otherwise, for example, plasmolysis spaces would develop at the poles of the cell, or the membrane would rip, leading to the leakage of cell constituents. But as described, the tensions would be uniform along the length of the cell as in the analogous case shown in Fig. 7.8 where the piece of rubber is pulled from its ends comparing A to B. On the other hand, if the force is exerted in the middle of each end, as in C, then some of the stress is diverted to the side regions, but not uniformly along the length if the sheet is made of a stretchable elastic material. D shows what may be a closer analog of the cell wall, where force is applied at one edge only. It shows a buildup of stress in the center of the free-hanging edge and a maximum on the primarily stressed end.

Even though the layers are continuous and are in close apposition in all regions of the cell, the stress distribution within the CM layer could reach a maximum at the cell center. This may well be the case because the cylinder wall enlarges in all of its layers during the cell cycle while the pole outer and murein layers are inert (see Chapter 13). If the phospholipids enter the CM at all regions then there will be a flow from the ends toward the center; this will decrease the tensions developing near the ends of the cell, but not those at the center unless the connection between CM and M is very weak. This is shown in Figs. 7.9 and 7.10.

The bipartite structure provides a mechanism to identify the center, but that is only a portion of what is needed. We, therefore, also proposed a specialized sensory system, similar in several respects to known two-component systems, that monitors the cytoplasmic membrane stress and at a critical value triggers localized constriction or septum formation leading to cell division. At a sufficient

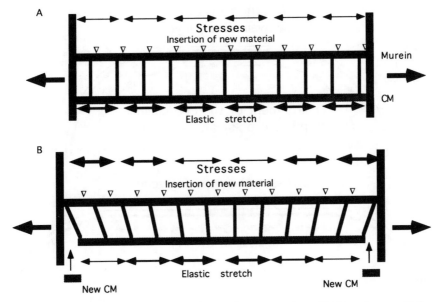

Figure 7.9. Development of localized stress in a composite wall of two layers. At the top, a body composed of an upper murein layer and a lower CM layer has become elongated due to the cell's turgor pressure forcing the ends apart. In response to the stretching at its ends the murein has enlarged by insertion of new material as shown by the ∇'s. Elastic stresses develop in the CM layer that are uniform over the length of the side wall. At the bottom a quite different situation is shown because the ends of CM layer are free-hanging. This leads to a maximum at the center. For this model to work for bacterial growth, new CM must be synthesized and recruited to the ends at least periodically during the cell cycle and joined to the old CM and attached to the murein layer.

value, this localized stress initiates cell division at that site. The basis of this model is thus quite different from that of any of the four models outlined above.

Triggering of Cell Division

The model assumes that once the stress threshold is locally exceeded the response occurs in a circumferential circle and causes an irreversible change in a membrane-bound molecule that is the sensor component of a two-component system (Fig. 7.11. Because of the tug-of-war process this happens in the middle of the cell, and the circumferential response occurs at the exact center. This membrane molecule then speeds the local wall growth and determines its inward direction. Through the enlargement of the wall the turgor pressure is decreased so that sensors in other regions of the CM are not subsequently triggered. This lowering

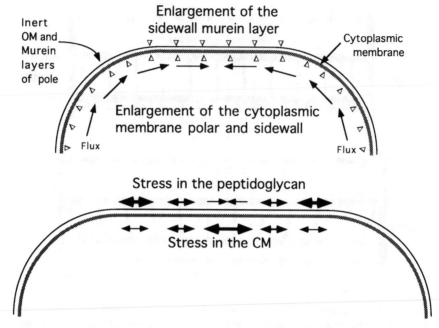

Figure 7.10. The Central stress model. Δ indicates incorporation of murein into the growing wall. The arrows show the flux of phospholipids towards the center of the cell wall. The model is based on the following assumptions: (i) the M and CM layers interacting physically with each other allowing the transfer of stress; (ii) the incorporation of murein components occurs diffusely but only in the side wall; (iii) the incorporation of CM material occurs all over the inner membrane, including the pole; (iv) there is a special two-component sensory system responding to tension in the CM that elicits cell division.

also prevents any additional site (such as the junctions of cylinder and old pole) from initiating a local division site. Thus, it is assumed that in all parts of the cell membrane there is a dedicated two-component system that monitors the tension in the phospholipid layer. Two-component systems exist for quite a few functions in prokaryotes, including osmoregulation (see Stock *et al.* 1987). The sensory elements, once stimulated, cause the effector elements to become phosphorylated to cause the synthesis of appropriate proteins.

The hypothesized model also attributes two additional functions to the sensory component responding to membrane tension. One is that the molecule changes irreversibly, possibly by autophosphorylation, and remains localized as a marker of the future cell division site. The second is that the sensor binds other components that lead to local constriction or septum formation. Local responses are not characteristic of known two-component systems, and therefore Höltje and I have assumed novel additional features for such systems. Several can be suggested for the trigger action. Perhaps the initiating cell division events are mediated by the

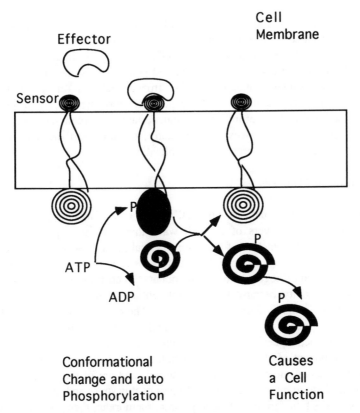

Figure 7.11. A two-component system for cell division. The now classical concept of the fundamental regulatory system involved in regulation of physiological function in prokaryotes is modified slightly to suggest how a cell can generate a division signal. Other than the basic concept, which certainly applies in many instances, this model is entirely hypothetical.

sensor in its conformationally altered shape causing changes in the phospholipid composition (Cullis and De Kruijff 1979; Gruner 1985; De Kruijff 1987; Bloom *et al.* 1991; Norris 1992). Perhaps it may result from the localized binding of the division proteins to the effector molecule that has initiated their synthesis. Perhaps it could be that the sensor portion of the hypothetical two-component system causes the *ftsZ* gene product to accumulate (Ward and Lutkenhaus 1985) as a ring localized circumferentially around the cell at the site where the threshold of the sensory system has been exceeded (Bi and Lutkenhaus 1991). Interestingly, filamentous cells blocked in cell division start dividing at pre-determined sites when returned to conditions that allow cell division (Donachie *et al.* 1984). This is consistent with the proposal that the sites had been marked and remembered, in spite of the failure of the immediate expression of the cell division event. As the postulated sensor is the trigger for cell division, there must be some way to "freeze" its immediate response, such as in an active conformation or to catalyze an irreversible process (a phosphorylation?). Although ordinarily wall growth directed toward cell division is initiated and persists for some time as a result, it is not unreasonable to imagine that the "frozen" state can remain for a longer period of time. Additionally, it must be noted that although normally chromosome replication is completed before cell division, certainly completion is not necessary in a variety of circumstances. Consequently, the proposed mechanism is not directly connected with chromosomal replication.

Formation of Minicells

In addition to the local maxima at the cell's center created as a result of differential growth of the two layers of the cell wall, there is also a permanent stress discontinuity at the junction of the hemispherical polar caps with the cylindrical part of the cell wall (see Fig. 7.12). At this location there is a 2-fold increase in one of the two orthogonal stress components (Koch 1988a). There is an abrupt 2-fold discontinuity because the circumferential, or hoop, component of the surface stress in the cylindrical region is Pr. The other stress—the meridional, or axial component—is mathematically equal to $Pr/2$ and therefore is the same as that in the hemispherical pole. Thus, there is no discontinuity at the junction in this component of stress. (These formulae assume that the two wall regions are homogeneous, made of the same material, and are of the same thickness.) Assuming that the murein supports only part of this stress and that some portion of the stress is transferred to the cytoplasmic membrane, then the tension would change significantly over a very small region of membrane. Thus, in a rod-shaped bacterium there are two different sites where there is a discontinuity in a component of the stress: the cell center where the rate of change of the axial stress changes discontinuously, and the junction of the pole with the cylindrical region where the hoop stress changes discontinuously.

There is another source of discontinuity at the pole junction. This arises because the murein and outer membrane are metabolically inactive, whereas the side wall is continuously enlarging. This fact and the flow of CM from the poles could also make this a special point.

Under a variety of conditions bacteria do divide at the inappropriate junction region and produce small, round, DNA-less cells, called minicells (Adler et al. 1967; Frazer and Curtiss 1992). This outcome can be favored by mutations in minCD or minE (de Boer et al. 1989), but it can also be engineered by overproduction of the ftsZ gene product (Ward and Lutkenhaus 1985). In the model proposed here (Fig. 7.13), this kind of error can be understood as the result of triggering the two-component system not only at the center of the cell but also at the cell poles because of a local discontinuity of stress. The latter event can happen when these sites are no longer under the veto power of the minCD cell division inhibitor.

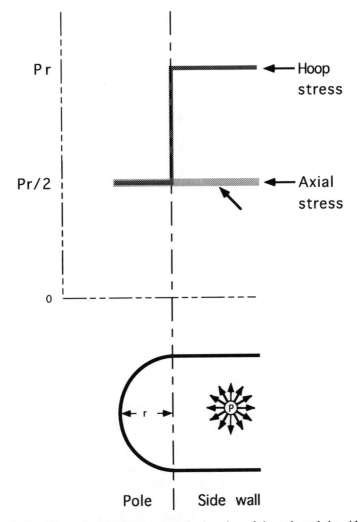

Figure 7.12. Discontinuity of stresses at the junction of the pole and the side wall. Below: the figure of an idealized cylindrical cell with a hemispherical pole is shown. Above: the stresses developing in the wall. The axial (or meridional) stress is shown by a thicker grey line; the hoop (or circumferential) stress is shown by the thinner grey line. Note that the hoop stress changes discontinuously at the junction of the pole and cylinder regions. The axial stress is the same in both regions of the cell, whereas the magnitude of hoop stress doubles abruptly in passing from the pole to the cylindrical region. Under certain circumstances when normal division in the center of the cell is blocked, the discontinuity in circumferential stress serves to trigger the hypothetical two-component system to initiate cell division, resulting in minicell formation.

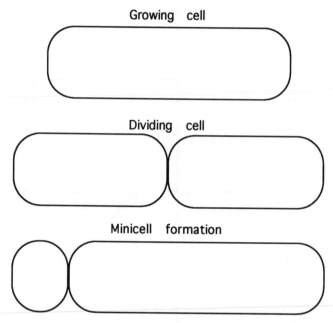

Figure 7.13. Minicell formation.

The Relationship Between Evenness of Division and Maintenance of Synchrony

After it was shown that bacterial cells divide more directly in response to their achieving a critical biomass than their achieving a particular age (Schaechter *et al.* 1962), it was quickly shown theoretically (Koch and Schaechter 1962) that if four criteria were obeyed, synchrony would persist, if developed. Although a portion of the synchrony would be lost after the first division, the remaining partial synchrony would be maintained indefinitely if: (i) division occurred quite precisely at a particular narrow size range, (ii) cell division always partitioned the cell into equal-sized daughters, (iii) biochemical processes proceeded at the same specific rate in all cells in the population, and (iv) full synchrony had been achieved initially. (These criteria have been further detailed in Chapter 6.) Maintenance of a partial synchrony results because cells at division that deviated by chance either above or below the mean division size would vary in the time to the next division, but in the subsequent division the deviation would be in the opposite direction; thus the total variation would not accumulate (Koch 1966; Koch and Schaechter 1962; and see Cooper 1991). Experimental confirmation was obtained (Koch 1966) through the analysis of the growth of single cells into microcolonies using the data of Hoffman and Franks (1965) and the cell pedigrees observed by Kubitschek (1962). Subsequently, other experiments were reported in which mass cultures maintained synchrony while growing in constant conditions in dilute suspension. This was shown by Marjorie Kelley (see Koch 1977) and by Cutler and Evans (1966) (see Fig. 7.14). Kepes and Kepes (1980) also prepared cultures that exhibited a large number of synchronous division steps. These reports of persistent synchrony are consistent with the idea that the four criteria mentioned above must have been rigorously obeyed in these cases. The important conclusion here is that the division process is normally extremely precise, and probably the average unevenness in division is even less than the experimental estimate drawn from electron microscopic measurements (Trueba 1981) in which there was only 4% *CV* in the ratio of the birth of mother sizes of dividing cells in an *Escherichia coli* culture. Consequently, even though the persistence of long-term synchrony has been demonstrated only rarely, it has been demonstrated widely enough that it can be considered an established fact,

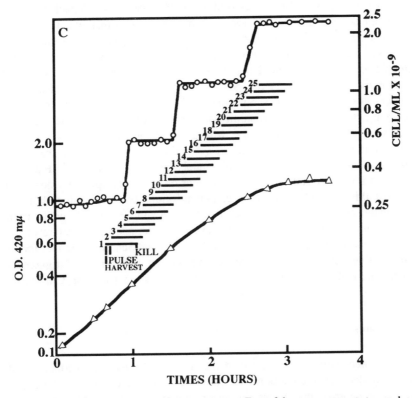

Figure 7.14. Experimentally observed synchrony. Few of the many attempts to synchronize bacterial species other than *Caulobacter* and *Rhodomicrobium* have succeeded. But occasionally and in quite diverse ways the attempts have succeeded with *Escherichia coli*. An example is shown here from the work of Cutler and Evans (1966).

and leads one to subscribe only to models in which division of the mother generates two very nearly equal-sized daughters.

Relationship of Cell Division to DNA Replication, Transcription, and Translation

Descriptions of the bacterial cell cycle are almost always based on the Helmstetter and Cooper (1968) analysis (Fig. 7.15). Their hypothesis posits that the time needed for completion of chromosome replication C and the subsequent period of time until cell divisions is complete, D are well controlled by cell physiological processes. But this formulation leaves unspecified the relationship of the doubling time T_2, also called I, the time between subsequent initiations of chromosome replication relative to the growth physiology of the bacteria. The importance of the growth of cell constituents not only is clear from the studies of Schaechter *et al.* (1962), but was also emphasized by the observations of Donachie (1968), who demonstrated that the cell seems to trigger some critical event, probably chromosome replication, at a particular cell mass, independently of the growth rate. This older work seems to be connected to *dnaA* binding sites in the *oriC* region and to sites of methylation of the adenine residues of DNA (see von Meyenburg and Hansen 1987). Studies with the flow cytometer (Boye and Løbner-Olesen 1991) have shown that initiation of sister chromosomes in a common cytoplasm occurs synchronously. In spite of these great successes, no studies so far have made a linkage with a biochemical measuring system that could estimate the cell's success in growth and cause cell division at the appropriate stage in the cell's growth. Höltje and I (1995) now suggest in Fig. 7.15 that the trigger responds directly to the tension in the cytoplasmic membrane and that this not only leads to a central position for cell division, but also coordinates cell division directly with cell length and less directly with current biomass and still less with the accumulated age. An additional feature of this model is that it may lead to revision of the Koch and Schaechter model for the control of cell division and replace the concept that cell division is triggered at a particular cell mass with the concept that it occurs at a particular CM tension and therefore a particular cell length.

Conclusions

Is there a necessity for evenness of division? Many bacteria and certain *E. coli* strains, under certain conditions, do not divide to create equal-sized daughters. So it is an anomaly that bacterial division in certain cases can be very precise. The observed equipartition led Koch and Höltje (1995) to suggest that this precision in the evenness of division during the creation of a pair of daughter cells reflects a fact that the method of the establishment of the cell center is

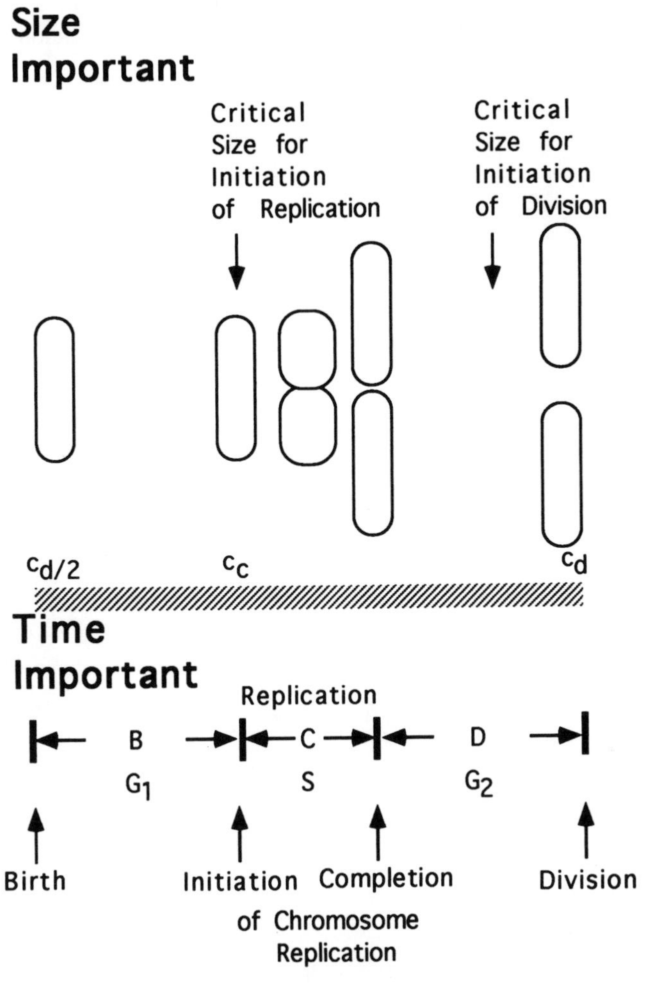

Size
Important

Critical
Size for
Initiation
of Replication

Critical
Size for
Initiation
of Division

$c_{d/2}$

c_c

c_d

Time
Important

Replication

|← B →|←C→|← D →|

G_1 S G_2

Birth   Initiation Completion   Division
          of Chromosome
          Replication

Figure 7.15. The Helmstetter-Cooper model retrofitted with the concept that the cell perceives something about its success in growth to trigger cell division events. Replacing earlier ideas about the cell "knowing" its age or its mass, the hypothesis exploited here is that the cell senses an extremum of tension in the cytoplasmic membrane, which then triggers chromosomal replication. At a more extreme value of the tension, either constriction in the Gram-negative organism or septation in the Gram-positive organism is initiated.

inherently precise because of the physical process of its generation. In this case, it may be easier to be precise than to be sloppy. It may be that the precision is not needed for evenness of the cell division process, but for the linkage of adequate cell growth with the initiation of chromosome replication and of cell division.

8

The Gram-Positive Coccus: *Enterococcus hirae*

KEY IDEAS

Enterococci have no cylinder part, but are made only of poles.
The poles are prolate, i.e., pointed like an American football.
Enterococci are adapted to a rich anaerobic environment such as the mammalian gut.
Enterococci grow zonally; i.e., two new poles form, starting at a wall band that
   surrounds the equator of the cell.
Growing poles achieve their ultimate shape as the septum is split and the wall
   exteriorized.
New growth zones are initiated when the ongoing ones cannot make wall fast enough
   to accommodate newly synthesized cytoplasm.
The wall of Enterococcus hirae *has aminated glutamic acid in γ-linkage.*
The wall of Enterococcus hirae *uses lysine and not diaminopimelic acid to form tail-to-*
   tail linkages.

The wall of Enterococcus hirae *has a polyamino aspartic bridge between the ε-amino group of lysine and the carboxyl group of the D-alanine of the other chain.*

The crossbridge of E. hirae *has only one negative charge, much less than for organisms such as* E. coli *or* B. subtilis.

Initiation of a growth zone probably controls chromosome replication, in a reverse way compared with the strategy of E. coli.

Enterococcal wall is extremely rigid after formation and crosslinking.

The zonal dome model predicts the pole shape on the assumption that much of the septal wall is incorporated into new external wall.

The split-and-splay model assumes that the septal wall area does not become enlarged on being split and exteriorized, but instead is twisted.

Split-and-splay requires replacement-synthesis that creates the additional wall needed to form the more pointed, prolate pole of E. hirae.

Whereas spherical shells have equivalent stresses in all directions, the prolate pole has a higher hoop stress that helps to form its prolate shape.

The prolate shape of the enterococcus has almost as favorable a surface-to-volume ratio as does a rod shape.

Because of the variability of pole volume, and especially because either one or two growth zones can become initiated, it is impossible to synchronize a culture.

For the study of streptococcal wall growth and morphology, the species that has become the standard organism was originally called *Streptococcus faecalis*. Then it was called *faecium,* then *Enterococcus,* then *hirae;* all changes were made for presumably good taxonomic reasons. Of the Gram-positive cocci, we will focus on this species because of the careful morphological studies with high-resolution electron microscopy. The two terms *Streptococcus* and *Enterococcus* will be used interchangeably except when naming particular organisms.

The Organism and Some History

Although the range of characteristics of the enterococci is quite broad, we can simplify and summarize them. When newly formed they are not spherical cells, but closer in shape to an American football. They grow by adding new wall material as an annulus at the equator that splits to form two new poles (Fig. 8.1). They divide in only one plane. This habit of growth sometimes leads to chains of cells, from which a Greek root, *streptos,* led to the name streptococcus. Streptos actually means twisted, but it also means chains. Typically, the entero-cocci have extensive nutritional requirements and therefore thrive in a complex, rich environment such as in the gut of animals or in milk. Part of the general defenses of the mammalian host against microorganisms is to deplete the environment of iron in a carefully orchestrated way so that the host has adequate iron, but microorganisms (whether they be symbiotic, commensal, antibiotic, or opportunistic pathogens) have difficulty in the iron-poor environment and fail to survive. The evolutionary counterstrategy of the enterococci to this challenge

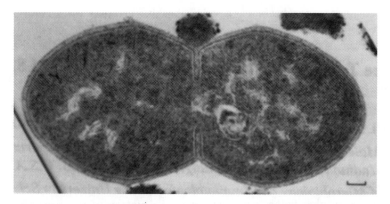

Figure 8.1. Electron microscope axial section of *Enterococcus hirae*. A thin section is shown, courtesy of Michael Higgins. The unit cell, as it arises through division of its mother, is shaped rather like an American football. At the later stage of the cell cycle, caught here, the central peripheral band has split and new wall has been added in between to create an annular septum that is partially split. In time, this will partition the cytoplasm into two independent cells. As the septum forms it becomes split externally by autolysin action. A key point in the development of the surface stress theory was the observation that the slope of a region in the developing pole is the same as the corresponding region of the completed older poles that are at least one generation older. This implies that the wall resulting from the splitting of the septum has a fixed slope that is not later changed as the cell division process continues. This is another bit of evidence that growth of the wall is zonal and that the wall is not reworked later. The figure is republished from Koch *et al.* (1981a).

was to perfect a metabolism independent of iron. Presumably, they were organisms that were basically fermentative and as animals evolved were able to occupy the gut habitat. This has meant that enterococci must forgo an oxidative metabolism and only transduce energy in a fermentative mode. This feature has (incidentally) made them of use in studies of energy metabolism and energy flow.

Lack of need for iron also allows them to be early invaders of the newborn mammal's intestinal tract. They grow unhindered by the paucity of iron in the mother's milk. In the normal succession process they grow luxuriantly in the intestine, but then are largely replaced by a sequence of other organisms. Some remain in the climax ecosystem of the adult mammalian intestine. From these facts it might have been deduced that they can function in cheese manufacture. Because their fermentative metabolism derives energy from rearranging the elements in sugars and some amino acids, thereby trapping some energy for their use, they produce lactic acid and other acidic molecules as waste products; these acids curdle and clot the milk. Thus, as they grow in milk they form acid abundantly. The resultant low pH prevents many other organisms, including the enterococci themselves, from further growth. Probably now that the name has been changed, somewhat hiding its original habitat, our type organism, *E. hirae*, may become more commonly used by the cheese industry than previously.

There are pathogenic enterococci such as the organisms responsible for hemolytic "strep throat," pneumonia, and endocarditis. Important understanding of the immune response was achieved by studying the "M" protein present in the wall of certain streptococci.

Our source of useful morphological information starts with the work of Toennies on the physiology of *E. hirae*. The study of the organism was handed over to a biochemist, Gerald Shockman, then to Lolita Daneo-Moore and Michael Higgins. These latter two are students of microbial physiology and ultrastructure, respectively. The team of Shockman, Daneo-Moore, and Higgins, working together in the Microbiology Department at Temple University, has been highly productive in elucidating many aspects of the biology of this organism, and incidentally produced the basis for the surface stress theory.

The wall chemistry is similar, but importantly different from the wall of *B. subtilis* or *E. coli* shown in Fig. 4.1; these differences probably are very important to the biology of the organism. First, there is an uncharged bridge inserted between the critical tail-to-tail peptide bond. Second, the α carboxyl group of the D-γ-glutamyl acid residues is aminated and the carboxyl group present in the diaminopimelic acid (DAP) is eliminated with the replacement of DAP by lysine; this leads to the important consequence of a much reduced net charge on the peptide. It follows, therefore, that the conformation and facility for changing conformations should be greatly different than that of the peptidoglycan of *E. coli* or *B. subtilis*.

E. hirae grows and divides in a way different from the rod-shaped organisms, either *B. subtilis* discussed in Chapter 9 or *E. coli* discussed in Chapter 10. Basically, it can grow only by forming poles. Very early studies with fluorescent antibodies established that wall growth in Gram-positive cocci was zonal. Growing cells were tagged with antibodies with a fluorescent probe against wall proteins and then allowed to grow during a chase and the samples periodically inspected in the fluorescence microscope. The finding was that the chain of cells acquired a zebra pattern because new zones of growth were formed that were separated by regions of old wall still labeled with the fluorescent antibody label.

The Morphology of *Enterococcus hirae*

Through careful electron microscopy it has been deduced that new growth occurs at the inside surface of the developing septum (Channel A); in addition, after splitting, the wall is thickened by addition to its surface (Channel B) (Fig. 8.2). Under some conditions the completed pole continues to be slowly thickened.

The seminal paper in this field was published by Higgins and Shockman in 1976. In this monumental and heroic work, cells from a growing culture were

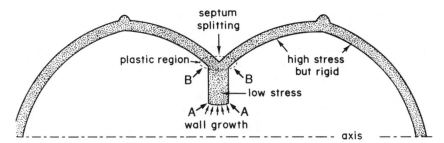

Figure 8.2. Growth pattern of *Enterococcus hirae*. Sites of wall growth (Channels A and B) are shown. Also shown is the major site of autolysis action, cleaving the septum to enlarge the developing poles.

fixed, stained, embedded, and sectioned. The sections were examined in the electron microscope and only those that by chance had been sectioned axially were used for analysis, such as, for example the cell shown in Fig. 8.1. Stringent criteria were used to reject off-axis sections. As mentioned, streptococci many times grow as chains. In order to avoid this complication, the level of tryptophan in the complex synthetic medium was augmented. This nutritional change somehow prevented chain formation so that most cells appeared as what would classically be called a diplococcal form, such as that exhibited by *Diplococcus pneumoniae,* now called *Enterococcus pneumoniae* (the common etiological agent of bacterial pneumonia).

A qualitative picture of the lifecycle was assembled from the many hundreds of pictures. Essential for the quantitative analysis of the cell cycle was the measurement with the aid of a computer of the various dimensions of each of the axial sections. The outlines of the inside and outside of the wall from the photographs were traced with a probe connected to the computer. Again with the aid of a computer, these dimensions were retrieved, analyzed, and combined to estimate surface areas and volumes of various parts of the developing and completed poles. In carrying out these calculations, the computer was programmed to reconstruct the third dimension on the hypothesis that the cell was truly axially (cylindrically) symmetric. For example, from the data on coordinates of points on the inner and outer surfaces of the wall, the volumes of wall and the volumes enclosed by each part of the cell could be calculated using the trapezoidal formula. The assumption that the organism remains axially symmetric after fixation is quite a reasonable hypothesis for Gram-positive organisms because the wall is so much thicker and more rigid than that of Gram-negative organisms. Moreover, the results from various control experiments were consistent with the assumption that there was little distortion of the cell by the fixation and sectioning process.

With the wealth of numerical data about the parts of each cell (old poles, primary developing poles, secondary developing poles, and their completed and incomplete septa) a choice had to be made to pick a suitable measure to be the indicator of progress of the cell through its growth cycle. Higgins and Shockman chose to use the surface area of the actually exteriorized pole, which they designated P_a. This allowed them to arrange the sections in increasing order of P_a. Then the other parameters they had calculated could be correlated with progress through the cycle. In this way they were able to establish the course of development. They were able to show that development of poles occurred in very nearly the same way from cell to cell. Fig. 8.3 shows that P_a correlates well with the pole volume.

The organism starts a wall growth zone by growing septal wall inwardly under a ridge or wall band that is present as a belt around the cell's equator. (In the laboratory slang in Philadelphia, this belt is called a "bibpee".) As the septum grows inward the ridge starts to split, and the splitting process bisects the septum. At some point the septum is completed, and at some later time the splitting is complete and the two poles separate. One could defend either of two points of view: that cell division is completed when the cytoplasm of the two daughters is no longer continuous, or that cell division is only complete when separation occurs so that the daughter cells are kinetically independent.

Triggering New Growth Zones

One aspect of cell division was not reproducible. This is the initiation of inward wall growth signaled by the splitting of the wall band; i.e., the initiation of new growth zones starts pole formation. This event, also, results in a variable relation of the new growth zone to the stage of other growth zones in progress. Sometimes two growth zones will develop almost simultaneously so that two secondary sites are competing with an older, primary site; sometimes only one develops (Fig. 8.4).

Variation from the E. coli Paradigm

The cell division process of *E. coli* as discussed in Chapters 4 and 10 seems to operate by a trigger event that occurs when the cell achieves a critical and nearly constant biomass. This event probably leads to chromosome replication, later to constriction, and finally to division. At different growth rates several ongoing rounds of chromosomal replication may be occurring simultaneously. Thus, the rod-shaped organism needs to perceive biomass prorated to each single completed chromosome.

The streptococci approach cell division quite differently. Instead of sensing biomass, the group at Temple University concluded (Gibson *et al.* 1983; Koch and Higgins 1984) that new growth zones appear to be initiated when the existing growth zones cannot increase the volume contained by the cell wall fast enough to support the ongoing protoplasmic growth. This hypothesis suggests that turgor

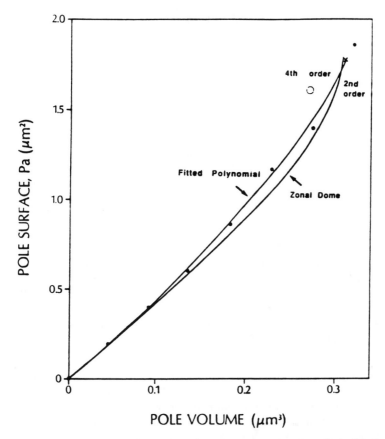

Figure 8.3. The relationship of the pole surface area to the pole volume in the developing pole of *Enterococcus hirae*. Theoretical curves and polynomial to fit the data of Higgins and Shockman (1976) are shown. For most of the process the newly created surface area is proportional to the new volume that it must contain. The computer program described in the final section of the chapter uses the second-order regression and its inverse to calculate surface areas from volumes in carrying out simulations of the growth of single cells and the properties of populations of cells. Reprinted from Koch and Higgins (1991).

pressure may play a direct role. When a zone is first initiated, the amount of new wall surface that can be made in a unit time accommodates a much larger cell volume than near the end of a zone's life when the septum has been completed and largely split. Then, deformation (bulging) of the remaining portion of the septum allows very little expansion of the cell. Although it is probable that turgor pressure serves a critical role, it has not been demonstrated directly. What is known is that the cellular control operates in a way dependent on the rate of

ASYMMETRIC

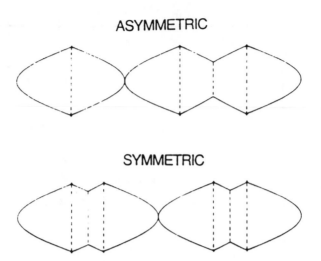

SYMMETRIC

Figure 8.4. Symmetrical and asymmetrical modes of new growth zone formation. When the primary zone can no longer enlarge at a rate to provide space for the newly created cell volume, then two patterns may be exhibited. Either one or two of the wall bands split, initiating one or two new growth zones. The variability of this process leads after only a few divisions to a population that is quite asynchronous when starting from a single cell. Reprinted from Koch (1992).

cytoplasm production relative to new surface area. This has been shown by Gibson *et al.* (1983) and by Koch and Higgins (1984) through the analysis of the distribution of growth zone initiating events as indicated by the formation of "baby" sites (small or early growth zones) in different phases of the cell cycle as a function of growth rate (Fig. 8.5).

At slow growth rates a primary zone can take care of the cell's needs much

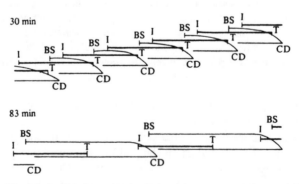

Figure 8.5. The relationship between the timing of pole development and chromosomal replication. BS stands for band splitting events and CD stands for cell division. Reprinted from Koch and Higgins (1984).

longer. Therefore, growth sites are likely to emerge shortly after division, and in most instances only one new primary site is created. In the short interval before a new site is created, such cells are called "unit cells." On the other hand, at more rapid growth rates new zones (usually two secondary sites) are formed before cell division of the primary site occurs that terminates the life of that growth zone.

Thus, *E. hirae* has a much different growth strategy than does *E. coli*. By sensing biomass *E. coli* has the best of all possible worlds. Under spartan conditions, the average cell size is small; thus, the cell can garner resources better because of a favorable surface-to-volume ratio and make more propagules per unit amount of biomass. On the other hand, under gourmet conditions the average cell size becomes bigger and fewer resources are wasted in the formation of cell envelope. Its system, because of its (presumed) ability to sense cellular biomass, allows free switching from one extreme of growth to the other. In between it has the ability to adopt suitable intermediate compromises. In the coliform, the diameter and length can vary based on secondary mechanisms without upsetting the basic control system.

The system for growth and replication of streptococci is apparently predicated on the maintenance of a constant wall band diameter. The measured diameter of the wall band at the equator of the cell has an experimental *CV* of only 5%. This is a very small variation for any biological parameter, and this observed 5% of course includes experimental error due to fixation, drying, and computer measurement. Evidently, each wall band serves as the template for its conversion into two wall bands. Copying of this structure can properly be claimed to be a second instance of a biological process of semiconservative replication; the only other case so far known is the replication of double-stranded DNA. As the wall band divides and the septum splits in half, each side acquires a new portion with exactly the same dimensions as its partner. If there were random fluctuations, any event enlarging the radius would have preference because the turgor pressure would tend to enlarge the cell. For a templating process to be successful, the wall material can have absolutely no ability to stretch. It may be that the *sine qua non* of the streptococcal strategy is to produce wall polymer that has an extremely high Young's modulus so that, on being joined to the split wall band, it forms a solid with dimensions that will not alter when it subsequently comes to feel the tension due to the cell's turgor pressure. The analogy is that streptococci are made of solid steel while the bacilli and coliforms are made of a network of steel springs.

The Special Wall Biochemistry of Streptococci

Several times in this chapter the inability of streptococcal wall to stretch has been alluded to. This is very unlike the elasticity of *B. subtilis, E. coli*, and

Streptomyces coelicolor (Chapters 9, 10, and 11). What is the possible chemical basis for this difference in stretching behavior? The basic carbohydrate chain of all these eubacteria is a repeating unit of (-β-*N*-acetyl-D-glucosamyl-(1→4)-β-*N*-acetyl-D-muramyl-(1→4)-). Each muramic acid residue contains as an integral part a D-lactyl group that is always linked to L-alanyl-Dγ-glutamic acid and then to other amino acids. The crossbridged peptidoglycan of the *Enterococcus* is of the Kandler and Schleifer peptidoglycan type A4α (L-alanyl-D-(α-amino)-γ-glutamyl-L-lysyl-D-alanyl-D-alanine) (see Rogers *et al.* 1980). Contrast this with the peptide chain of *B. subtilis* and *E. coli,* which is L-alanyl-Dγ-glutamyl-*meso*-diaminopimelyl-D-alanyl-D-alanine (classified as the A1γ type). For the physical properties there is a key difference: the enterococcal wall crossbridging peptide has a single negative charge at neutral pH. This is because the α-carboxyl of the glutamic acid residue is aminated and there is no free carboxyl group on the L-lysine in comparison to the diaminopimelic acid present in *B. subtilis* and *E. coli.* The enterococcal wall has an additional peculiarity, *viz.,* while the rod-shaped organisms have no bridge amino acid or peptide, enterococcus has a crossbridge consisting of β-amidated aspartyl residues connecting the lysine of one chain with the D-alanine of another.

In the *B. subtilis* and *E. coli* type of wall the crosslinked tail-to-tail pair of tetrapeptides has five ionized carboxyl groups and one amino group at the pH of neutral culture medium; in *E. hirae* there is one. These facts suggest, but do not prove, that this dramatic difference in net charge could lead to repulsion between the chains of highly negatively charged sacculi of *B. subtilis* and potentially might result in the greater elasticity of the wall fabric, whereas the intramolecular forces might keep the enterococcal wall always in a compact configuration.

The Chromosome Cycle

The other problem that must be solved for stable growth over an indefinitely large number of generations is to couple DNA replication and the pole cycle. This mechanism must be quite different from the *E. coli* paradigm where chromosomal replication is the trigger for the completion of cell division. The exact mechanism for the streptococci has not been established, although a start has been made by using autoradiography with tritiated thymidine (Higgins *et al.* 1986). Analysis for this organism is more difficult than for *E. coli.* The experimental difficulty results from the fact that within a single cytoplasm there are several poles forming, and consequently the pattern of DNA synthesis, at the level of the whole cell, is quite complex. A reasonable, but unproven, model is that the strategy is exactly the reverse case of *E. coli;* i.e., DNA synthesis starts after an interval timed from the start of pole formation. It follows that in a cell with a primary growth zone and two secondary zones that were all initiated at different times there could be three genomes in which DNA synthesis is ongoing, each replicating a different portion of the genome at the same time. The conclusion from these thymidine studies when considered in conjunction with the parallel autoradiographic study using labeled leucine and uridine (Higgins *et al.* (1986) and the earlier morphological studies is clear: although the initiation and completion of the poles proceeds in the same way in all the growth zones present in a cell, the timing does not follow a rigid schedule. The completion of a new secondary zone slows the completion of a primary zone, and two secondaries compete with a primary even more. An additional source of variability is that the completed pole heights are poorly controlled. Even though the wall band's diameter is rigidly controlled, the volume of completed poles has a *CV* of 18%. This higher variability may result from the action of the split-and-splay mechanism for achieving a more pointed pole (described below), but it may be part of a mechanism that ensures that chromosome replication and segregation of complete nucleoids to both the new and old poles of the cell occur before completion of the septum.

It should be mentioned in passing that the growth strategy of this organism makes the formation and maintenance of a synchronous culture impossible due to the variability of cell volumes at cell separation. An even more important factor is variability in the number of initiation events of the wall band splitting. The timing will be quickly upset if sometimes one or sometimes two sister wall bands split when the primary growth zone cannot keep up the pace of cytoplasm synthesis. Also, when only one wall band splits, before the time when the primary zone is completed, then two cells are produced, one will already have a partially completed primary growth zone and the other will not (Koch and Higgins 1991).

Surface Stress Theory

I learned about the morphological growth process of the streptococcal pole while attending a conference on bacterial cell walls in 1978. I was struck by the shape of the half-completed pair of poles of the streptococcus such as shown in Fig. 8.1. It reminded me of the shapes that soap bubbles can achieve under certain conditions. Could the shape of a streptococcus be determined by physical processes such as those determining the shape of a soap bubble?

As a youth I had read the book by Boys (1890) entitled *Soap Bubbles and the Forces Which Mould Them*. This book, written before the turn of the century, had later been reprinted many times; it had been read by many adults and adolescents simply because it is a fascinating area of popular science. Although Boys really wrote for the amateur scientist and for a general audience, for the professionals, the physics and mathematics of soap bubble films had been serious business for one-hundred years before that. The guiding principle is that the film will adjust to the shape that minimizes the area (and therefore the work of surface tension). The simplest case is film surfaces that are formed by the dipping of a wire frame into a soap solution. If the shape is a circle, a rectangle, or any regular or irregular planar shape, then the resultant film will be planar and tautly cover the wire form. If the wire frame is not planar, nonplanar surfaces will form that generate the minimal surface possible. In addition to physical constraints, in these examples due solely to the wire frame itself, if there are other constraints the total surface energy is still minimized. We have already had the example of the physical situation that gave rise to the *Young-LaPlace equation* of Chapter 6; there the additional constraint is that there is a specified pressure difference across the membrane adding the $P\Delta V$ work term to the equation.

Remembrances from my reading of Boys' book included the fact that if two initially spherical soap bubbles of equal size happened to fuse while they were floating freely in the air, they would form a planar septum between them and the remaining region of each bubble would be spherical and make an angle of 120° with the septum and with the other bubble (Fig. 8.6A). This minimizes the total surface of the system for the volume contained. Such a shape has some resemblance to Fig. 8.1, although it would be a better approximation for a staphylococcal cell, which at some stage of the cell cycle may look exactly like the dual soap bubble (Giesbrecht *et al.* 1985). These ideas occurred to me while I was returning from the Gordon Conference where I had learned about Higgins' work. The return path to Indiana was not simple; first a car ride from New Hampshire to Philadelphia with Ronald Doyle in which the conversation was totally devoted to prokaryote wall biochemistry and physiology, then a day of learning about the streptococcal system from Lolita Daneo-Moore and Michael Higgins. I think that the conjunction of ideas led to what Pasteur in 1854 called "nature favoring the prepared mind." ("Dans les champs de l'observation le hasard ne favorise que les esprits préparés.") In any case, as I flew from Philadelphia to Louisville, it jumped into my mind that what was needed was an understanding of the additional constraint(s)—perhaps only one was needed—to achieve a football-shaped streptococcal pole. Arriving in Louisville, I immediately went

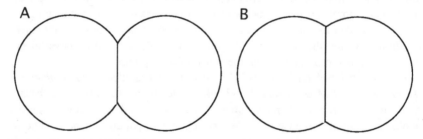

Figure 8.6. The stable shape formed when two soap bubbles of the same size fuse. A. Note that the interface between the two is planar and that the angle made by the septum and the outer surfaces is 120°. This fact was first popularized in the book by Boys in 1890. B. The equilibrium shape when two soap bubbles fuse, on the assumption that the surface tension of the planar surface is 0.6 of that of the spherical surfaces of the same bubble.

to a toy store for soap bubble pipes and soap solution, then to Doyle's laboratory to borrow other apparatus to see if I could blow soap bubbles in the shapes assumed by a diplococcus as it grew and divided. I could (Fig. 8.7).

Formation of a shape similar to *E. hirae* was very encouraging, but a false start. The sequence of shapes as the poles developed required that I gradually pull the pipes apart and blow just the right amount of air into the system. If the spacing between the spheres was self-adjusting in a way to minimize the work of surface tension, as shown for a freely suspended sphere in Fig. 8.8, then pointed poles would not be produced. It therefore follows that if the forming poles were a fluid membrane and had a shape determined by physical forces, like the soap bubbles, then the appropriate constraint had to act to control the separation of the tips of the old poles and the internal pressure in a very special way. When the zone started the distance would be very short, and as the poles were finished and separating the distance must be greater than twice the equatorial radius. See Fig. 8.8 for the calculated shape for two half-soap bubbles where pressure and surface tension are constant and the spacing is mechanically fixed at a distance too short and too long compared to a distance without constraints.

The alternative was simply that most of the new pole was rigid, as we had already concluded the old pole to be. Therefore, the slope of the older parts of the new pole did not matter, but once formed it retained its slope indefinitely. The only thing that does matter for this model is the physics of the surface forces where the septum is being split. Then the surface tension forces act on the still-plastic material of the wall as it was split and before it is hardened by the addition of material from Channel B and possibly by the formation of additional crosslinks. Because there is a septum present, however, the situation is different from the case of cylindrical elongation considered in Chapter 6.

As the septum splits, its material becomes part of the externalized pole. The

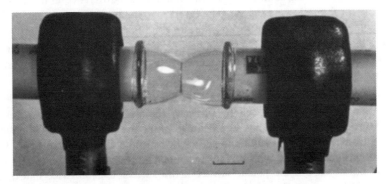

Figure 8.7. Soap bubble simulation of pole shapes similar to those formed during streptococcal growth. By judicious manipulation of two soap bubble pipes connected together, it is possible to achieve a simulation of the shape of *Enterococcus hirae*. Reprinted from Koch *et al.* (1981a).

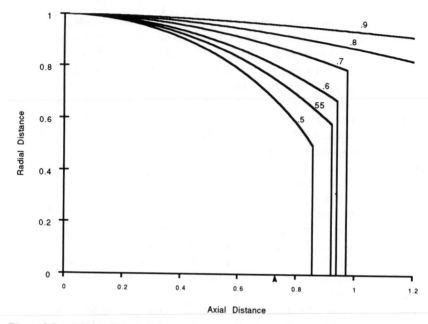

Figure 8.8. Effect of constraints on the shape of fused soap bubbles. The curve marked .5 corresponds to Fig. 8.6A and is just the spherical left-half and the planar septum and corresponds to two fused soap bubbles. The other curves correspond to the longer poles that would arise if there were extra axial tension or if the curved portion of the wall had the indicated surface tension higher than $T = 0.5$. The other vertical lines connecting with curves correspond to planar septa that would be observed. Calculation of these shapes required the theory presented in Chapter 6.

amount of the contribution of septal material makes a difference in the shape of the pole for the following reason. The fused soap bubbles (Fig. 8.6A) with their 120° intersections represent the equilibrium condition. If the septal disk were to split a little more, then its area would be smaller and the external surface larger. If the septal disk were a little larger, then the external surface would be smaller. For the soap bubble the total surface area is minimal for the specification that the surfaces meet at an angle of 120°. This only happens because the contribution of a unit of surface is the same no matter which part of the pole or septum it is. On the other hand, if the septal material during pole formation contributed more or less to the total surface than just twice the decrement in septal surface, then the energetics would be different and the angle would be other than 120°. This would be the case if the planar surface had a surface tension twice that of the curved surfaces, see Fig. 8.6B. The contribution has to be different from that of the still-adjacent plastic peripheral wall because the septal wall is thicker. This in turn could determine the angle of the forming pole.

The Zonal Dome Model

If the wall of the streptococcus was rigid once formed, then one could assume that the older wall served as a platform for new wall to be added in the formation of the developing pole. Then a little bit of algebra would lead to the calculation of a pole shape; this theoretical prediction was in close agreement with the observed shape (Koch *et al*. 1981b). This derivation will be reproduced here with a fuller explanation. The logic is straightforward. We start again from $P\Delta V = T\Delta A$, as we did several times in Chapter 6, but what is different now is that we have some septal wall that, due to its splitting, becomes external wall. The application of the formula $P\Delta V = T\Delta A$ to bacterial wall growth implies that the pressure forces volume enlargement and that this work is converted to the work favoring autolysis that then enlarges the surface area. The growth zone is shown diagrammatically in Fig. 8.9. These ideas led to the following balance equation. (Note that S is the slope of the wall and that it and dr are negative as the wall grows inward.) Then the combined amount of new external wall area created by wall growth on both sides of the septum is:

$$2\pi[r + (r + dr)][dr^2 + dz^2]^{1/2} = 4\pi r[1 + S^2]^{1/2}dz.$$

In words: this equation calculates the circumference at the midpoint of the trapezoidal new wall element and multiplies it by the axial length of the new segment. The factor of 2 on the left side is really $2 \cdot 2/2$. One 2 comes from the formula for calculating the circumference from the radius, one 2 is because

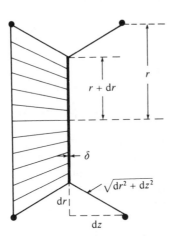

Figure 8.9. Geometric relationships during the formation of new wall. The figure is modified from Fig. 3 of Koch *et al*. (1981a). See text. By convention the slope S is defined as $-dr/dz$.

there are two segments of new wall, and the third 2 is needed to calculate the average radius from the radius at either end of the growth zone. The right-hand expression arises when dr is neglected in comparison to r and the equation is expressed in terms of the slope. The amount of external wall area produced from previous septal wall is:

$$\delta\pi[r^2 - (r + dr)^2] = -\delta 2\pi r dr = -\delta 2\pi S dz,$$

where δ is the thickness of the septal wall expressed in terms of the thickness of the adjacent externalized wall. The difference between the two expressions is the net area dA forced by the pressure-volume work, given by:

$$\Delta A = 4\pi r[1 + S^2]^{1/2}dz + \delta 2\pi r S dz.$$

Note that this is the difference between two positive quantities because S is negative. On the other hand, the volume increment of space enclosed by the new wall is given by:

$$\Delta V = 2\pi[r + (r + dr)]^2 dz/4 = 2\pi r^2 dz,$$

where again the right-hand side has been simplified by neglecting dr in comparison to r. Now substituting these values into $P\Delta V = T\Delta A$, we obtain:

$$Pr/T = 2\sqrt{1 + S^2} + \delta S.$$

For the case where $\delta = 2$ (where the septum is twice the thickness of the initially externalized wall) this equation becomes:

$$S = (Pr/4T) - (T/Pr) = dr/dz.$$

All this work was done to get an expression for the slope that then can be integrated to determine the pole shape. The methods as given in Chapter 3 to integrate this expression are sufficient. The answer is:

$$z = z_{max} + (2T/P) \ln[1 - (Pr/2T)^2], \quad \textit{Zonal dome model}$$

where $z_{max} = -(2T/P) \ln[1 - (Pa/2T)^2]$ and a is the radius of the base of the dome-shaped pole as well as the radius of the wall band at the initiation of the growth zone. Actually, from measured values of a and Z_{max}, the single parameter P/T can be determined. Fig. 8.10 shows the match between the formula and the measured shape of the pole.

The Amended Theory: Split-and-Splay

In the last section, the case in which δ had a constant value of 2 was called the "zonal dome model." On the other hand, the actual measured values of the ratio

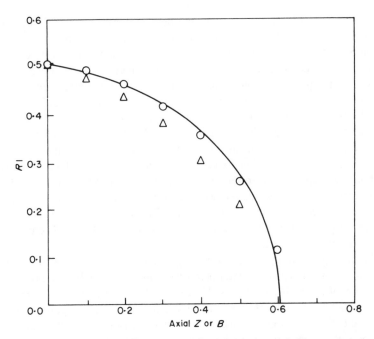

Figure 8.10. Fit of cell measurements to the zonal dome model. The open circles and the open triangles represent the measurement of Higgins and Shockman (1976); the solid dark line represents the predictions of the zonal dome given two parameters, the radius of the wall band and the maximal axial height of the completed pole. Taken from Koch (1992).

of the septal wall thickness to the adjacent externalized pole wall were found to be near $\delta = 1$ for newly initiated poles and to approach $\delta = 2$ only close to the completion of pole formation. The theoretical prediction for the shape of poles developed by a process in which δ varied in accordance with the experimental findings was found to be much poorer than for the zonal dome model (see Fig. 4 of Koch *et al.* 1981b). Thus, there was a contradiction between the fact that the zonal dome model fitted so well and the numerical balance of the volumes of cell wall measured by Higgins and Shockman (1976), which required that considerably more volume of peptidoglycan was needed that could not be accounted for by the exteriorization of previously formed septal wall. Consequently, wall material had to come from someplace else. Although in 1981 we noted this discrepancy, we did not pursue the matter further because the discrepancy could have been accounted for by some combination of stretching or the addition of Channel B material that was involved in thickening the exteriorized poles. On more recent reconsideration of the problem, I (Koch 1992) felt that the surface stress theory as applied to *Streptococcus* needed to be revised, not with respect to the energetic considerations, but with regard to the quantity and site of synthesis or to the possibility of expansion of the peptidoglycan volume by the surface stresses. There were three possibilities that had to be considered.

One possibility is that either a special part of Channel B or a third channel secretes additional wall in the narrow region where septal splitting is taking place. Such a process would have to intercalate new material throughout the thickness of the septum in the zone where the septum is splitting. This process would have to be in addition to the wall accretion on the inside surface of the annulus of the forming septum (Channel A) and to the gradual thickening of the externalized pole wall (Channel B). There is no precedent, however, for a biochemical mechanism whereby peptidoglycan can be inserted into the existing wall at any place other than the outside surface of the cytoplasmic membrane. Only formation of nascent wall adjacent to the cytoplasmic membrane is expected, because the penicillin-binding proteins (PBPs) are membrane bound and the nascent oligopeptidoglycan is secreted through the membrane. Therefore, this possibility was discarded.

A second possibility, mentioned by Higgins and Shockman (1976), is that the wall becomes stretched on exteriorization. This is certainly the case of pole formation of B. subtilis as discussed in Chapter 9; however, for the streptococcal pole it is not a candidate. Higgins and Shockman reasoned that significant stretching of the E. hirae wall is unlikely from their measurements. It is improbable also because of the relative invariance of pole shape during formation pointed out above, the relative pointedness of the pole, and the quantitative analysis (Koch 1992) that demonstrated that little expansion takes place after the septum closes.

Formulation of the "Split-and-Splay" Model

Because neither of these two possibilities seemed reasonable, a third possibility was considered to account for the contrasts of pole formation of E. hirae relative to B. subtilis. Some calculations concerning the geometric relationships of pole formation [which will not be detailed here but are presented in Koch 1992)] and the above considerations suggested the following model. As the septum closes and the external wall is split, the annular surface of murein (to which new wall is being added) enlarges in radius due to the associated splaying motion. This in turn requires that extra material be added to replace the displaced wall. The extra replacement material results in the completed pole's having a larger surface area than otherwise expected (see Fig. 8.11). The essential premise of this model is that the wall does not, and cannot, expand once formed. It is assumed that although the wall is negligibly stretched upon being exteriorized, each ring of bisected septal peptidoglycan is, however, reoriented; i.e., the septal peptidoglycan of E. hirae is splayed. This displaces the inside of each half-ring outwardly as the wall to which it has just been attached yields to the radial and axial tension.

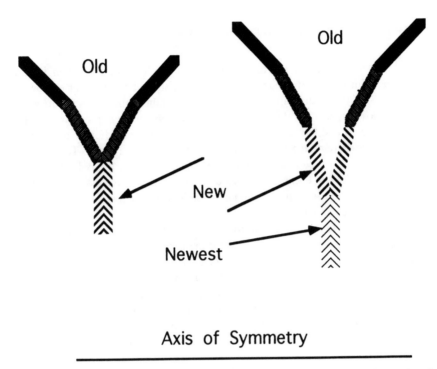

Axis of Symmetry

Figure 8.11. The "Split-and-splay" model for the formation of the prolate pole of *Enterococcus*. It shows that as the septum splits the newly exteriorized wall twists (splays), thus enlarging the diameter at the splitting surface and requiring extra wall to be added to the inside of the closing septum. The figure is republished from Koch (1992).

Thus, although the surface area covered by a ring element remains nearly constant, its circumference becomes larger, and the width of the ring narrower. At the same time, the orientation of the wall switches from being perpendicular to the cell axis, to being oblique to the cell axis. Consequently, the component in the direction perpendicular to the cell axis of the new wall surface becomes smaller. Therefore, new septal peptidoglycan must be added, not as an additional channel, but by additional output from Higgins and Shockman's Channel A to replace peptidoglycan that had moved outward away from the cell's axis due to the splaying process. Then additional wall must be added from Channel B to give the wall its final thickness. Although a lamella of unsplit wall is not capable of much splaying, any extension is magnified in enlarging the radius of the region at the junction of the septum and the external wall, particularly in the initial phases of pole formation when the wall bends outwardly the most. This model was implemented in a computer program that satisfactorily matched the observed shape (Koch 1992).

Stresses on the Forming Septum and the Shape of the Pole

What is the difference in mechanics of the process of pole formation in the organisms with differently shaped poles? Physicists and mechanical engineers have developed the mathematics for the stress and strain relationships for pressure vessels. Those especially relevant to the present problem are expressions for the meridional stress and the hoop stress (remember that the meridional stress tends to elongate the cell and the hoop stress tends to enlarge the radius). The stresses in a vessel of any shape can be computed; however, when the pole shape can be approximated as an ellipsoid of revolution, the stresses are easy to calculate. Formulas for the *ellipsoidal meridional stress* and the *ellipsoid hoop stress* are given on page 161 in Chapter 6.

The important feature relevant to our microbiological application is that whereas the expression for the meridional stress is always positive, the expression for hoop stress can be either positive or negative depending on the angle γ and the ratio a/b, because of the first negative sign in the numerator of the expression for N_Θ. In the extreme oblate case where a is much smaller than b, the hoop stress is less tensile as the pole starts to form and may be even compressive (negative). Consequently, if the pole starts to invaginate strongly, it will not experience large tensile (expansive) hoop stresses and will develop a flattened

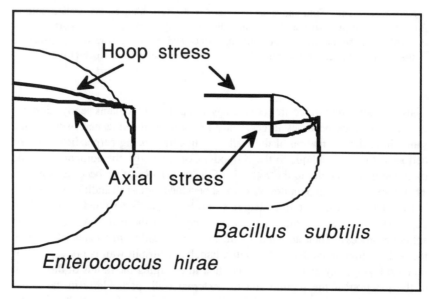

Figure 8.12. Stresses in the pole of *Enterococcus hirae* and *Bacillus subtilis*. These have been calculated using the shape of the poles approximated as elliptical shells of revolution, but using mean measurements of the radius at the base of the poles and their height. See text.

oblate pole, as in the *B. subtilis* case (Fig. 8.12). Because of the elasticity of *B. subtilis,* the split-and-splay mechanism never starts to act. Conversely, because of its pole's larger radius and prolate shape, the force stretching the still-patent septum of *E. hirae* is always tensile (see Fig. 8.12).

Surface-to-Volume Considerations

Microbes have no internal circulatory systems; thus, if a bacterium cannot move as fast as small molecules, such as glucose, can diffuse to them, there is reason for them to have a high surface-to-volume ratio. Different shaped organisms do have different S/V ratios. It has been often argued that rod-shaped bacteria are constructed that way in order to have a high S/V. But the argument can be applied almost as well for the pointed cocci. Fig. 9.2 shows the S/V for hypothetical organisms of simple shapes all of the same volume. A rod-shaped organism with flat poles and a ratio of length to width of 3:1 was taken as the prototype. It can be seen that as the rod gets thinner, its S/V increases, and reciprocally, the fatter the organism becomes, the poorer the S/V gets. Using the same abscissal scale, (Fig. 8.13) i.e., the total length of the cell, the same is done for elliptically

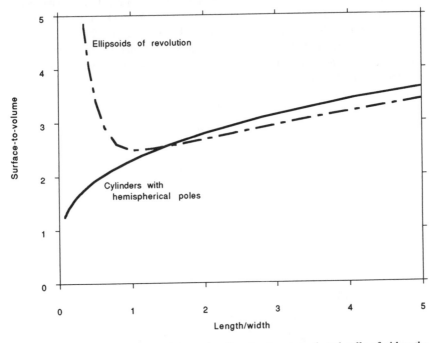

Figure 8.13. The surface-to-volume ratio of variously proportioned cells of either the rod shape or of the elliptical shape. All points correspond to the ratio of the length to width of the volume as a cylindrical cell with hemispherical poles whose length is 3 μm and width is 1 μm.

shaped cells. Although the pole wall of *Enterococcus hirae* is not precisely an elliptical shell of revolution, it is easily close enough to make the point clear that the more pointed the prolate pole, the greater the S/V. For comparison a cell that has the standard volume but has hemispherical poles is also shown.

Of course as mentioned earlier, surface is less important for slow-growing organisms or organisms that remain together in an aggregate. But they may, as part of the culture cycle or their evolutionary history, need a high S/V, although it probably is not needed under the conditions used for our studies in the laboratory.

Population Distributions

There is much analytical work left unfinished regarding the cell cycle of *Enterococcus hirae*. The available electron micrographic data from Higgins' lab contain a treasure trove from which the controls for wall growth may be deduced as well as how the controls are altered during shifts to better and poorer nutritional media and into antibiotic-containing media. There are data from experiments with tritiated precursors. The real problem is that, although exponential balanced-growing populations can be studied, synchronized cells cannot be maintained. A strong start to the further analysis of these problems is that a massive computer program has been constructed and is capable of simulating large populations and keeping track of the progress of many cells through their cycles (Koch and Higgins 1991). Because the model for the growth of the population is firmly based, the program can be, but has not been, used to test a variety of models to see in what ways *E. hirae* is different from the better-studied organism *E. coli*.

9

Gram-Positive Rod-Shaped Organisms: *Bacillus subtilis*

KEY IDEAS

*A Gram-positive rod is characterized by thick wall, constant diameter, new septa
  precisely spaced between old, and variable splitting of septa.*

*It elongates side wall by laying down new layers of peptidoglycan and teichoic acid
  immediately outside the cytoplasmic membrane.*

As a layer moves out it stretches because of its great elasticity.

When a region of wall is stretched to its "elastic limit," autolysins cleave the wall.

The cleavages occur most often in the portion of the wall that are most stressed.

*Because of the elongation of the wall, the elasticity of the peptidoglycan is expended
  in this mode and none is left for expansion in the hoop direction.*

*The cleavage pattern results in helical cracks that partially support the intact cylinder
  region.*

The cleavage pattern of growth causes rotation of one end of the cell relative to the
  other during growth.
The shape of the pole is the simple consequence of splitting a planar septum composed
  of material that can expand by 50%.
There are four main mechanisms and several more that may function to prevent the
  autolysins from destroying the cell.
Sometimes the autolysins do destroy the cell.
Small molecules can diffuse through the peptidoglycan fabric, but large ones cannot
  and can only move through the wall as it turns over in its inside-to-outside growth
  process.

The Organism and Some History

Bacillus subtilis is a microorganism found generally in soil. It is not one of the
opportunistic consumers of living, freshly dead, or dying biomass, but rather it
waits near the end of the chain of consumers. Although its metabolic capabilities
are nowhere as great as the collective abilities of pseudomonads, it makes up
for this by being more patient. To do so it depends heavily on its ability to
sporulate and to germinate quickly.

B. *subtilis* has been well studied because of two factors. The first was that
John Spizizen found a strain and conditions so that genes could be transformed
from one line to another. This made it possible to make *B. subtilis* into the
workhorse for the Gram-positive organisms as *E. coli* is for the Gram-negatives.

The second reason for the extensive study of this organism is its ability to
sporulate. Intensive studies have been carried out and are ongoing. Genes have
been mapped, sigma factors identified, and new and novel regulatory systems
unearthed.

Morphology

The organism has a rod shape and maintains the same diameter throughout the
cell cycle, no matter whether grown in rich or poor medium (Fig. 9.1). It has
been generally assumed that the rod shape provides a better surface-to-volume
ratio than a sphere of the same volume. Of course, it provides a poorer surface-
to-volume ratio compared with a spherical cell of the same diameter (Fig. 9.2).
So instead the logic must be phrased differently: Given the constraint that the
cell must have at least a certain volume, then a spherical shape is poorest; a bi-
conical shape is better; a rod-shape is better still; a branching system of filaments
still better; and a flat leaf-like structure just as good. A choice between the last
two shapes depends on how small the diameter of the branches or how thin a
leaf-like structure can be made and still function.

The bacillus seems only to know how to grow with a particular dimension for
its diameter and does not know how to branch. Consequently, the rest of the

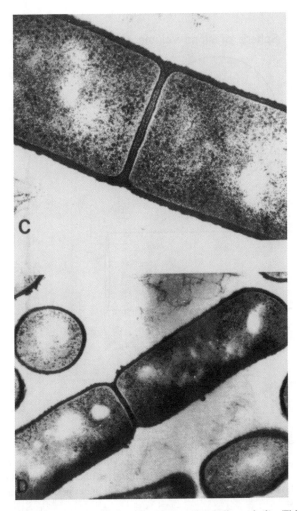

Figure 9.1. Electron micrograph of a thin section of *Bacillus subtilis*. This organism's side walls are very closely the shape of a right cylinder. The poles are flattened compared with a hemisphere. Photograph courtesy of Dr. Terrence Beveridge.

cell's physiology must conform to these constraints. In terms of its growth habit, the wild type cell can adjust how long a length is devoted for each nucleus and how many μm of length there are between initiating crosswalls and between the ends of morphological units as they separate from one another. Different growth conditions influence the number of nuclei and the number of septa per morphological unit (Sargent 1975). The morphological unit is not, however, the entity of reasonable interest because we can paraphrase and extend the argument given in the last chapter for a chain of streptococcal cells. This argument is that each

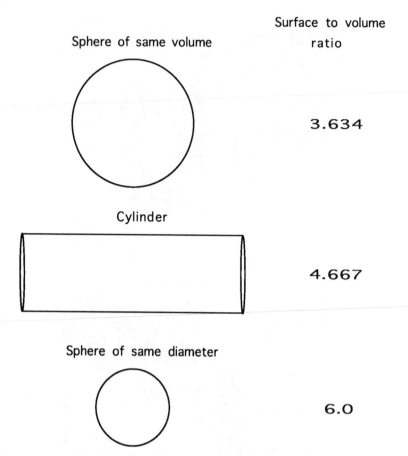

Figure 9.2. Surface-to-volume ratio of a closed cylinder (center), a sphere of the same volume (top), and a sphere of the same diameter as the cylinder (bottom). For the calculations shown in the figure, the diameter of the cylinder has been taken as unity and its length as 3.

cell present in the chain is the true physiological and reproductive unit and that the chain itself is not. The aggregate of cells is of interest from the ecological viewpoint; i.e., adhesion to surfaces is firmer for chains of cells. From the physiological viewpoint, although the chain is composed of physiological separate cells, the chain's length is just the trivial balance between growth and the final splitting of completed septa. The chain length probably does not matter to organisms growing in suspension when nutrition is adequate and dispersal is not an issue. Timing from closure of a septum to the closure of the next septum, in some sense, is the more basic clock for the organism rather than the length of the chains and the timing of cell separation.

Fein and Rogers (1976) selected organisms resistant to methicillin. This led to the isolation of mutants that grow as filaments during ordinary culture. That is, when observed at equal culture densities, the mutant organism existed as longer filaments (undivided chains of bacteria) than are seen in the wild type control culture. Originally, this was thought to be because they have reduced levels, compared to the parental strain, of both of the *N*-acetyl-muramoyl-L-alanine amidase and *N*-acetyl-D-glucosaminidase autolysins. Recent work by Vitkovic (1987) has suggested that they may form as much of these autolysins, but may rebind liberated autolysins less readily so that there are fewer associated with the wall.

It was noted in our laboratory that the wild type strain grew as filaments if grown as a very dilute suspension (Koch *et al.*, unpublished). This phenomenon had been noted as well by others, but we devised an explanation that seems to bring together many ecological, physiological, and biochemical facts. Our hypothesis was that *B. subtilis* simply does not have a mechanism dedicated to cell division *per se,* but it does have a mechanism to create septa. Clearly, this organism grows as a rod by the inside-to-outside strategy and can create septa at fixed intervals. In very dilute suspension, we hypothesized, there is no mechanism to split the septa, and therefore long filaments develop. Modulation by mutations that decrease the production and avidity of the autolysins for the wall results in cells remaining as filaments in cultures at higher cell densities. In wild type strains, therefore, there is a cooperative, cell density-dependent process that controls septal splitting and cell division. This process, which can hardly be called a mechanism, exploits the exoenzyme nature of the autolysins that have fundamentally been made for a different purpose.

Autolysins are secreted through the cytoplasmic membrane and have the essential role of cleaving stress-bearing bonds of the sacculus, a process necessary for elongation. In a dispensable role, they cause the dissolution of the shreds of peripheral wall. Third, the autolysins serve in a way, responsive to environmental conditions, to lead to division. The system works this way: during elongation, autolysins are formed and move through the wall as it grows. Once they reach the outside of the wall, their action on the wall liberates them from the cell. However, the autolysins have an affinity for the wall and, consequently, they rebind from the medium to the cell wall, including regions above septa. In the regions where septal splitting is actively occurring, they act to bisect the septum. The result is that septal splitting depends on the amount of autolysin rebound to peripheral walls (see Fig. 9.3). Because the autolysins equilibrate between the medium and the cell surface, the fraction bound depends on the affinity and on the total concentration of autolysins. Although the autolysins rebind, they do not have an extremely high affinity for the wall, and thus the rate of splitting and separation of daughter cells depends on the environmental conditions. When the culture is very dilute, a large fraction of the autolysins that had migrated to the surface are liberated, and not rebound. Instead, the concentration of enzyme

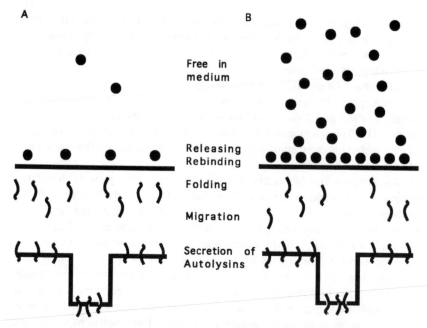

Figure 9.3. Difference of autolysin at low and high density. The figure shows autolysin production, movement through the wall, liberation, and rebinding from the medium. The wall is about 25 nm thick. Autolysins are secreted through the cell membrane, progress through the wall, at some point become native, and become liberated as a result of their function. Depending on the concentration of autolysin in the culture, the fraction of autolysins bound on the outer aspects of the wall will vary. A. Cell in an early sparse culture in which very few of the autolysin molecules had been liberated from earlier growth of the cells and have been rebound on the cell surface. B. Cell from a dense culture grown from a small inoculum. As the culture grows, more and more autolysins have migrated to the outside surface of the cells, hydrolyzed wall, dissociated, and then become available to be rebound to the wall; this yields a higher surface concentration.

molecules is higher in the medium than on the cell surface (Koch 1988d) and does not foster cell splitting. On the other hand, as the culture grows and its density increases, the total concentration of autolysins also increases and, consequently, the fraction bound increases and favors an increased rate of cell division.

This accumulation of old autolysin molecules is a simple mechanism that favors cell splitting when the cell titer has become high. It serves an anticipatory function; i.e., "making a guess" of how soon the environment can be expected to be depleted; when recession is anticipated it speeds the division rate to create more propagules. ("Propagule" is the ecologists' word for a unit that can grow by itself.) Near to the stationary phase, a smaller amount of biomass per kinetically independent unit is appropriate. This is also the stage in which motility would be most useful and, indeed, the dissolution of filaments (no matter how they were produced) is necessary to cause cells to become motile. Of course, there are other additional mechanisms controlling motility whose genetic mechanisms are only now becoming understood.

It is of interest to note that the mutants isolated by Fein and Rogers (1976) are not devoid of autolysins, but they, and other similar mutants, have reduced levels of both autolysins. Mutants completely lacking autolysins have never been isolated, presumably because some low level is essential for the inside-to-outside growth process. Consequently, it is probable that *B. subtilis* has evolved genetic mechanisms to modulate the level of both autolysins without totally repressing either. According to Vitkovic (1987) this is due to changes in the cell wall that prevent their rebinding, but it has not been excluded that these *lyt* mutants are regulatory mutations at a hypermutable locus. From the cell's point of view, either arrangement allows the easy and reversible generation of mutants that have down-regulated both autolytic enzymes, while retaining a basal level that leaves enough activity for elongation growth. There are environmental circumstances where down regulation and a persistent habit of filamentous growth are beneficial. In flowing aqueous systems or water-saturated soils, adhesion to rocks and macrophytes favored by a filamentous habit would be advantageous; it would also give more favorable access to nutrients in an oligotrophic environment. In other environments where filamentous growth is not advantageous, the normal wild type level would be selected for and replace the slowly dividing filamentous variety. It is certainly the case that many normally dividing cells possess an array of physiological and genetic responses that lead to filamentation. These may have important ecological as well as metabolic advantages to the organism when it is challenged under natural conditions.

Exoenzymes

Besides the autolysins, *B. subtilis* secretes a number of exoenzymes, some of which are so large that they cannot diffuse through the wall fabric. Some, such as β-lactamases, protect against antibiotics, and others, such as levansucrase, act to provide carbon substrates for growth when the substrate is available. The production of an exoenzyme is a financial gamble for the cell. This is because diffusion of either the enzyme or of the degradation products from the large substrate molecule away from an isolated cell would obviate their value to the exoenzyme-producing cell. Consequently, quickly liberated, small-sized exoenzymes are only useful as cooperative phenomena when a large population of exoenzyme-producing cells is concentrated close together. Then the enzyme secreted by one organism may produce growth substrates for another, and vice versa, and the enzyme produced at one time may be helpful at a later time. I would expect, but do not know, that the induction of levansucrase (a protein of 55 kD molecular weight) by sucrose would be density dependent.

Permeability of the Wall

For the autolysins and the other exoenzymes, secretion of the nascent protein through the membrane is carried out by the signal peptide mechanism. Then the proteins must diffuse or be carried through the thick wall fabric by the inside-to-outside growth process in either the unfolded or folded state. For many exoenzymes, only when they reach the outer surface would they serve a useful role for the organism.

Paul Demchick, in my laboratory, has recently extended early work of others to try to measure the porosity of the wall containing its peptidoglycan and teichoic acid components (but with the few wall proteins removed) to determine how much the wall structure impedes the penetration of macromolecules. The details of his studies will be given because they present novel biophysics and are another application of the mathematics developed in Chapter 3. He started with commercially obtainable dextrans that had been produced by *Leuconostoc mesenteroides* and then partially degraded. He labeled partially purified size fractions with fluorescein and fractionated them further to obtain preparations that were more uniform in size. He also prepared sacculi from growing cultures of *B. subtilis* and *E. coli* by treating them with very hot SDS (sodium dodecyl sulfate) to kill the organisms, denature proteins, and virtually instantaneously prevent autolytic activity. These preparations were then treated with proteinases and nucleases and again extensively washed.

The actual experiments utilized a fluorescence microscope and special depression slides created by treating an ordinary microscope slide with hydrofluoric acid for a particular time at a particular temperature. In this way he could create depressions that were 4 μm deep, a depth at which the sacculi are not distorted by compression when the cover slip is put in place, but still are shallow enough to allow visualization of the sacculi by negative staining. When the dextrans used were large, then the background appeared green-yellow and the nonfluorescent sacculi stood out in relief. On the other hand, if the dextrans were small then it was very difficult to visualize the sacculi at all, i.e., there was no negative staining. In a critical size range of the dextrans the sacculi had an intermediate character. We were initially surprised that in these cases their negative staining changed little with longer incubation times. This lack of time dependence suggests that the wall preparations behaved very much like a sieve, and that even with the purified fractions, there was enough size heterogeneity so that the smaller sizes would pass through the wall, equalizing the concentration of smaller species on both sides, whereas the larger species would not penetrate even when given a much longer time. By quantitating the sizes of the fragments, Demchick was able to show that dextrans smaller than 25 kD equilibrated almost instantaneously, those of about 70 kD were barely able to pass through the wall, and bigger ones could not penetrate at all.

Kinetics of Equilibration into Sacculi

The process of penetration to the point where efflux equals influx is relevant at this point and provides a quick review of the mathematics developed in Chapter 3. Let us define symbols as follows. First, the quantities that vary during an experiment: q = the quantity of dextran transported, t = time, C = concentration inside the sacculus, and ∂C = the change in concentration from outside to inside. We then define the quantities that are constant during a particular experiment: V = sacculus volume, A = area of the sacculus, P = the fraction of that area that the dextran can go through, D = diffusion constant of the dextran in water, ∂x = thickness of the sacculus, and C_∞ = concentration in the medium. Diffusion through the wall should follow Fick's law, which we can write with these symbols as follows:

$$dq/dt = (dC/dt)V = PDA\partial C/\partial x. \qquad \textit{Fick's diffusion law}$$

The quantity q is equal to $C \cdot V$ and because V is constant, dq/dt has been replaced by $(dC/dt)V$ in the second expression. The two differential quantities, dC and ∂C, are not interchangeable: one defines change in time, and the other, change in space across the saccular wall. The latter can be replaced with $(C_\infty - C)$. With these two changes done, we now have an equation with only two variables instead of four. Now the equation can be rewritten so that the concentration variable is on one side and the time variable is on the other side:

$$dC/(C_\infty - C) = [PDA/V\partial x)]dt = Gdt.$$

The items within the square brackets are all constant during the diffusion process with a given monomolecular dextran preparation. Let us call this constant $G = PDA/(V\partial x)$; it is formally equivalent to a first-order reaction rate constant. Because G depends on P and D, it depends on the size of the probe and the nature of the wall. For the mathematical details, we substitute X for $(C_\infty - C)$ and dX for $-dC$, and we end up with almost exactly the same mathematics that we had in the case of bacterial growth, but with a negative sign.

$$dX/X = -Gdt \qquad \text{or} \qquad X = X_\infty e^{-Gt}.$$

Now translating back in order to use C instead of X, we have:

$$C = C_\infty(1-e^{-Gt}). \qquad \textit{Equilibration kinetics}$$

This expression is usually referred to as an exponential saturation curve. No matter what the value of G, net flow always terminates when $C = C_\infty$. Curves with different values of G just differ in the time course to get there (Fig. 9.4). If we could accurately measure the time course we could find the value of G and hence calculate the effective pore size of the wall. But we do not have the time course data, and the expressions for P and D are complex and depend on the pore size for either of the models we will use below. But because Demchick has a number of measurements with different-sized probes followed to different endpoints, this equation allowed him to estimate pore size. This is possible because we can rewrite the *Equilibrium kinetics* equation as follows:

$$(C_\infty - C)/C_\infty = \ln[-Gt].$$

He always used the same concentration, so that C_∞ is kept constant. For pre-set times of observation (3 min, 1 hour, and 1 day), he searched for a probe of a

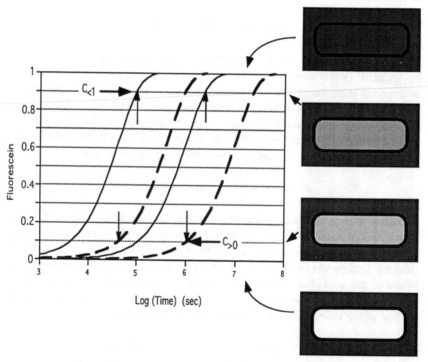

Figure 9.4. Time course of the filling of sacculi. The graph shows four time courses for the entry of labelled dextran into sacculi. Initially the sacculus appears to be negatively stained. See lower right hand image. At some latter time the fluorescein-labeled dextran penetrates enough to just slightly decrease the negative staining. This condition is the lower criterion. Subsequently enough penetration occurs as in the second from the top curve so that one can barely distinguish that the negative staining has not been entirely lost. This is the upper criterion. Finally after a long time, equilibration is complete and the negative staining has been lost.

size that was big, but not so big that there was no entry it just allows the detection of a slight decrease in amount of negative staining, which we designate as $C_{>0}$. He also searched for the probe of a size that was just large enough that it just barely has a detectable amount of negative staining, which we designate as $C_{<1}$; this yielded a total of six pairs of values for estimation of pore size. With either of the two criteria, $C_{>0}$ or $C_{<1}$, C_∞ would be the same for both pairs of times and probe size so that $(C_\infty - C_{>0})/C_\infty$ or $(C_\infty - C_{<1})/C_\infty$ would also be a constant. Thus, algebra leads to:

$$[G_{<1}t]_1 = [G_{<1}t]_2 \qquad \textit{Criterion for perceptible negative staining}$$
$$[G_{>0}t]_1 = [G_{>0}t]_2. \qquad \textit{Criterion perceptible loss of negative staining probe}$$

$G_{<1}$ is the corresponding rate constant to $C_{<1}$, which refers to the concentration of fluorescein-dextran for probes that only barely lower the fluorescein before that of this medium $G_{>0}$ is the corresponding rate constant to $C_{>0}$, which refers to the concentration that only allows a small amount to enter during the test period. The subscripts 1 and 2 refer to the paired values of the G and the time that gave the particular criterion.

Now we will examine the factors that enter the expression for G; i.e., $PDA/(V\partial x)$. P is the proportion of the surface area through which the dextran can penetrate. Dextrans with different sizes that are smaller than the pore can penetrate through it, but bigger ones are impeded by hitting the walls. We will take this into account in two ways. First we will discuss a computer simulation that assumes that the dextran is a random coil and only will enter if it glides smoothly through the pore. This strategy is developed in Demchick's Ph.D. dissertation (Indiana University, 1994). The results will be compared with those of a different model developed below where it is assumed that the probe molecules are rigid spherical molecules of radius R and molecular weight M and the pores are cylindrical and of radius r. For this model, the chance that an approaching probe would enter is simply the ratio of the area that the probe can go through relative to the total pore area (see Fig. 9.5A), i.e.:

$$P = N(1 - R/r)^2,$$

where N is the number of pores on an average sacculus. A second factor must also be introduced (Fig. 9.5B). This has to do with the hydrodynamic retardation of a particle passing through a fluid-filled hole. The factor is part of the well-known Stokes' law and is:

$$[1 - 2.104(R/r) + 2.09(R/r)^3 - 0.95(R/r)^5].$$

This factor is unity if the particle is small compared with the radius of the hole; i.e., R/r approaches zero for small R. The factor becomes zero when the particle is the same size or greater than the hole. One might think that penetration would

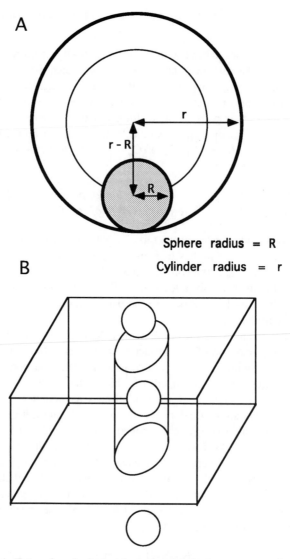

A

r

$r - R$

R

Sphere radius = R

Cylinder radius = r

B

Figure 9.5. A. Entry of a spherical probe molecule into a pore—a very simplistic model that assumes that the cell wall contains circular cylindrical pores of radius r. Probe molecules of radius R diffuse to the surface and pass through only if they enter a pore without hitting its wall. The diagram shows that the probability of passing through the pore depends on $(r - R)$ being less than r. If the center of the probe is anywhere in a circle of radius the quantity $(r - R)$ concentric with the cylinder, it will go through the pore. Thus, the probability of passage is $(1 - R/r)^2$. The probability of passage is unity if the probe is very small and approaches zero as R approaches r. B. Passage of a spherical object through a cylindrical pore. As a sphere passes through a fluid-filled cylinder, it is retarded because the viscous water must flow around the probe but is restricted by the walls of the pore. Stokes showed that slowing can be large and depends on the radius of the sphere relative to the radius of the cylinder. The formula is given in the text.

be an all-or-none process, but the resistance to the fluid flowing around an object passing through a channel greatly slows the passage even for particles that are sufficiently small so that they can penetrate.

D is the diffusion constant of the probe, which is inversely proportional to the square root of the molecular weight of the probe M. For each dextran we can calculate an effective particle radius R from the expression:

$$R = [3 \cdot M/(4 \cdot \pi 6.023 \cdot 10^{23} \cdot 10^{-21} \cdot 1.3)]^{1/3},$$

where 6.028×10^{23} is Avogadro's number, 10^{-21} is the number of cc in a cubic nm, and 1.3 is the assumed density of the material. The mean area of the sacculi A and the thickness of the wall, ∂x are both constant properties of a given batch of walls, and no doubt are characteristic of the species, strain, and growth conditions. Consequently, either constant for the *Criterion* equations can be replaced by a new constant (that will be designated by a capital letter $C_{>0}$, and $C_{<1}$) that equals an expression with only quantities of interest; i.e., the unknown value of r and the experimental parameters R and t:

$$C_{>0} \text{ or } C_{<1} = (1 - R/r)^2[1 - 2.104R/r + 2.09\,(R/r)^3 - 0.95(R/r)^5]R^{-1.5}t,$$

where $C_{<1}$ refers to the size that is almost fully penetrant and is barely prevented at the observation from entering and $C_{>0}$ refers to the sized probe that is almost fully prevented from entering.

Then by substituting a data pair for the same criterion for any two experimental times, two expressions are obtained. They can be used to eliminate $C_{>0}$ or $C_{<1}$ and allow a value of the pore size r to be calculated. There are six pairwise combinations by which this can be done. These data are shown in Table 9.1 for both the random coil and the rigid sphere model for both *B. subtilis* and *E. coli*. For *B. subtilis* when the values computed for both models were averaged, the pores had an apparent radius of the hole of 2.12 nm that should barely let a particle of molecular weight of 45 kD squeeze through. Little difference results from the two criteria or for the choice of model. It is clear that the entire wall of both organisms studied is a sieve with very few large imperfections or cracks because the observed values are about the radius expected for a fully crosslinked stretched fabric of about 2 nm.

Biological Reasons for Exoenzymes of Different Sizes

Many of the exoenzymes secreted by bacilli are smaller than 45 kD, but some are larger. There may be a role for both small and big exoenzymes. Presumably, the former can quickly penetrate the wall of living bacteria and serve a useful function for a large population of bacteria. For a sparse population, this might be counterproductive because the enzyme and its products would diffuse away. A large exoenzyme would be more useful; it would take longer to be liberated

Table 9.1. Dextran sizes satisfying the criteria and calculated pore radii[a]

| | Probe Size Meeting Criteria (kD) | | Calculated Pore Radius (nm) | | | |
| | | | Random Coil Model[b] | | Rigid Sphere Model[c] | |
	$C_{>0}$	$C_{<1}$	$C_{>0}$	$C_{<1}$	$C_{>0}$	$C_{<1}$
E. coli						
3 min	46.6	26.8	2.82^d	1.93^d	1.96^d	1.61^d
60 min	57.0	32.8	2.67^e	1.83^e	1.84^e	1.53^e
1,440 min	69.6	40.1	2.95^f	2.01^f	1.89^f	1.63^f
			2.37		1.74	
				(2.06)		
B. subtilis						
3 min	29.7	20.8	2.51^d	2.72^d	1.71^d	1.62^d
60 min	40.1	31.2	2.73^e	2.74^e	1.60^e	1.47^e
1,440 min	49.0	42.2	2.32^f	2.70^f	1.71^f	1.62^f
			2.62		1.62	
				(2.12)		

[a]Two criteria were used: $C_{>0}$ is the minimally detectible fluorescence that barely decreases the negative staining; and $C_{<1}$ is the fluorescence level that is barely lower than that of the surrounding medium and the sacculus is barely visible.

[b]Estimation of the pore radius for the model that assumes that the dextran is ideal random coil.

[c]Estimation of the pore radius for the model that assumes that the dextran hydrogen bonds to itself to act as a hard rigid sphere.

[d]Calculated from the 3-min and 1,440-min observations.

[e]Calculated from the 3-min and 60-min observations.

[f]Calculated from the 60-min and 1,440-min observations.

as it passed through the cell wall and thus could serve the individual cell that had produced it. The fraction of enzyme associated with the cells for this case can be calculated easily by assuming that the enzyme is produced in parallel with biomass. If one started with a very small inoculum, at any later time the amount of enzyme formed would be proportional to the biomass. One generation is sufficient to liberate into the medium all the enzyme present in the side wall at the beginning of that generation. The biomass has doubled in this time and the amount of enzyme formed during that generation and still associated with the wall will equal the amount that has been liberated up to this time. Thus, 50% of the enzyme that has ever been made by the cell and all the cell's antecedents will be present within the cell wall. Consequently, when the culture is sparse and if the protein has little inherent affinity for the wall, then large exoenzymes serve the role of allowing growth of a sparse culture on a substrate small enough to diffuse into the interstices of the wall where there is active, but trapped, enzyme. Some workers have used the slang expression, "Gram-positive periplasmic space." The periplasmic space is defined as the region between the inner and outer membrane of Gram-negative organisms. But the idea behind this misnomer is that some proteins in Gram-positive organisms are outside the cytoplasmic membrane but still or currently retained by the cell, and that these behave effectively like the periplasmic proteins of Gram-negative organisms. Consonant with these concepts is the finding that certain enzymes may be attached on the outside of the cytoplasmic membrane via linkages to the lipid.

Morphogenesis

Zonal Versus Diffuse Growth

The replicon model of Jacob *et al.* (1963) has had a very strong impact on modern biology and has left its mark in microbial genetics as well as in eukaryotic biology. It was based on genetic studies with *E. coli* and on morphological studies with *B. subtilis*. The idea is that the energetics of cell membrane growth provide the mechanical work to separate daughter chromosomes and to place a single chromosome into each daughter cell; this idea came from the studies with *B. subtilis*. Such a process would work if the chromosome were bound to the membrane near the center of the newborn cell and if new membrane were incorporated only between the two sites binding the sister chromosome origins to the membrane. The morphological studies of the original paper were not conclusive, but the morphological studies on *B. subtilis* by Ryter (1967), and later autoradiographic studies from Karamata's laboratory (Schaeppi *et al.* 1982) at first seemed strong support for the original model in which the attachment site of an unreplicated chromosome was in the middle of the cell. Ryter's studies depended on using tellurite as a labeling agent. TeO_3^{2-} is reduced to Te on the outer surface of the cellular membrane. Crystals of Te remain in the region between the wall and the cytoplasmic membrane and can be viewed in the electron microscope. After the tellurite in the growth medium was removed, the new wall was formed that was unlabeled with tellurium crystals in the central region of the cell. These experimental studies indicated to the authors that zonal growth took place in the center of the cell, but these claims did not hold up. This is partly due to the toxic action of tellurite, which confused interpretation, but mostly due to the different turnover of the side and polar wall discussed below in this chapter. This also confused the studies from Karamata's laboratory (Schaeppi *et al.* 1982) of autoradiography by pulse-chase labeling with a wall precursor showing alternating patterns of labeled and nonlabeled regions; they presented an interpretation consistent with the original replicon model. It is now clear that these observations are a reflection of the slow turnover of cell poles relative to the side walls and are due to the retention of old outer non-stress-bearing wall in the *lyt* mutants that were used in the later studies.

Today the evidence is consistent with growth in three cellular regions. One zone constitutes the entire cylindrical portion growing from inside to outside and not just a central part thereof, as in the Jacob, Brenner, and Cuzins model, and is strictly zonal only from the radial point of view and strictly continuous and uniform from the longitudinal one. The second is the zonal formation of the crosswall as successively smaller annuli from the outside to the cell's axis (Fig. 9.6), and the third is the slow turnover of completed poles. In none of the

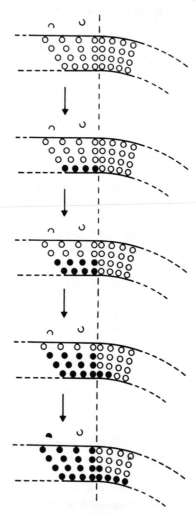

Figure 9.6. Turnover of the side walls and poles of *B. subtilis*. The systematic inside-to-outside turnover of the cell is shown by the rapid passage of cohorts of peptidoglycan units through the side wall and the slower passage through a mature pole. This diagram is based on the results from the laboratories of Archibald and is taken with permission from Clarke-Sturman *et al.* (1989).

processes is wall material inserted throughout the thickness of the wall, but instead new wall is added only inside of existing wall and immediately outside the cell membrane.

The first process is the inside-to-outside growth process that allows the cylindrical wall to elongate safely. Cleavage of the stress-bearing part of the wall is an essential feature for growth. In many, but not all, strains of *B. subtillis,* the older side wall is digested eventually and is sloughed or turned over into the medium. In filamentous strains the no longer functional side wall remains attached near the polar or septal regions of the cell. The second process is the repetitive addition of layers of murein through the cytoplasmic membrane beneath a centrally located region to form a septum. The third process is the slow turnover of the poles, with a gradient of synthetic activity starting at the junction with the cylindrical wall and proceeding towards the tip.

The first indications of these three processes came from the work of Archibald and of Doyle. Starting in the mid-1970's, Archibald's group used a virus that adsorbed to the teichoic acid present in the wall of the bacterium. They used this as a probe to study wall metabolism. If the accessible surface had teichoic acids then the virus would adsorb, but if the surface had the phosphate-less teichuronic acid, it would not. Bacteria were studied that had been grown in chemostat culture under first one and then another ratio of phosphate to potassium. When potassium was growth limiting, the cells possessed the phosphate-containing form (teichoic acid); when phosphate was limiting, the wall contained the phosphate-free form (teichuronic acid). In the potassium-limited steady states viruses absorbed over the entire cell surface, and in the phosphate-limited state the viruses failed to bind. Thus, the attachment of viruses to the cell surfaces allowed aspects of wall growth and turnover to be established. During the transient between the two states it could be observed that the poles retained the previous state for a much longer time than did the side walls. Another trick was used in Doyle's laboratory (Mobley *et al.* 1984); it employed a mutant that was temperature dependent for the glucosylation of teichoic acids. Thus, the glucosylated wall formed at low temperature could be distinguished from wall formed after the temperature was increased. The state of glucosylation of the teichoic acid could be measured with concanavalin A, which is a lectin that binds to macromolecules containing glucose; the process could be visualized with fluorescein-tagged concanavalin A. This study led to the same conclusion that the side wall turned over rapidly and the poles turned over slowly.

The most critical set of experiments are those of Merad *et al.* (1989) and Clark-Sturman *et al.* (1989). In these experiments, as above, chemostat cultures were established with levels of potassium and phosphate that were either limiting or nearly limiting. By shifting so that the limitation was changed and by staining with lead and uranium (which allowed the phosphate in the teichoic acid of the walls to be visualized in the electron microscope), it was possible to examine the mode of growth of both side wall, pole formation, and turnover. The progressive movement of new wall displacing the old throughout the cylindrical region as the inside-to-outside process functioned could clearly be shown and critically excluded the original version of the replicon model. These experiments confirmed the earlier work and showed definitively that completed poles turn over very slowly. In addition they showed that the tip of the completed poles turn over very slowly indeed and that there is more turnover activity adjacent to the cylindrical region (see Fig. 9.5).

Localization of Septation

Growth of septa takes place in the same way that streptococcal wall formation occurs: i.e., by the addition of extra wall to the peripheral wall at the cell center to form a septum. It is necessary to gloss over the process of crosswall formation, because nothing is understood about the process, so only the positioning of the septum can be considered. *B. subtilis* has an exquisitely precise mechanism for locating its own center. Although several mechanisms have been suggested (see Chapters 7 and 13) only one mechanism for the precise localization of the site of septation in *B. subtilis* had been put forward (Koch *et al.* 1981a). This paper pointed out the experimentally observed precision of the division process, and examined a number of possible measuring sticks that the cell might use for determining its own middle. It noted that the only reasonable candidate is the cell's own chromosome; other possible molecules were much too short. It was suggested that the binding sites for chromosomal origins and termini occur at the junction of the pole and side wall. Koch (1988c) later modified this and suggested that a single point at the very tip of the poles was the chromosomal attachment site, based on the finding of Sonnenfeld *et al.* (1985) that the origin DNA is retained in the pole. The newer version of the replicon model proposes the following. When replication starts (at O in Fig. 9.7), one of the two sister origins will migrate and displace the terminus DNA then situated at O'. This results in a symmetrical structure. The displaced terminus DNA, designated T, may then become free or remain bound in a nonlocalized way to the membrane. The complex holoenzyme functioning in the replication of the DNA, called the "replisome" and designated by R, which was initially at O, gradually moves to the precise middle of the cell. Similarly, the terminus locus of the chromosome, indicated by T, which may also be associated with the replisome during most of the replication, becomes located adjacent to R at the completion of replication. The model functions so that as the chromosome is completed its terminus is in the precise center of the cell due to the tensions generated in the symmetrical structure of the chromosome during the replication process. This so-called theta structure, named because it topologically is like the Greek letter Θ, is necessarily perfectly symmetric and if symmetrically attached to the two ends of the cell would accurately center the location of new septum. For the model to function there would need to be the formation of two new binding sites that initially

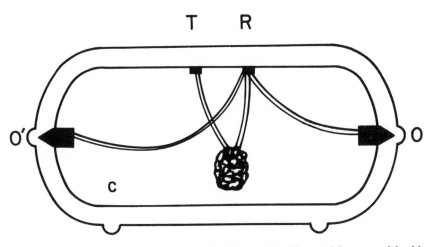

Figure 9.7. A mechanism for a rod to find its middle. The model presented in this diagram was presented by Koch *et al.* (1981a) to account for the precision with which the rod-shaped cell finds its own middle to produce nearly identically sized daughters. A step-by-step version of this process is shown in Fig. 7.4. It depends on the theta mechanism of the chromosome replication. Starting from the origin, bidirectional replication proceeds until the terminus DNA is reached. A proposed mechanism for segregating chromosomes and precisely dividing the cell in halves depends on this. The elements of the centering are shown here. At the poles of the cell there are special sites, labeled O and O', that can bind either the origin, *oriC*, or terminus portions of the chromosome. After a new round of replication starts, a new copy of *oriC* is formed, one copy of *oriC* remains attached (or dissociates and reattaches to the original site), and the other copy of *oriC* finds the equivalent site at the other end of the cell and displaces the terminus region of DNA previously bound there. Then the terminus DNA (indicated by T), probably with other factors, probably attaches to the cytoplasmic membrane, but not in a fixed position. In the stage after that shown here it becomes associated with the region of the DNA where replication is taking place (indicated by R for replisome). Because the replicating chromosome is in a symmetrical "theta" structure, as the DNA is synthesized, the proteins associated with it and the secondary structure of DNA create forces that lead, at least by the time of completion of chromosome replication, to the centering of the terminus in the middle of the cell cylinder. In this model, completion of replication triggers septation in the Gram-positive organisms or the formation of a constriction in Gram-negative organisms. In both kinds of organisms it leads to the formation of two new membrane-bound binding sites for either the origin or terminus. This model is similar to the replicon model of Jacob *et al.* (1963) but can operate when elongation takes place over the entire cylindrical part of the cell.

would hold the two newly replicated termini;—each of which would be on opposite sides of the septum and end after division in different daughter cells. Probably crosswall formation in between would take place by accreting new wall to form the septum, and the splitting would be started before the crosswall closed.

For septal wall formation, this pair of sites must stimulate additional wall formation (and probably a cessation of the inside-to-outside growth process). Because the side wall is 25 nm thick and turns over once per generation and the radius of the cell is 800 nm, a cross-sectional area of wall $2 \cdot \pi \cdot 25 \cdot 800$ or 125,600 nm^2 is formed per generation. Septa may be completed in about a third of a generation, because their cross-sectional area is generated at $\pi \cdot 800^2 \cdot 3 = 6,028,800$ nm^2/per doubling time or 48 times faster than the inside-to-outside process. The average length of the cell is about 48 times that of the thickness of the unsplit septum; therefore, the area engaging in inside-to-out side growth is about 48 times that of a developing septum. Consequently, in an asynchronous culture there would be, on the average, a nearly 50/50 division of wall synthesis between future pole and future side wall material.

Kinetics of Wall Turnover

Part of the evidence for the inside-to-outside mechanism of wall growth for the Gram-positive rod-shaped bacteria is the kinetics of wall turnover. A number of laboratories have published numerous tracer experiments; these have been reviewed and analyzed by Koch and Doyle (1985). For a variety of strains and under different experimental conditions, it is found by pulse-labeling with radioactive tracers that tag the wall, followed by a chase, that there are three phases of turnover. During the first there is total retention of label; this lasts for about one cell generation (Fig. 9.8). After this time the radioactivity is released by the second process which is approximately first order (Fig. 9.9). After two-thirds of the label is lost, there is a slower, approximately exponential turnover of the smaller amount of label. This was label that was introduced into forming septa at the time of the pulse. This third process liberating residual radioactivity has been followed for eight or nine generations. Koch and Doyle (1985) simulated the inside-to-outside model and fitted the body of experimental data in the literature and data from unpublished experiments. (Fig. 9.10) It was shown that the lag time of the first process and the half time for turnover of the side wall component in the second process were generally equal to a doubling time of the organisms (Fig. 9.9). These conclusions provided strong support for the inside-to-outside side wall growth mechanism that demands this close coupling of growth with turnover.

ORIENTATION OF THE GLYCAN AND THE PATTERN OF FORMATION OF CRACKS IN THE SURFACE

Little more need be said about the basic inside-to-outside process, to complement the exposition given in Chapter 6 and above. More discussion, however,

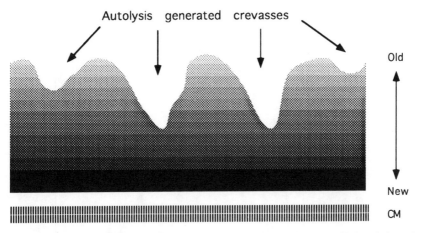

Figure 9.8. Age and autolysis of wall in the side walls of a Gram-positive rod-shaped bacterium. The position of wall formed at earlier time is shown. Older wall has been spread over a longer length of the cell because of elongation growth. This is indicated by the lighter shading. Older, more peripheral wall has a greater chance of being cleaved and released from the cell by the turnover process.

is necessary concerning the orientation and order of the peptidoglycan chains and crossbridges. Arguments have been made above to show that the insertion of new chains is not done by a one-to-one templating process, but that conclusion does not prohibit a degree of order in the wall structure. The kind of partial order possible is different for the thick-walled Gram-positive rod and for the thin-walled Gram-negative rod. In the latter case the new wall is inserted into a wall structure that has been partially oriented by the differential stresses (see Chapter 10). Of present concern is the Gram-positive case, where the new wall is added to the slightly older layer that bears only minimal stress. Thus, a new layer has few cues for ordering itself. Can it be oriented in any other way than in a random direction as there is no templating or crystallization characteristic of the polymerization locally available? The more peripheral layers can have little effect on orientation at the time of synthesis. Consequently, the linkage of new oligopeptidoglycan chains with other new chains and to the older wall of the next adjacent outer layer has little possibility to be other than random.

Because there is a good deal of elasticity of the wall fabric, considerable elongation is possible before stress becomes so intense that autolytic cleavage events are favored. As the growth process continues, the layers would have the properties of a rubber sheet that is pulled in one direction, corresponding to the elongation process in the axial direction. Expansion in the other plane, corresponding to the hoop direction, due to the radial expansion of the layer as it moves to the periphery is much smaller. This situation persists until a layer progresses outward to a position where the stress is so great that the first autolytic events occur.

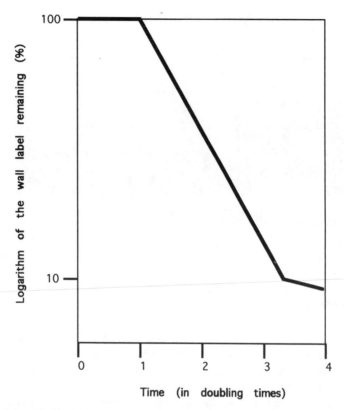

Figure 9.9. Idealized kinetics of the turnover of side walls of a Gram-positive rod-shaped bacteria. The amount of radioactivity retained in the cylindrical part of the wall after a short pulse of radioactive precursor has been added is shown. The lag before release of radioactivity is the time for new wall to migrate out to the region where enough cleavages have taken place to liberate some wall material. If stress due to growth is the prime factor favoring enzymatic cleavage of wall material by the autolysins, the radioactivity retained by the walls decreases exponentially with a half-life that equals the doubling time of the culture. Figure redrawn from Koch and Doyle (1985). The beginning of the third phase of turnover is shown by the decrease in slope at the bottom of the right-hand side. In this phase only the slow turnover of the poles is evident.

Once splitting starts, there is a differential factor between the hoop and the meridional direction. As shown in Chapter 6, the stress in the hoop direction is twice that in the axial direction. Therefore, autolysin action would be more rapid in the axial direction than in the hoop direction. On the other hand, the elongation process in which the cylindrical region lengthens 2-fold every generation would tend to favor cleavage of bonds oriented in the axial direction. The results of this conflict are that autolysis should create strips of more peripheral intact wall that would give partial support to underlying intact wall. The stress would be oriented so that a system of helical cracks is created (see later in this chapter

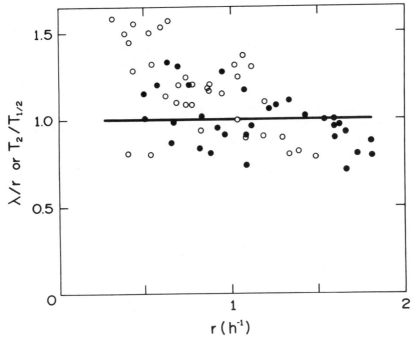

Figure 9.10. Turnover of the side walls of Gram-positive rod-shaped bacteria relative to growth rate. Turnover of the poles is much slower than that of the side walls, so the initial loss of radioactivity depends more on the side wall turnover than that of the poles. The graph shows the measured rate of growth rate relative to the degradative rate as a function of the growth rate. Specific growth rate is designated with r instead of the symbol μ used elsewhere in the book. Other symbols: first order degradation rate, λ; doubling time of culture, T_2; and half-time for turnover, $T_{1/2}$. Republished by permission from Koch and Doyle (1985).

and Chapter 13). With further growth, intact wall at the bottom of cracks would tend to split at an obtuse angle to extend and perpetuate each helical crack. Wall that lies between the cracks would survive longer and only later be subject to increased stress and increased autolytic activity.

Consider an analogous situation, a geological uplift where a flat ocean bed is becoming a high plateau. Water erosion will create a series of gullies and valleys. Finally, a series of sharp ridges will form. If the uplift and erosion processes continue at the same rate, the ridges will persist in form although their substance is eroded but replaced by a continuing uplift.

Rotation During Growth

The idea that formation of helical cracks would be the mode of wall cleavage permitting safe elongation of the Gram-positive cylinder came only belatedly though the combination of new experimental results, the development of the surface stress theory, the attempt to generalize the theory for nonradially symmetrical cases, and attempts to understand the observations of Mendelson.

The experimental results of Mendelson (1976, 1978) were crucial. He had constructed a double mutant whose spores, when germinated, grew as a filament that remained attached to the spore coat at both ends. As these filaments elongated, a closed circular structure at first developed. It then changed into a wound-up structure with the same topology as supercoiled covalently closed plasmid DNA (Fig. 9.11). That is, the structure was wound into double-stranded helices with the two strands at each end connected to each other. Mendelson correctly interpreted this to mean that as a cellular unit elongates, one end rotates relative to the other. He, incorrectly I believe, interpreted that to mean that the wall grew by the helical insinuation of new material.

In the second paper, he studied entities that he called macrofibers. These are formed by further development of the filaments connected as a circle, but also from a variety of strains that grew as filaments. His later work showed that cells with the same genotype growing in the same medium under almost all conditions would achieve the same handedness of growth. The handedness could be altered genetically and physiologically by altering the medium in various ways. The

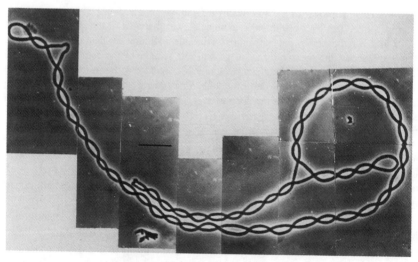

Figure 9.11. Structure produced by germination of spores of Mendelson's double *B. subtilis* mutant. Microphotograph taken from Mendelson (1976). It shows the spore of a mutant that grows as a filament and remains attached to the spore coat. This creates a circular structure that with further growth forms a twisted structure. This demonstrates that the individual cells of the filament rotate, one end relative to the other, as they grow.

theory proposed in the last section predicts helical cracks and perpetuation of such cracks and their handedness, but it does not determine the choice of screw axis. It can be imagined that very weak forces initially establish the direction of twist as the spore germinates and then the strong forces discussed above perpetuate them. Irregularity of handedness results when single-stranded spirals are generated by treatment with various antibiotics (Tilbey 1981).

The ability of shifts of medium or temperature to lead to alteration of the handedness of rotation in a few generations suggests that the system of cracks is metastable. Initially, I wondered whether other physical forces were involved (Koch 1988b, 1990a). Yet, the convincing evidence that cylindrical growth of Gram-positive organisms proceeds by a process associated with rotation fundamental to the inside-to-outside process came only recently and excluded other forces playing a crucial role. The evidence came from my experiments (Koch 1990e); filaments growing in depression slide culture were followed visually and photographed, and drawings were prepared (Fig. 9.12). Many filaments, each hundreds of cell units long, were observed, but for the critical class of experiments care was taken to follow only those that initially were neither attached to the glass surfaces of the depression slide nor folded back upon themselves. The unambiguous conclusion was drawn that rotation was an inherent part of the *B. subtilis* growth pattern. (The experiments were carried out at sufficient dilution so that only one filament was in the microscopic field and interaction between filaments was not a factor.) The long filaments writhed as they grew, second by second. Microphotographs were taken, but it should be emphasized that the pictures were not sufficient in themselves to demonstrate relative rotation of the two ends of the filament. Because of the limited depth of focus of the microscope, the fine focus adjustment could be used to make the conclusion concerning rotation unambiguous. From observations that noted whether particular portions of the filament were above or below the reference plane of focus, the only possible interpretation was that rotation was generated by the growth process itself. It could be concluded that, at least while growth was occurring at 37 °C in an undepleted rich medium, the ends were rotating relative to each other at about one revolution for every two or three cell's length increase.

Individual cells in a filament have no helical stress (torque) during the above experiment because the filament is free to rotate. This is not so if a portion of the filament becomes attached to some object. For example, if the filament attached to the glass surface of the depression culture or the cover slip, stresses would develop and at some point a portion of the filament could be seen to detach from the glass, rotate, and then reattach. Later, another portion would come loose and repeat this gyration, always with the same handedness. Stress also would develop if a free-floating filament folded back on itself. Once this happened a double-stranded helical region would form as rotation of the ends took place as growth proceeded. But the helical shape was not part of the inherent structure of the cells, as could be established by compressing such

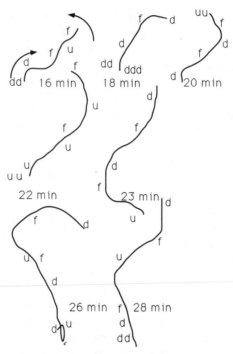

Figure 9.12. Rotation of freely suspended single-stranded filaments of wild type *B. subtilis*. An 18 h culture of *Bacillus subtilis* strain 168 was diluted 2 × 10⁴-fold in fresh aerated Luria Broth, transferred to a depression slide culture, and incubated at 37 °C for 180 min before the zero time photograph was taken. The numbers indicate the time in min thereafter. Photographs were taken when large parts of the filament happened to be momentarily in the plane of focus. Additional sketches and photographs at intervening times and a fine focus adjustment were used to construct this drawing. Regions of the filament not in focus are marked with u and d (up and down) for above and below the plane of focus. The region in the plane focus are indicated, where appropriate, by the symbol f.

helical structures by pushing the cover slip or sucking them back and forth in a Pasteur pipet. These procedures would break the filaments, and the parts that were no longer parts of double strands then became straight.

The B. subtilis *Pole Shape*

As the septum starts to form, the future pole wall is not under stress. Stress appears only after splitting starts. Ian Burdett and I (Koch and Burdett 1986a,b) wondered whether the establishment of the convex pole shape was due simply and solely to stresses caused by the pressure differential that appeared after splitting of the septum took place, which externalized the pole wall.

Our approach was simple. Starting with filaments formed from an autolysin-deficient mutant, we caused the existing crosswalls to split by the addition of an exogenous enzyme (lysozyme), observed the shape of poles that were produced, and compared the surface area with that characteristic of completed poles produced during the rare natural divisions of the mutant or the typical pole of the mutant's parental strain. The reader may question why the lytic enzyme made by a hen to protect its eggs against bacterial infection would not destroy the bacteria by degrading the cylindrical wall. Of course this happens, but the rates of that process turn out to be slower than its function in causing septal splitting. We did not have the native autolysin as a purified reagent, so we chose to use hen egg white lysozyme, because it is cheap, very pure, and readily obtainable. A preliminary experiment showed that 10 μg/ml was the appropriate level because it did not cause lysis in 10 minutes at 35 °C, but did so at longer times. When filaments were treated for this brief period and then fixed, stained, embedded, sectioned, and examined in the electron microscope, it was seen that the new poles created were of the shape and had the area of the poles of normally growing, wild type organisms.

Because the conversion of septum to pole took place so quickly, apparently no new material is needed during this conversion through bisection of the planar septum into two poles. The pole shape was thus determined by a mechanism that we dubbed "split-and-stretch." Our measurements showed that the surface area had increased by 50%. A doubting Thomas, however, might say that much wall could have been formed and inserted in the 10 min of lysozyme treatment. So this experiment was not definitive.

We carried out a critical experiment that excluded new growth as the source of the new surface area. For this we used vancomycin, an antibiotic that blocks peptidoglycan formation (see Chapter 4). This particular antibiotic was chosen because it is known to have a fast mode of action (see Rogers *et al.* 1980). With it we could be confident that wall synthesis was blocked in much less than 20 sec. We added a thousand times the median lethal dose of vancomycin (appropriate for our bacteria) to a growing culture of bacteria; we immediately followed this with 10 min of lysozyme treatment. Electron microscopy showed the same result as before. As a result, it is a "hard" fact that stresses due to turgor pressure acting on the newly split septum convert it to the convex pole and this is done without the addition of any new wall material. Further, the change in shape occurs so fast that probably little enzymatic remodeling of the existent wall is involved.

If a soap film is formed over a bubble pipe, the film will be planar when no air is blown into the stem. As air is blown in, the film expands and the resultant bubble has a shape that is a segment of a sphere. When the appropriate amount of air is blown into it, a hemisphere is generated with a surface area of $2\pi r^2$, where r is the radius of the pipe's bowl. Thus, the area at this point is 100% more than that of the original planar septum. The increase we observed was 50%; the difference is relevant because in many studies the authors assume that the pole is a hemisphere. This is definitely wrong. Our findings are that the pole shape of *B. subtilis* is not hemispherical, in fact it is not even a portion of a spherical surface, but is more flattened. So it seems that a soap bubble is not a good model for the shape of this pole.

More realistically, imagine that a flat sheet of rubber were affixed to a circular pipe and then a source of pressure applied. Let us assume that the mechanical properties of the rubber sheet are the same in both directions; i.e., the material is isotropic and Young's modulus is the same in all directions in the plane of the sheet. Whether the material is isotropic or not, the septal region near the cylinder part of the pipe must exhibit at least a different kind of expansion than that near the center (the apex of the presumptive pole). At the center of the septum, the stresses will be equal in all directions within the plane. This is a case that has already been discussed in Chapter 6. Near the edges of the septal disk stresses will appear only in the axial direction and not in the hoop direction. So the pole shape will depend in a complex way on how one elastic modulus will be influenced by stresses imposed in other directions.

It is not yet clear how expansion of the wall fabric takes place in terms of how often rupturing of covalent bonds is necessary. In the likely case where the forces due to stresses are adequate to cause all weak bonds to rupture and rearrange, but not at all sufficient to rupture covalent bonds mechanically, the values of Young's moduli are irrelevant and the elastic limit (extensibility) is the important parameter. Consider a fabric such as that diagrammed in Fig. 9.13. Imagine that the unit elemental area of this fabric is composed of a cartesian network of springs that are easily elongated to a degree and then can exhibit no more stretch, but are strong enough so that they do not break. When the stress is constant in all directions, then the square unit will become a bigger square unit. If the stress is only in one direction, then some units will expand in one direction, but if the units are obliquely oriented with respect to the force, then the square will stretch but become a narrow elongated rhombus occupying little surface area, possibly less than it did when unstressed. So, near the margins of the pole there will be little expansion of the area when stress is applied.

Evidently, the rule for the relationship of the ratio of the two perpendicular forces to the factor of area expansion depends critically on the properties of the actual peptidoglycan fabric, no matter whether the fabric is one or many layers thick. Still, it is easy to show that the observed shape of the pole could be accurately generated by a simple abstract model for of this kind of fabric (Koch and Burdett 1986b).

To this discussion should be appended a consideration of the pole of one other bacillus. It is a diagnostic criterion, but not always observed in every cell, that the pole shape of *Bacillus anthracis* is flat or planar. How may this be interpreted? Possibilities are that the wall of this organism's pole (at least) cannot expand when stressed, because the wall fabric in the pole is chemically different from that in *B. subtilis,* because it is much thicker than the side wall, or because the

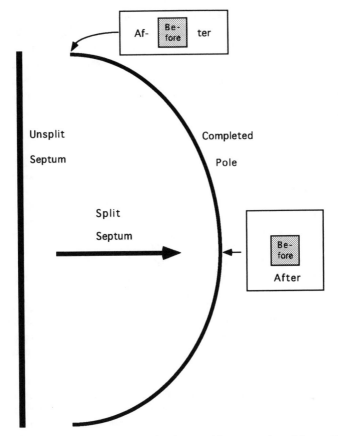

Figure 9.13. Pole shape determination by the stretching properties of the peptidoglycan fabric of the unstressed septum. When the septum is split and exposed over its entire surface to the turgor pressure of the cell, it will bow out. The stretchability is different near the center and near the periphery of the septum. At the center the wall fabric can stretch evenly in two dimensions. A small square would be distorted to a larger square. Near the periphery a small square would be expanded more in the axial (meridional) direction than in the circumferential (hoop) direction.

pressure differential is not sufficient. Another (unlikely) interpretation is that the pole wall is reinforced as splitting takes place.

Summary of Mechanisms That Allow the Autolysins Safely to Enable the Elongation and Cell Division Processes

Fig. 9.14 combines the various suggested factors that might protect the cells against their autolytic enzymes. First, the stress is less at the membrane and greater at greater distances away from the cytoplasmic membrane, because of expansion. Second, the autolysins, when secreted, must fold up into a native configuration to become active enzymes; this may not be possible in the interstices of the peptidoglycan. Third, autolysins originating from the earlier generation of growth coat the external surface favoring septal splitting, but not cell lysis. Fourth, the extrusion of protons as part of chemiosmotic energy generation makes the pH unfavorable near the cytoplasmic membrane, maintaining intact layers of peptidoglycan enclosing the cell.

Figure 9.14. Factors that may protect the Gram-positive rod from self-destruction and allow its growth and division.

A. The distribution of stress is one of the factors that protects the cell from self-destruction. The peptidoglycan is not under tension as secreted through the membrane and initially as crosslinked to the older wall and to other nascent wall. A layer, however, comes to feel tension as new layers are added underneath and the cell elongates. Eventually, a layer is maximally stretched. At this point it may be easily breached by autolysins, relieving some of the stress and relaxing a portion of the wall. Autolysin action is localized to regions under stress because enzyme cleavage activity depends on the stress in the sensitive bond. A second factor protecting the cell from self-destruction is the kinetics of folding of the autolysin. The autolysins are secreted through the cytoplasmic membrane and probably move through the wall with the inside-to-outside growth process. At some point, they fold into a native config-uration. If they cannot fold to a native configuration while surrounded by wall network, then native autolysins will form only at the outer surface of the wall. The third factor protecting the cell is that the bulk of the autolytic activity comes from or via the medium and acts predominantly on wall that is not intimately involved in the structural integrity of the cell. On reaching the outer surface the autolysins then act to cleave wall and, in that act, they are liberated for an instant from the cell of origin. Depending on the concentration of cells and whether growth has been from a small inoculum, the concentration of native autolysin molecules in solution and the amount bound on the outside of a cell will be different. This protects cells in dilute culture because most of the autolysin is not bound to cells. The fourth factor is the energy metabolism of the cell, which functions to extrude protons that lower the pH of wall in the near vicinity of the membrane (Joliffe *et al.* 1981). The pH lowering extends further as the cell grows and exchanges protons for cations (mostly potassium) (Koch 1986a). Probably autolysins on liberation rebind to the same cell, but not necessarily in the same location. Thus, enzymes from previous generations (and to a lesser degree from other cells) may coat a developing crack or crevice, particularly in the region where the septum is being split. The figure shows these movements of the autolysins. Thus, cracks in the wall continue to be deepened and complete dissolution of an original layer takes several generations. Splitting of a completed septum would be faster because the forces at this site are greater.

B. Progression of autolysins through the wall, accumulation on the outer surface, and the formation of cracks or crevices in the surface. Again, it has been assumed that most of the autolysins do not fold into their native configuration until they reach the surface.

C. Postulated helical grooves on the surface of cylindrical Gram-positive rod-shaped bacteria. These results because the stresses in different levels of the wall mean that a system of helical cracks or fissures will propagate itself as the cell grows.

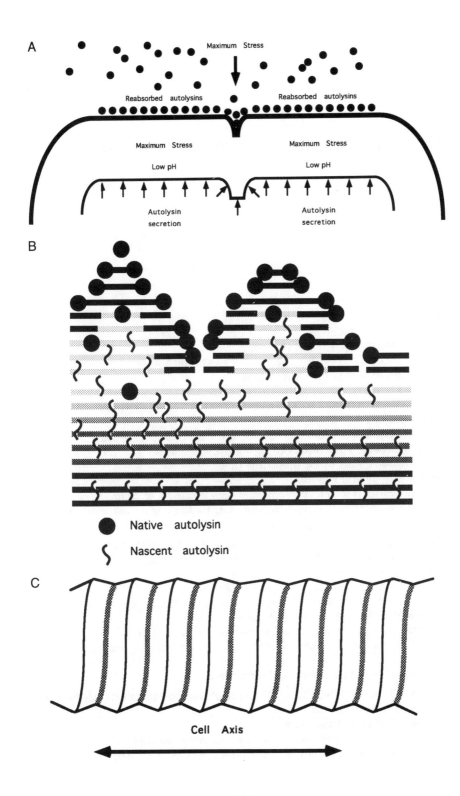

Native autolysin

Nascent autolysin

Cell Axis

10

The Gram-Negative Rod: *Escherichia coli*

GENERATION OF THE ROD SHAPE AND DEVELOPMENT OF THE POLE
  The Problems Caused by Being Thin-Walled
  The Contrast with the Gram-Positive Strategy
  The Gram-Negative Strategy
  The Variable-T model for Gram-Negative Pole Formation
    Potential cellular mechanisms that could alter the surface tension parameter.
    Calculation of the shapes of the developing pole
  New Models for Growth of the Gram-Negative Wall
    Allosteric model and "smart" autolysin
    Holoenzyme model
    Höltje's three-for-one model
    Minor modification of the three-for-one model
CONCLUSION ABOUT THE NATURE OF THE PEPTIDOGLYCAN FABRIC OF
  THE GRAM-NEGATIVE ORGANISM AND THE GROWTH PROCESS
  The Unit of Gram-Negative Wall Architecture: The Tessera Implications of the *in
    vivo* Structure not being Maximally Extended

KEY IDEAS

Practical and esoteric reasons led to E. coli *being the object of study for molecular
  biology.*
E. coli *is well adapted to a specialized habitat.*
Its niche in its habitat depends on its elegant coping strategies.
Permease systems were a new kind of "enzyme" in 1956.
*Although the Monod equation is key to understanding modern ecology, it has
  deficiencies.*
*The "Best" formulation includes the Monod and Blackman models; and it is more
  realistic.*
*The wall envelope includes the inner and outer membranes, the murein, and a
  periplasmic space that sometimes contains MDOs.*
E. coli *changes its length and width depending on nutritional conditions, but keeps the
  ratio fairly constant.*
E. coli *divides very exactly in the middle.*
*High-resolution autoradiography is a very important tool, but it has limitations of
  resolution.*
The distribution of label can be calculated from the distribution of grains.
*It was thought earlier that incorporation of murein occurred predominantly in the cell
  center.*
*Now it is known that the side wall is enlarged by diffuse insertion of peptidoglycan all
  over its length.*
*The incorporation of label in a developing constriction is intense, but is not all at the
  exact center as posited by the "leading edge" model.*
The working (stress-bearing) part of the sacculus of E. coli *is one molecular layer
  thick.*
*There are portions where appended new wall is waiting to become structural and
  other portions where turning over wall has been only partially released.*
The sacculus is quite elastic and can be expanded over 4-fold range of area.

The state of ionization of acidic and basic groups is very important for cellular functions.

Light scattering at low angles is a very powerful tool to study the wall fabric.

The pole of the Gram-negative rod is formed in a completely different way than that of the Gram-positive rod.

The Variable-T model and the "three-for-one" model combined together is the most likely mechanism for cell division and pole formation.

The unit of the wall architecture is the "tessera," which is composed of two octasaccharide chains, crosslinked at the ends with tail-to-tail tetrapeptides.

The Organism and Some History

A century ago Dr. Escherich was searching for the organism causing cholera. He found an organism that today is called *Escherichia coli*. Later it was believed that this organism was the major one present in mammalian fecal contents. Because it was rapidly lost in soil and water, its presence was used as a measure of fecal contamination in water quality assessment. The fact that it could utilize lactose reinforced the idea that it was associated with mammals that fed their young with lactose-containing milk. It was for these reasons that selective and enrichment methods were developed. In addition to the ability to ferment lactose, some tests depended on the ability of the organism to grow at 45°C and some on its ability to resist the action of bile salts. Both qualities would be expected for an organism whose major habitat was the mammalian intestine. Technical improvements, like the development of eosin-methylene blue (EMB) agar, were made for these public health purposes. Later these techniques became major tools in the development of molecular biology.

Molecular biology, in fact, arose through studies of this organism in two ways. First, Delbrück started a school fifty years ago that had the goal of understanding the biophysics of viruses. The basic idea of "the phage group was to study the viruses and use the bacterium as a constant incubator supplying housekeeping functions for the viruses. Delbrück felt that these unimportant features should be kept constant in order to permit the essential genetic function of the viruses to be dissected. *E. coli* was chosen as the host organism in part because it was believed that *E. coli* was nonpathogenic. In that assumption the Delbrück school was wrong; it was a major killer of women during childbirth a century ago, and even today certain strains cause economically important disease in newborn farm animals. It also can cause septicemia in adult humans. Fortunately, no serious medical problems related to *E. coli* developed among the phage workers. The studies on the bacteriophages did lead, incidentally, to a data base about the host organism. They also led to studies of the nature of mutagenesis. The random basis of mutation (as opposed to the theory of direct mutation) came from the experiments of Luria and Delbrück (1943) and the indirect selection method by Lederberg and Lederberg (1952), both with *E. coli*. The finding of genetic recombination by Lederberg depended on the chance selection of a pair of appropriate, compatible mating strains of *E. coli.*

The other impetus for the development of molecular biology came from the Pasteur Institute in Paris. On one floor of the building, Jacques Monod and colleagues studied the basis of lactose metabolism of *E. coli* and discovered operator mutants, the operon, and the repressor. On another floor of the same building, these same concepts were developed through the studies by François Jacob on the lysogenic virus, lambda, a virus that grows on *E. coli* (Jacob and Monod 1961). Obviously, there were many other workers who played critical parts in the development of the origins of modern molecular biology, but here we want to focus on the organism.

E. coli is mainly a commensal organism of the mammal, usually isolated from the mammalian feces or intestine. It does supply vitamin K, so to some degree it is a symbiont. However, it can kill neonatal animals (some strains are particularly dangerous to piglets); it was also a major cause of childbirth fever, an important human killer until modern medical techniques and hygiene became common. So it is a pathogen, but an opportunistic one.

Another of our early ideas was wrong: it is not the major organism of the large intestine, as was once believed. It usually constitutes less than 1% of the organisms in the large bowel; its numbers are insignificant compared with the strict anaerobes. Only forty years ago techniques were developed to grow strict anaerobes; before this we did not know how abundant they were. There are times when *E. coli* cannot be cultured from the intestine of hygienic adult humans of the first world. So it may be thought of as a wayfarer with oases in the available gastrointestinal tract. It has one feature needed for this peripatetic life style: the ability to grow with or without oxygen, i.e., it is a facultative anaerobe. The point is that it can cope and grow using oxygen and grow also in an oxygen-free environment by fermentation or anaerobic respiration with nitrate, nitrite, or fumarate as an electron sink. Under all these conditions it can grow rapidly and efficiently. Although found in the colon (whence its name) and feces, it must move into the environment and back into another mammalian's intestinal tract. It must do this at least once a host generation for its continued propagation and, in fact, it does so many times during the lifetime of any mammal. Although not the first occupant of the newborn's intestine, normally it arrives shortly after birth and persists unless very hygienic conditions are maintained or unless there is medical intervention with antibiotics. It has been argued that different strains remain sequestered in the gut, presumably in the ileum, and that occasional organisms from there are the founders of most of the waves of coliforms that pass through and out the colon (see Koch 1987b).

E. coli has been used as the prototype of the Gram-negative rod. Its nutrition is limited compared with pseudomonads, but the regulation of its metabolism and anabolism is exquisite. Its ability to salvage certain nutrients from the environ-ment is very elegant and precise; moreover, its regulation of that ability is a continuing source of amazement to the average microbiologist. This salvage ability suggests that it is very well adapted to the various parts of the mammalian (and other) gastrointestinal tracts. In these environments it is fine-tuned and, like an Olympic racer (Koch 1971, 1979, 1980b), is able to start growth very quickly when the opportunity presents itself. It does not form spores as such, but it responds to a worsening environment by becoming smaller and does survive very well in poor environments or even during full starvation.

The organism has the hallmarks of having had hundreds of million of years to adapt to life in the intestine. (Mammals started to evolve in the Jurassic period, 180 million years ago.) The ecosystem in the gut has some very constant properties, such as constant temperature. There is variability of some properties, like the input of nutrients, but even this variability can have a constancy during a sojourn in one host. Thus, some hosts take three meals a day, some eat almost continuously while awake, some have dinner only when they capture a quarry. Consequently, cyclical adjustments are necessary and their nature is dependent on the host. The niche in which this bacterium specializes is to be able to maintain a foothold in a flowing gut ecosystem and to survive in the other environments through which it must pass at least well enough to colonize yet other intestinal tracts. Its ancestors have been challenged to respond to starvation conditions and to compete in very rich environments. It is an optimistic organism that "believes" that no matter how bad things become they will always get better. It has reason to be optimistic because that has always been the lucky fate of its ancestors. Although these last two sentences express rank teleology and anthropomorphism, they do summarize a great deal of this organism's regulatory biology. For example, the organism must be, and is, prepared to make the most of the opportunity when there is a "shift-up" and the environment improves, making life temporarily easy.

Understanding some of the biology of E. coli and its viruses led to the revolution that created molecular biology. Now the organism is used as a tool for the study of other living things, and our knowledge of it has grown enormously as a spin-off simply because there has been such a general and widespread effort. Much of this is outlined in the two editions of the compendium by Neidhardt et al. (1987, 1995). From a global biological point of view the important reason for focusing on this single organism is that it is an organism that can grow without help from other organisms given a mineral salts medium and a carbon source, so that in some sense it contains the full set of answers to "the problems of life." It has 4.7×10^6 base pairs of DNA. It needs only a fraction of that, perhaps only 10%, for growth under balanced conditions. We now know the sequence of 43% of its genome. Before the millennium is over, we can expect to understand the meaning of these 3,000 kb of DNA comprised of housekeeping DNA, home-keeping DNA (=DNA needed for the physiological processes of morphology, replication, and division), DNA of genes in the act of evolving, junk DNA, emergency DNA, and contingency DNA.

In this chapter, some of the problems of morphology, replication, and division will be examined. The topics considered here are peripheral to the view of many microbiologists, but they are important and the exposition given here balances the treatments found elsewhere in biochemistry texts and in books that are labeled as microbial physiology.

Garnering Outside Resources—Uptake Mechanisms

At the supply level, the organism must have uptake systems on its external surface to bring necessary substances inside in sufficient amounts for growth. This chapter on *E. coli* is an appropriate place to develop the subject of "uptake" because this was the organism that yielded the critical insight into active transport. How this field developed is well worth retelling. The workers in Monod's laboratory in Paris (Rickenberg *et al.* 1956) were studying this organism's ability to use lactose. Not unexpectedly, it was found that mutants could be found that could not grow on lactose as a sole carbon source. Actually, they could grow on a medium that supplied an additional carbon source, such as succinate or glycerol, but were called "*lac*" negative because they neither metabolized or fermented lactose; they also yielded differently colored colonies on suitable indicator agars in petri plates. Such mutants, however, fell in several categories; the major one comprised those that had lost β-galactosidase. This is the enzyme that splits lactose into the two utilizable sugars, glucose and galactose. The minor class of mutants behaved the same, except that when grown in the presence of a galactoside, β-galactosidase was present in the total culture. The uninteresting finding was that the supernate of this class of lac-negative cultures, like the major class and like the wild type organisms, had very little β-galactosidase. The interesting finding was that if the cell membrane was ruptured with toluene, the injured cell suspension of this subclass of mutants had developed the ability to split lactose. Such cells were designated as "cryptic because of their hidden ability to split lactose. They needed to be grown in the presence of a galactoside for them to form the enzyme, so they still had the normal regulation of induced enzyme formation. In the key paper of Rickenberg *et al.* (1956), the logical puzzle was solved: there were two, instead of one, gene products unique to lactose metabolism—β-galactosidase and a new one that today is called "galactoside permease." "Permease" was a coined word. The word suggests that the function was to aid the permeation through the lipid bilayer. The ending "ase is that used in signifying enzymes. Here was a strange enzyme: it does not alter its substrate chemically, and therefore its name caused some biochemists to become furious. This is nearly as bad as calling it the "here-to-there-ase." But the epithet stuck and permeases of many kinds are important to the growth of bacteria and to development of the field of bacterial physiology.

The original justification for the name "permease" was that it exhibited enzyme-like kinetics. The rate of entry into the cell had a saturable substrate dependence like other enzymes. The kinetics were nearly hyperbolic, as in the case of the first enzyme ever analyzed; i.e., the action of yeast invertase on sucrose. That study led to the well-known Michaelis-Menten formulation of enzyme kinetics in 1913. The formulation for the permease case can be made as follows: the substrate outside the cell, at concentration G_0, binds to the cytoplasmic-membrane bound permease P to form a complex that "resolves" to become the product, which is the substrate inside the cell. The concentration of galactoside inside the cell will be designated by G_i. The reaction could be studied because the workers had in hand radioactive galactosides that were substrates for the permease, but were not cleaved by the β-galactosidase (they were analogues containing sulfur at the critical glycoside bond). The Pasteur Institute workers also had ways (by membrane filtration) to separate the cells from the medium faster than substrates could permeate the cell membrane. The uptake process was reversible because a chase with unlabeled galactosides caused the rapid efflux of the internal pool of labeled galactoside. That was forty years ago and much more is known today.

The gene for permease, *lac*Y, has been cloned, sequenced, mutated, and the version in other organisms studied. It has also been the subject of many other kinds of studies. Today we understand that the protein is a membrane protein that coils through the bilayer 10 times with its sequence of 200 amino acids. This structure has the properties of binding galactosides with a range of aglycone residues in either the α or the β configuration; it stoichiometrically binds protons on the same face of the cytoplasmic membrane as a galactoside molecule has bound. This form interconverts to an identical structure; i.e., one with both the proton and the galactoside still in the cis arrangement but on the opposite side of the inner membrane. Neither a proton by itself nor a galactoside molecule by itself can traverse the membrane with the aid of the permease. In effect the galactoside permease is a machine that normally functions to pump galactosides from outside to inside, powered by the protonmotive force, i.e., by the excess of protons pumped outside the cell by the electrogenic process of proton extrusion (chemiosmosis). In an energized cell, the kinetics appeared to Monod and his colleagues as follows:

$$G_0 + P \underset{k_2}{\overset{k_1}{\Leftrightarrow}} GP \overset{k_3}{\Rightarrow} P + G_i. \qquad \textit{Monod permease mechanism}$$

Just for practice and to remind the reader of Chapter 3 and a first course in biochemistry, this mechanism translates into mathematics as:

$$d[GP]/dt = k_1[G_0][P] - (k_2 + k_3)[GP].$$

In the steady state where the rate of entry is balanced by outflow, the two terms on the right-hand side are equal and this expression can be set to zero, yielding:

$$(k_2 + k_3)/k_1 = K = [G_0][P]/[GP],$$

where K by definition is $(k_2 + k_3)/k_1$. By calling the total amount of permease "p"; i.e., $p = [P] + [GP]$, we can then add $[G_0]$ to both sides of the second form of the equation and write:

$$K + [G_0] = [G_0]([P]/[GP] + 1) = [G_0]([P] + [GP])/[GP] = [G_0]p/[GP].$$

The rate of the entry process is equal to the rate of dissociation of the complex to yield the internal galactoside as a product, therefore:

$$v = k_3[GP] = k_3 p\ [G_0]/([G_0] + K).$$

Then by calling $k_3 p\ V_{max}$, we obtain:

$$v = k_3[GP] = V_{max}\ [G_0]/([G_0] + K). \qquad \textit{Monod permease}$$

In this way, we have rederived the classical 1913 Michaelis-Menten equation in the 1924 Briggs and Haldane elaboration but in the context of bacterial uptake as proposed in the 1956 derivation by Monod (Cohen and Monod 1957). K (frequently designated by the symbol K_m) is the Michaelis constant and is the substrate concentration that gives the half maximal rate; i.e., when $[G_0] = K$ then $v = \frac{1}{2} V_{max}$ (see Fig. 10.1).

The point of the 1913, 1924, and 1956 derivations is that useful results are obtained even though we have no idea of actual concentrations of enzymes or membrane-bound components. This is in spite of the fact that we have used square brackets, such as $[G_i]$, to indicate molar concentrations as is conventional with chemists. Actually today in many cases we can know the concentrations for soluble enzymes of known molecular weight and known weight concentrations, but still this is actually a presumptuous symbol to be used for a material embedded in the cytoplasmic membrane. That was the point: our ignorance does not matter because we end up with an expression in which all the quantities are measurable, the concentration of substrates and rates of reaction, so that the useful and significant quantities, V_{max} and K_m, can be extracted from well-designed experiments.

The extensive use of this simple model has been particularly important in the fields of microbial ecology and biotechnology. The hyperbolic relationship for transport when transport limits growth is called the *Monod permease* equation, and there must be many thousands of publications describing complex ecosystems or considerations of the design and operation of commercial fermentations that do use this formalism to dissect the multiple processes involved. But it is wrong!

It is right in the limited case in which proton extrusion is never limiting, the permease is always limiting, and the permease molecules do not compete with each other on the surface of the cell. But imagine that it was once right in an

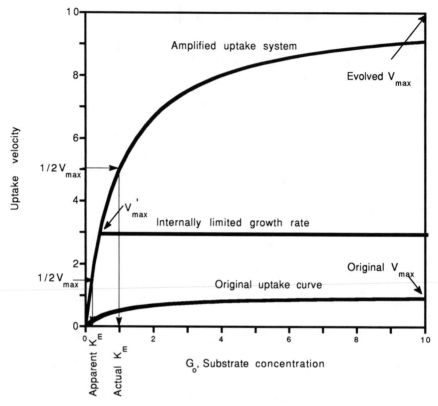

Figure 10.1. Evolution of an uptake process. The hyperbolic curve is characteristic of many enzyme reactions and also some transport processes across membranes; in this latter case the data appear to satisfy the *Monod permease* equation. The curve shows how the V_{max} and K_m can be picked off the rate versus substrate concentration curve. There are more elaborate ways to treat kinetic data for the simple hyperbolic case and there are far more elaborate enzyme mechanisms. The equation was derived in the text for the case where the catalytic activity is a membrane-bound permease that favors the interchange of galactosides (together with protons) from outside to inside the cell. Actually, a much more elaborate treatment than that given in the text is required to treat the interaction of both protons and galactoside. As formulated in the text, there is no explicit mechanism to explain how the substrate could be accumulated to high concentrations by an energy-consuming process. In fact, this process is powered as the result of active extrusion of protons due to catabolic processes; the protons can return to the inside via the permease, if they "drag" along a galactoside.

Two curves are shown for different values of V_{max} and the same value of K_m for the uptake of nutrients into a growing cell. It is imagined that the curve with the lager V_{max} actually represents the cell taking in nutrients faster than it can use them. Maximum ability to use the substrate is denoted by the horizontal line marked V_{max}'. Now the uptake will not be hyperbolic but follow the dark horizontal line and yield a much different apparent K_m. This case is an example of *Blackman* kinetics.

earlier eon; how would evolutionary forces respond? Whenever the permease's substrate was limiting, selection would favor better or more numerous permeases. Eventually, the transport capability would become so great that the entry rate would be magnified and entry at low substrate concentration be very steeply proportional to substrate concentration (compare the original lower hyperbola to the upper curve corresponding to increased transport ability in Fig. 10.1). Then even at quite low concentrations, the substrate would be taken into the cell at a rate that more than suffices for all the cell's needs (Koch 1967). The cell would need ways to shut off the uptake or let the excess leak out of the cells. These ideas fit into a model devised by Blackman in 1905 (even before the concept of enzyme kinetics had been formulated). For the Blackman type of limitation, uptake is characterized by a second-order rate constant for uptake; i.e., k_1 in the formulation above. But then there is a discontinuity and uptake is squared off when $v = V_{max}'$, where V_{max}' has a prime symbol to refer to the maximum ability of the organism to grow given an unlimited internal supply; i.e., no longer a supply-sided limitation, but a limitation on the demand side (see Fig. 10.1). V_{max}' must be less than V_{max}, but otherwise is unrelated to it. The equations of the two branches of uptake are then:

$$v = k_1 S; \quad S \le V_{max}'/k_1 \qquad \text{\textit{Blackman, low S}}$$
$$v = V_{max}'; \quad S > V_{max}'/k_1. \qquad \text{\textit{Blackman, high S}}$$

The first branch is the substrate limited case (supply side limitation), whereas in the second branch growth is limited by the cell's internal physiology (demand side limitation). Of course, we do not know how far the evolutionary process for any given case in any given organism has gone, so one has to resort to refined experiments to establish the kinetics in particular cases.

But before we examine experimental results, first consider an intermediate formulation, which had been developed a year before the *Monod permease* equation by Jay Boyd Best (1955). It considers an enzymatic process in tandem with diffusion. I'll not go through the algebra, but the Best solution for the two processes working in concert is:

$$v = V_{max}' (S + K + J)\{1 - [1 - 4SJ/(S + K + J)^2]^{1/2}\}/2J.$$
$$\text{\textit{Best diffusion-enzyme}}$$

This is more complex than the *Monod permease* uptake equation. The extra parameter is J, which is defined as $J = V_{max}'/(AP)$, where A is the surface area of the bacterium and P is the permeability constant, i.e., the diffusion constant in the membrane divided by the thickness of the membrane. [Additionally, I have worked out the case for two enzyme-like processes in tandem (Koch 1967, 1982b), but it does not fit the data any better and it does have one more parameter, so we will only consider further the *Best diffusion-enzyme* equation.] When J is very small the equation reduces to the Monod equation and when J is very large it becomes the two limbs of the Blackman equation (see Fig. 10.2).

Sample curves for the gradation between the extreme models are shown in Fig. 10.2 and actual data for glucose consumption by *E. coli* are shown in Fig. 10.3. This set of experimental results suggests that the Best equation is indeed needed because an intermediate result between the Monod and Blackman predictions was observed. The value of J fitted from such data, because of its meaning of $V_{max}'/(AP)$, can be useful in dissecting the value of the three factors that make it up. More extensive use of the *Best diffusion-enzyme* equation will be invaluable to the next generation of microbial physiologists and microbial ecologists.

From a wider point of view, L. von Bertalanffy introduced general systems theory a generation ago to give a deeper understanding of the biology of growth; his modern counterpart is S.A.L.M. Kooijman who recently published a book (1993) which generalizes these ideas further to embrace "dynamic energy budgets in biological systems." These ideas are relevant here because the basic tenet of the various application of both general theories is based on the idea that uptake

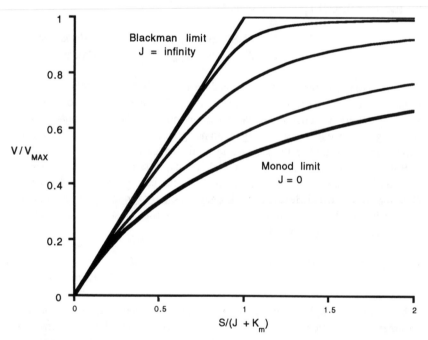

Figure 10.2 Dependence of uptake rate on concentration for different mechanisms. Sample kinetic curves for the *Monod permease, Blackman,* and *Best* equation for the uptake of a substrate by a bacterium are shown. The equations are given in the text. The *Best* equation is the more general and includes the other two when the parameter J has extreme values.

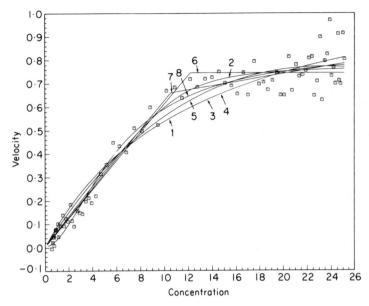

Figure 10.3. The growth rate of *E. coli* as a function of glucose concentration. A culture growing in a very low concentration of glucose was followed in a computer-linked spectrophotometer using a long light path and a flowing system so that the culture was aerated and maintained at constant temperature. The data have been fitted to the *Best* formula. The data do not fit the *Monod permease* equation, rather they are closer to the *Blackman* equation. The results actually were intermediate. Figure reprinted from Koch (1982b).

of nutrient is generally dependent on the surface area of the organism. Here we have been focusing not on the surface area but on the concentration of transport entities on the surface. In this case surface limitation is critical at very low substrate concentrations because then every molecule that hits the surface is taken up and consumed and therefore the surface area is limiting rather than the permease assemblies (Koch, 1971, 1982b, 1986b, 1990b, 1995d).

The Periplasmic Space and MDOs

Thirty-five years ago Heppel made an important discovery. If Gram-negative bacteria were given a strong osmotic down-shock (i.e., the culture diluted into distilled water) some proteins were released into the medium, but others species of protein were not. Why not all kinds? The riddle was solved when it was realized that the Gram-negative wall was a sandwich in which an inner membrane and outer membrane surrounded the thin, but strong, peptidoglycan layer. Also sandwiched in this space with the peptidoglycan was a special class of proteins.

During an abrupt osmotic down-shock, water from the environment rushes into the cell and causes it to expand. Under most conditions the combination of the inner membrane and the peptidoglycan fabric is strong enough so that the cytoplasmic membrane is pushed more strongly against the peptidoglycan screen-like network and not ruptured as the outer membrane is, and therefore cytoplasmic proteins do not leak out. On the other hand, water rushes into the periplasmic space, possibly because only water can enter faster than can salts and the other small molecules characteristic of the initial medium. Consequently, the outer membrane swells and ruptures, ejecting into the medium the proteins present in the periplasmic space. The outer membrane then usually reseals.

It is now appreciated what an important organelle the cell envelope is, i.e., the inner membrane, periplasmic space, sacculus, and the outer membrane. The outer leaflet of the outer membrane generally keeps out water-insoluble materials, whereas the inner leaflet keep out water-soluble materials. The porins traversing the entire outer membrane make exceptions permitting transit for the molecules of molecular weight less than 600-700 (depending on shape and charge). There are other exceptions to the outer membrane serving as an impenetrable barrier; these are due to the special outer membrane components facilitating the uptake of iron and of short glucose polymers (maltose, amylose, and degradation products of starch and glycogen). The outer membrane affords protection against many antibiotics and against the bile acids present in the gastrointestinal fluids (see Fig. 10.4). The periplasmic space contains digestive enzymes and binding proteins that aid in the uptake of molecules into the cytoplasm. Without the outer membrane these substance would escape, so it may be better to view the periplasm and the outer membrane as a single physiological entity that acts as a "selector" of which environmental components are allow into the atrium of the cell and as the "server" of these to the cytoplasmic membrane. Moreover, it is the cell's first line of defense.

Much work has been done to study various aspects of the cell envelope. The outer membrane lipid, carbohydrate, and protein components have been analyzed, and studied, and their roles identified. The periplasmic proteins were isolated and their biochemical and physiological properties identified. Many of them have important roles in speeding transport to the cytoplasmic membrane or as enzymes to hydrolyze molecules, such as nucleotides, to moieties that can be taken up and utilized. Later, it was discovered that small phosphate-containing polysaccharide materials, called membrane-derived oligosaccharides or MDOs, are formed from the inner membrane and accumulate in the periplasmic space. They were identified and shown to have a role in maintaining the osmotic pressure of the periplasm when cells were grown in low osmotic strength medium. They were large enough so they did not pass through the porins in the outer membrane. The MDOs, being fixed polyanions, also create a Donnan potential across the outer membrane (see Chapter 5).

There is something of a contradiction here. The outer membrane must be

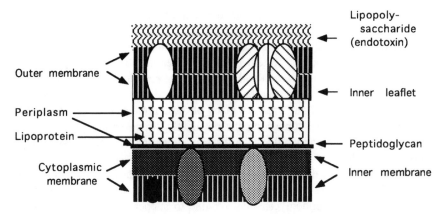

Figure 10.4. The cell envelope comprises a very important organelle. For a Gram-negative cell envelope, we should include the outer and inner leaflet of the OM bilayer, the lipoprotein, the stress-bearing M layer, the periplasmic space and its constituents, and the cytoplasmic membrane. The periplasm and the outer membrane effectively combine as one key entity that protects the cell from external chemicals and feeds the cell. The murein layer and the cytoplasmic membrane combine to resist the osmotic pressure difference with the environment and provide selective permeability. The OM contains porins, depicted as ovals, and the outer leaflet contains lipid A (the endotoxin) linked to strain specific oligosaccharides.

strong to resist the osmotic pressure difference. On the other hand, the outer membrane can evidently be ruptured by a sudden decrease in the osmotic pressure of the suspending medium, suggesting that it is not very strong. That idea that the outer membrane is weak is also supported by studies of model membranes and the very well-known phenomenon of the sloughing or blebbing of the outer membrane into the surrounding medium. The material from the outer leaflet of many Gram-negative outer membrane material is more commonly known as "endotoxin." When it enters into the circulation of a mammalian host it is the source of inflammation, fever, and shock. Based on a knowledge of its composition, one has no reason to assume that the outer membrane is especially strong. The inner leaflet of the lipid bilayer is made of the usual phospholipids. Phospholipid membranes are not very strong. Consider that this is the material that stabilizes the foam on a glass of beer; the foam has only limited keeping qualities though much better than the keeping qualities of foam on a glass of champagne. There is a special structure of the outer leaflet; it is composed of a special substance, lipid A, and a complex of specific carbohydrate chains that determines the specific subtype or strain of the species of Gram-negative organism. But these chemical features are not expected to give the outer membrane an exceptional resistance to tension. The ability to resist tension of the outer membrane must be many orders of magnitude less than that of the covalently linked peptidoglycan sacculus. The stress that the outer membrane can support quite evidently depends on the membrane's being able to transmit the stress to the peptidoglycan fabric underneath it.

Transfer of stress is done by a special protein, called Braun's lipoprotein (see Braun and Wu 1995); this protein holds the outer membrane to the sacculus. One end of this molecule is covalently linked to the peptidoglycan, and the other has three fatty acid residues that become an integral part of the inner leaflet of the outer membrane. There are very many of these proteins (100,000 per cell). Evidently, these "ties" connecting the covalently strong peptidoglycan to the weaker outer membrane would make the later more able to resist tension. This means that the higher the concentration of the lipoprotein ties per unit area of outer membrane, the larger the osmotic pressure difference that the outer membrane can resist without rupture. The area between attachment sites, if small enough, when supported by the peptidoglycan may be able to support a significant osmotic pressure differential. This may be the basis of the resolution of this apparent contradiction.

Because neutral salts and solute can equilibrate across the outer membrane and therefore do not contribute to the osmotic pressure. In the absence of the MDOs, the periplasmic proteins would constitute the main contributor under steady-state conditions. This osmotic pressure should be quite slight, however, because the number of protein molecules is small even though their mass, because of their molecular weight, is considerable. On the other hand, when grown in a low osmotic strength medium the MDOs are formed, and they and their counterions contribute a good deal to the osmotic strength of the periplasm (see Chapter 5 and also Chapter 13). As noted in the former, MDOs are formed only under conditions where they can effectively function to maintain an osmotic pressure. So even though the porins permit small cations and anions to permeate, an osmotic pressure and a "periplasmic turgor pressure" must develop across the outer membrane. This could be important because it may be necessary to maintain the periplasm's osmotic pressure higher than that of the cytoplasm, and in a growth medium with very few small solutes this would cause a problem. Also, when the organisms are growing in their normal habitat, the gut, there may be abundant lower molecular weight polysaccharides that cannot penetrate the outer membrane, but are sufficiently numerous to cause an osmotic differential between the environment and the periplasm, sucking water from the entire cell.

Morphogenesis

Although the shape of E. coli has been idealized by many workers as a cylinder with hemispherical caps, that is only a rough approximation. A more realistic description of its shape and its morphogenesis follows. Quite precisely in the middle of the cell a constriction starts to develop. The constriction, although it may start at a point on the wall, quickly spreads around the cell circumference so that the developing furrow is soon cylindrically symmetric. To begin with, the groove is very narrow. Even when the constriction is almost complete, the sides are nearly flat and perpendicular to the cell axis leaving very little gap between the two sides (see Fig. 10.5). Subsequently, as the developing poles fill out, the shape changes to a more gradual curvature, and at the time of separation the poles have bulged out to almost hemispherical proportions. On treatment with certain penicillins the separation can be delayed or aborted and the cell can bulge out at these regions. Under other conditions with certain penicillins, so-called blunt constrictions develop in which the constriction is U-shaped and not V-shaped (Nanninga *et al.* 1982).

After cell division is completed, the two daughter cells may adhere together and be counted as one or two, depending on the analytical technique. They

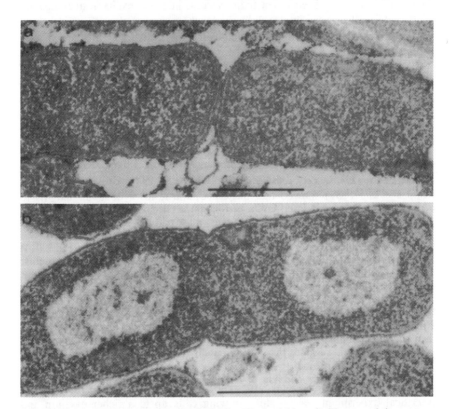

Figure 10.5. Electron micrograph of dividing *Escherichia coli.* Photograph is courtesy of Conrad Woldringh from the University of Amsterdam. The initially sharp constriction will alter shape and become more gradual as it progresses to form the two new poles, then the average shape of a pole will approximate a hemisphere. This change in shape can be directly observed in growing bacteria with the phase microscope. The hemisphere approximation is often used, but is too unrepresentative for many purposes.

almost always can be seen as two cells in the electron microscope and usually as two in the phase microscope, if high enough power is used. This poses a problem when a hemocytometer, such as a Petroff-Hauser chamber, is used to enumerate the cells. These chambers have a 20-μm depth and the highest power objective (100X) can critically focus only on the bottom of the chamber when a thinner cover slip than those supplied with the chamber is used. The problem is that the thinner cover slips are not as flat, thereby causing an error. However, this is only a minor error. The doublet problem is exacerbated if a Coulter counter type device is used; then two cells still connected together will be counted as one larger cell. The individual daughters increase in mass continuously as far as can be told (see Chapter 3). The new biomass in some small part may be accommodated by further bulging of existing poles, but not very much additional volume is created in this way. The major volume increase results from elongation of the side wall. By looking at fields under the phase or electron microscope, one may clearly see that the side walls of some individuals are very nearly cylindrical with constant width, but there are also individuals that taper significantly towards the ends. The statistical data of Trueba and Woldringh (1980) demonstrated that the diameter at the broadest part varies to a small, but significant, degree, as the newborn cell grows to become a cell that is just about to divide.

During balanced growth a new constriction starts the cell division process. The morphometric results from many papers from the Amsterdam laboratories of Nanninga and Woldringh showed that cell division is very precisely in the middle of the cell. One of these studies (see Trueba 1981; Grover et al. 1987; Koppes et al. 1987) showed that the CV of the ratio of daughter lengths to maternal lengths for certain strains is as small as 4%, and because experimental variation contributes to this value, the actual variation in the division process may be very small indeed. However, other strains may have a CV as high as 14% (Grover et al. 1987). Obviously, only the first type of strains are suitable for the study of the kinetics of cell division and the growth of individual cells. This range of behaviors of strains means that usually, but not always, division accurately bisects the cell. Clearly, however, division can be much more precise than seemingly it need be.

When cells in balanced growth are shifted into media supporting more rapid growth, the cells become wider and have a longer mean length. The cells do not increase in all directions simultaneously as happens upon blowing more air into a cylindrical balloon. Instead, the central part of the cylinder becomes wider. A critical experiment was to shift growing cells up to a richer medium and examine samples in the electron microscope. At the time of the first division event subsequent to the shift, the newly arisen cells have one fat end and one narrow end (Woldringh et al. 1980) (see Fig. 10.6). This implies that growth immediately after the shift has caused the cells to bulge outward in the middle. Thus, the cause of this peculiar morphology is that the old poles do not, or

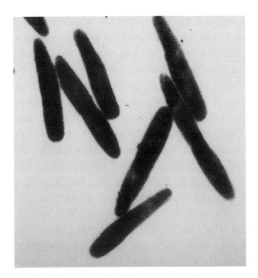

Figure 10.6. Electron micrograph of a "shifted-up" culture of *Escherichia coli*. The electron microscope photograph is reprinted from Woldringh *et al.* (1980). Before the shift, the cells were narrow and short. In the rich medium they start to bulge in the middle, so at the first division they have a narrow and a broad end. See text for a more detailed interpretation.

cannot, enlarge in radius, at least not very rapidly, and consequently the cell bulges. When a new constriction event occurs, the radius of the nascent pole is larger and is the definitive diameter for cells in balanced growth in the new medium. This process is reversible and permits widening and narrowing of the cell controlled by the nutritional status of the medium and is probably associated with a change in turgor pressure and/or the effective surface tension parameter T, which then defines a new width for the cells in the new balanced growth state. It must be noted that Cooper (1989) has proposed a constrained hoop model that glosses over the bulging in the middle of the recently shifted-up cell and instead focuses on the measurement of length and width. He proposes that the cells do not become wider during the first generation, but that because the critical size (amount) of cytoplasm at division shifts immediately, the cells in the first division have to grow to be longer.

High-Resolution Autoradiography with the Electron Microscope

This subsection is a methodological aside to allow the appreciation of the principles and methods of a very important and useful technique. Tritium is ordinary hydrogen in which there are two extra neutrons present within the nucleus, it is designated as T or ^3H. Because of its extra content of neutrons, it is radioactive and decays with a half-life of 12.3 years. It is valuable in biology because it serves as a tracer, not only for hydrogen, but also for carbon and other chemical species that have stably bound hydrogen atoms. Its special property is that it decays by emitting a very weak electron, or β-particle. Its energy is comparable to that of the electrons emitted by the electron gun in a black-and-white television set (maximum energy of ^{3}H is 18,600 electron volts). The β-particle can traverse a vacuum, but only a very short distance in biological tissue—no more than several microns. Thus, its site of incorporation into a biological structure is marked by nearby disposition of electronic energy causing biological, physical, and chemical actions. The method of detection that gives the most information on structure is autoradiography. For this method one prepares a very thin sample of the radiolabeled material. (Sectioning is done, in many cases.) The samples are covered with a thin film of photographic emulsion or a coat of liquid emulsion, which then solidifies. The sandwich of sample and film are then stored in a dark, dry, cool environment. When adequate exposure to the electrons has activated a sufficient number of silver atoms, the film is developed and then the sandwich examined. High resolution is achieved by using a very thin photographic emulsion and examination of the film and the underlying cell section in the electron microscope. Because of the thin emulsion, the efficiency is very low, but it yields photographic grains only near the site where the tritium atom decayed, becoming a helium atom and emitting the electron. The electron, being so light relative to atomic nuclei, takes a very erratic path through the forest of atoms in the biological sample; its movement is very close to that of a random "walk." It interacts with matter and loses energy by causing atoms in its path to become excited or ionized and in so doing transfers some of its energy. As a small part of these processes the electron loses energy to the silver atoms in the emulsion and activates them so that they can later form grains of silver during the photographic development process.

For high-resolution work the film must be thin, because then the β-ray has a good chance of activating only very nearby silver atoms. It then passes from the other side of the emulsion, and after leaving the emulsion it has almost no chance of bouncing back into the emulsion at some place more remote and generating adventitious silver grains there. With very thin layers the absolute best resolution can be 100 nm; with the technique used by the microbiologists in Amsterdam the resolution is nearly this good. The development and the electron microscopy are demanding procedures, but are now quite well defined and routine.

The Correspondence Between the Source Shape and Grain Distribution

The labeled isotope sources in which we are interested (rod-shaped cells) all have cylindrical symmetry. Universally, the position of grains of silver has been recorded based on distances measured along the length of the cell, while the position of the grain normal to the central axis of the rod-shaped objects has been neglected. This reduces the problem of interpretation to a one-dimensional problem. The β-particles from a point tritium source in biological material will radiate in all directions but will activate silver atoms only if the electron happens to enter the emulsion. Thus, most of the decays do not cause any silver grains to become activated. The blackening of the thin emulsion will be most intense immediately above the source. From the way that the intensity falls off with distance in the plane of the film from this spot, it is possible to calculate the distribution of grains around any biological source by considering the actual source as a collection of point sources.

Because the position of each grain is usually recorded as its distance along the X-axis, we need only consider the contribution of all point sources at one X-position on the number of grains at another X-position and can correct for the distance of the source or the grain normal to the axis. Thus, the function that describes the concentration of photographic grains with distance above a point source must be integrated (summed) over all distances perpendicular to the axis of the symmetry at a given distance along the axis from the point source. Once this is done, the expression must be integrated again to account for the contribution of all the decays emanating from points in a uniformly labeled, narrow, rod-shaped source of half-length h lying on the axis. This integration results in a formula for the number of grains at a chosen position along the length of the object contributed by decays emanating from the entire cylindrical source. The mathematics will not be given here (see Koch 1982d, 1983 for more details), but the final formula will. Let g_c, the numbers of grains from a cylindrical source, be the collective grain count at position X from a line source of half-length h, centered at position C. K is a constant that is a combination of many constants including the specific activity of the source, photographic efficiency, etc, and d is the half-distance of density of the blackening of the emulsion; i.e., the distance from "point zero" that reduces the grain density to half maximal. The equation is:

$$g_c = 2d^2K[\tan^{-1}\{(X - C + h)/d\} - \tan^{-1}\{(X - C - h)/d\}].$$
Grains from a cylindrically symmetric source

This equation solves the basic problem if the tritium atoms are distributed uniformly on a line source, a swath, or the cylindrical part of the bacterium. It does not solve the problem for more complex cases, such as when some radioactivity is present in the poles or when additional radioactive atoms are incorporated in the region where a pole is to develop. However, it allows the solution for more complex geometries to be gained by building them into a computer program. What must be done is use the *Grains from a cylindrically symmetric source* equation over and over again for different short cylindrical sources of half-length h; centered at different positions C; with different specific activities K; arranged in such a way as to simulate the source object and thus predict the distribution of grains. Usually we must work backward from the distribution of grains to see what distribution of tritium atoms in the source object would generate the observed pattern. Then, if we can relate the tritium content to the amount of biomolecules, we have accomplished our goal.

Early Autoradiographic Studies on Peptidoglycan Synthesis

Important autoradiographic studies came from an international cooperation between the laboratories of Uli Schwarz at the Max Planck Institute in Tübingen and Antoinette Ryter at the Pasteur Institute in Paris. With the mutant strain from Germany and the autoradiography done in France at the electron microscope level, it was possible to pulse-label cells with tritium-labeled diaminopimelic acid and thereby uniquely tag the peptidoglycan and see the relationship of wall growth with the cell cycle. An important fact emerged: zones in the middle of the side wall of many cells in the population were extremely active in wall formation during growth in both poor and rich medium (Ryter et al. 1973; Schwarz et al. 1975). The zones of darkening of the photographic emulsion by the radioactive decays were so intense that the contribution of synthesis in the rest of the wall seemed insignificant. These results, initially, seemed in accord with the replicon model of Jacob, Cuzins, and Brenner (1963).

As the surface stress theory was developed and the engineering considerations recapitulated in Chapter 6 became clear, I came to doubt this aspect of the replicon model. My doubts arose because the only elongation process that permits stable growth demands diffuse incorporation throughout the cylinder. Actually, these and other studies had noted that grains appeared other than at division sites and Schwarz et al. (1975) had noted that there was a "second channel." The conclusion that there was substantial murein added to the side wall was supported in the experiments of Verwer and Nanninga (1980) and later papers from the Amsterdam laboratory and by the lower resolution, light microscope, autoradiography of Burman et al. (1983). A reanalysis of all of the available data (Koch 1982c, 1983) led to replacing the idea that the middle of the cell was a "staging" area for later lateral movement with the idea that two types of growth were occurring independently. Actually, the surface stress theory demanded this because its second postulate is that the wall growth is forced by increases in cell cytoplasm. This idea has recently been stressed and elaborated by Cooper (1989b, 1991). Before constriction starts, the addition of side wall leading to elongation is the only way to accommodate the new cytoplasm, but because such wall growth does not accommodate as much cytoplasm per unit area of wall when constriction starts, it is necessary that additional wall be incorporated. This could be as a separate process where a constriction was just about to take place, or a constriction was actually forming. There, wall must be rapidly laid down. My published (Koch 1983) and unpublished calculations showed that in all the reported experiments the diffuse addition process was sufficient for the accommodating cytoplasmic synthesis through cell elongation. Consequently, cell division involves periodic, rapid, intense, localized wall synthesis at the site of constriction independent of the other, separate, more gradual and diffuse synthesis associated with cell elongation. Both processes, but mainly the latter, contribute to enlarging the sacculus to accommodate the new cytoplasm. Table 10.1 shows this point, but made again utilizing more recent experiments of Woldringh *et al.* (1987) (see Fig. 10.7) that carried out the autoradiographic analysis both on pulse-labeled cells and on sacculi. Because the sacculi are so thin, there is no problem with absorption of the tritium energy within the biological sample and thus the two types of samples complement each other.

More Recent Studies by the Amsterdam Group

Since the late 1970s the Molecular Cytology group at the University of Amsterdam has continued the work started by the French and German groups. Important technical advances in terms of bacterial strains, data collection, and analysis have been made. They were able to launch a massive effort in order to gain an

Table 10.1. Division of synthesis between the murein of polar versus side wall[a]

Class length	Number of grains		
	Total	Side wall	Constricting wall
Nonconstricting			
1.54–1.97 μm	393	393	
1.97–2.26 μm	487	487	
2.26–2.69 μm	397	397	
Constricting			
2.28–2.74 μm	301	50	251
2.74–3.04 μm	283	120	163
3.04–3.50 μm	138	38	100
Total Grains	1,999	1,485	514

[a]The silver grain distribution over sacculi prepared from a steady-state population of MC4100 grown at 28° C in a minimal medium. The cells were pulsed with ^{3}H-diaminopimelic acid for 15 min. Data taken from Woldringh *et al.* (1987) and recalculated. All the grains over sacculi without an evident construction were allotted to the side wall category. The grains over sacculi with constrictions were partitioned subjectively into side wall and constricting wall categories. The constricting wall category includes only the nascent poles. The previously formed poles have little label (Koch and Woldringh 1995). It can be seen that approximately 25% of the new synthesis is for the wall of the pole. This is expected from the average dimensions of the intact cells and the approximate assumption that the completed poles are hemispherical. For pulse-labeled cells it was shown (Koch 1983) that $\%P_p$, the percentage of the pulse in polar wall, is given by $100 \times r/L_b$, where r is the radius of the cell and L_b is the length of the newborn cell. The latter can be calculated from the formula given in Chapter 3 by dividing the mean of 2.45 μm by 1.4 yielding 1.75 μm. Therefore, $\%P_p$ is $100 \times 0.395/1.75$ or μ/1.75 μm or 22%.

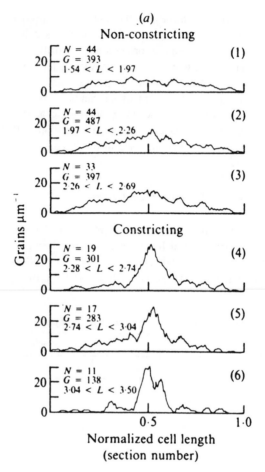

Figure 10.7. Autoradiographic analysis of the synthesis of new wall. Data from Wold-ringh *et al.* (1987) of silver grains as a function of normalized cell length. Pulse wall label yields more grains of silver over the central portion of the cell where constriction and division are to take place. The cylindrical wall grows all over but not at the high intensity of the nascent poles. The old poles incorporate label only very slowly and are essentially inert.

amount of data sufficient to be meaningful and interpretable. I will deal with three key papers.

SYMMETRY OF WALL SYNTHESIS

Verwer and Nanninga (1980) asked the question: Is wall growth asymmetrical? The underlying question was whether growth was apical (see Chapter 11). For this they pulse-labeled the same carefully chosen mutant strain used by the French and German workers, fixed and washed the cells, coated them with photographic emulsion, exposed and developed the silver grains, and then examined and photographed them in the electron microscope. This is the usual procedure, but their analytical procedure involved counting separately the grains over the two halves of the cell, calling the one with the most grains the "left half." They were interested in testing the unit model. This is a model that imagines that the cell elongated at only one end. There were clearly grains over both halves of the cell. But it was possible that the Poisson statistics of radioactive decay obscured the interpretation. Therefore, an indirect approach was needed because the distribution of grains is not so clear-cut that simple inspection of the autoradiographs suffices. So they tested the number of grains over the left member versus the right member against the prediction of random chance under the hypothesis that the two halves of cells actually had equal amounts of radioactivity and the data had been sorted to put the larger number in the left-hand bin. They concluded that the unit model failed and synthesis of wall occurred equally in both halves. I later reanalyzed their data (Koch 1983). My more refined analysis of the statistics indicated that there were 1.81 times more grains on one side than on the other; this was not enough to resuscitate the unit model. In that publication, I speculated that this might be caused as the result of off-center initiation of the site of constriction. I tentatively suggested that the initiation site was initially not precisely in the middle, that it might become more accurately centered as the cell cycle progressed. However, no studies to test this point have been reported. So while the unit model is rigorously excluded and uniform diffuse symmetric growth is likely, it is possible, but not proved, that constriction is initially started further off-center than the final division point of the mother cell.

HOW NARROW IS THE SITE OF INCORPORATION INTO THE MUREIN OF THE DIVIDING CELL?

Wientjes and Nanninga (1989) in their high-resolution autoradiographic studies took special care to measure the distribution of grains from a short pulse experiment around the site of highest radioactivity that marked where the next division was to occur; they found that the distribution was very narrow. This led them to support the "leading edge" model, previously suggested by MacAlister *et al.* (1987), and conclude that incorporation of new glycan into the growing wall was localized at the deepest part of the forming constriction. This is very much in accord with a narrow zonal model that I had investigated (Koch 1990g). Further calculations showed that the "leading edge" (also called "finite growth zone") model was not a possible alternative to actual variants of the Variable-T model (Chapter 6).

Let us ask: Do the autoradiographic data show that the zone of wall growth is restricted to the tip of the constriction? The answer to this question is No. In fact, insertion of new material could be taking place over most of the developing constriction. This can be deduced from the published graphs of Wientjes and Nanninga (1989) combined with the *Grains from a cylindrically symmetric source* equation. Wientjes and Nanninga published three distributions of grains from slightly constricted, medium constricted, and deeply constricted cell populations. Although there were grains all along the cells, this background could be subtracted from the region of the central intense zone and the width at half-height read off a xerox enlarged version of the published figure. At half-height the half-distance values were 109 nm, 100 nm, and 133 nm for the three types of cells. If the source were a point, then these values would be estimates of d. These values are very small as studies by physicists (Salpeter *et al.* 1978) indicate that a value of d of 100 nm is the smallest (the highest resolution) obtainable by this type of high-resolution technique. So two conclusions can be drawn: (i) that the Amsterdam workers have nearly reached the maximum resolution for tritium autoradiography, and (ii) that the zone of incorporation is indeed narrow. But is it narrow enough? Use of the computer program indicated that the zone had to be more than 80 nm (0.08 μm) for significant broadening of the apparent half-distance beyond the maximum resolution. In the last phase of cell division, if incorporation were taking place uniformly over two hemispherical poles, the zone of incorporation would be equal to the diameter, or twice the radius of the cell, or 0.8 μm. Thus, the zone of incorporation is about 10 times smaller than the dimension of the structure being formed. This would seem to support Wientjes and Nanninga's conclusion.

But there is a seeming contradiction between the narrowness of the incorporation zone and the electron microscope pictures and phase images that show that the shape of the pole alters over its surface as it is formed. The changing shape would be consistent with diffuse incorporation over the entire polar surface and the Variable-T model, whereas the "leading edge" would seem to be compatible with a narrow zonal model and no change once a region of the pole has been formed. There appears to be a way to resolve the paradox, which is that the constriction (consistent with the electron micrographs of Burdett and Murray (1974), Woldringh (1973), and other papers) is very narrow as most of the new wall is inserted, so that all the pole is the leading edge and not just the tip of the constriction, because it is almost a "leading flat surface."

THE RATE OF INSERTION OF PEPTIDOGLYCAN ALONG THE LENGTH OF A NONCONSTRICTING CELL

Perhaps the most definitive high-resolution autoradiographic studies are those reported by Woldringh *et al.* (1987). This careful work used a mutant that could not use diaminopimelic acid (DAP) for lysine manufacture, but could make DAP and also would use the tritium-labeled DAP from the medium. Thus, wall synthesis in this organism could be studied by short pulses during balanced growth without the perturbation imposed by washing the culture to remove the nonradioactive diaminopimelic acid, which had to be done with the mutant used in earlier studies. The published paper cited experiments to follow growth, etc., but importantly, it presented studies on whole cells as well as on sacculi prepared from the cells. The studies included length distributions and width measurements. Although the same conclusions were reached with both cells and sacculi, the combination of both types of experiments eliminated a series of possible artifacts. The studies with sacculi are particularly important because they are so thin that they are not subject to the criticism that radiographic efficiency varied depending on the location of the tritium atom. Another advance in these experiments was the use of very thin sheets of emulsion that were transferred to the top of the biological specimen. These sheets have a uniform thickness, whereas the alternative dipping process gives an uneven layer of emulsion. In the analysis and presentation of the data, they normalized the lengths of the cells. The results for the sacculi are shown in Fig. 10.7, where they have been fitted to the Variable-T theory as discussed below.

How Thick is the Gram-Negative Wall?

Analytical measurements of the amount of peptidoglycan per unit surface area are difficult to obtain. The first work was reported from Braun's laboratory (Braun *et al.* 1973; Braun 1975) utilizing both careful analytical techniques and well-controlled electron microscopy; they reported that the murein was three layers thick. However, there was a numerical error. At that time many biological journals were shifting their formats to standardize the units for small distances from angstroms (10^{-10} m) to nanometers (10^{-9} m). The paper as printed stated that the wall thickness was three layers, but when the error in units was corrected, the estimate turned out be 0.3 of a molecular layer, which would imply that the layer was incomplete. Various data obtained later, when expressed in these terms, vary a good deal up to a high of three layers, but I see no reason to doubt the original measurements (of course, after the numerical error correction) of 0.3-layer as valid as any of the later chemical determinations. It would correspond to a monolayer, but one in which the peptide chains are in an extended conformation and in which not all of the possible crosslinks have been formed. On the other hand, I see no reason to doubt the 3-layer estimates. They could accommodate a few newly attached chains that were not yet fully linked or stress-bearing and also accommodate glycan units that are no longer stress-bearing and eventually will be turning over. The best current analytical data are those of Wientjes *et al.* (1991), who conclude that the murein is basically monolayered.

The workers in Tübingen have put an interestingly different interpretation on the analytical data. Höltje and Schwarz and their colleagues chose the most generally accepted measurements at the time, which estimated the wall to be two to three molecular layers thick (at least in fixed preparations in which the wall is bearing no stress). They suggested that the inside-to-outside model such as I had proposed for *B. subtilis* (and described in the last chapter) would hold as well for *E. coli* with the modification that there are only a few, instead of many, layers. They further make the clever suggestion that an important aspect of Gram-negative wall growth is that the autolysins responsible for the cleavages permitting enlargement may be located in the inner face of the outer membrane (Höltje and Keck 1988). This hypothesis allows a conceivable solution for safe elongation of the Gram-negative cylinder without bulging or constriction. It is not a full solution because it is not evident how the enzymes would recognize only the stress-bearing wall that has stress-free wall fully linked to it.

Recently, there has been an important advance using neutron scattering to measure the thickness of the Gram-negative wall. Labischinski *et al.* (1991) employed neutrons for scattering studies. These studies are not trivial because a strong source of neutrons is needed, which required carrying out the studies at a reactor at the Brookhaven National Laboratory. Such a source was necessary because neutron beams have the resolving power to analyze atomic dimensions. Neutrons have an advantage over electrons or X rays in that they can be used to study suspensions of bacterial walls. This is because the sacculi can be suspended in D_2O instead of ordinary water. The deuterium, having twice the mass of the hydrogen atom or the neutron, causes much less neutron-scattering than do the hydrogen atoms of the peptidoglycan as the neutron passes through the sample. This is simply a matter of elementary physics: an object loses a maximum amount of its kinetic energy when it collides colinearly with another object of the same mass. Less energy is transmitted from the beam to either a lighter object, which is easily accelerated removing little energy, or a heavier object, which suffers little acceleration in the elastic collision. Thus, the D_2O solvent is almost transparent to the neutrons, whereas the peptidoglycan gives a strong scattering signal. The results of the critical experiments show the wall is largely (75–80%) a single layer. The remainder is most probably three layers thick. This experiment is very much in accord with the idea that the stress-bearing wall of Gram-negative organisms is fundamentally single layered, with non-stress-bearing regions in which new wall is being attached and regions in which older wall has yet to be excised.

Turnover of the Wall of Gram-Negative Organisms

The *Neisseria gonorrhoeae* wall is made and then degraded and fragments of peptidoglycan are excreted into the medium in large amounts. This wall turnover may be true of Gram-negative organisms generally. Goodell, working for the most part at the Max Planck Institut in Tübingen, presented experiments that are consistent with the turnover of the wall of *E. coli* (Goodell 1985; Goodell and Schwarz 1983, 1985). This work has been supported by recent work of Driehuis

(1989) in the Netherlands and newer work from Glauner and Höltje (1990) in Tübingen, but has been contradicted by Cooper (1988a) in Ann Arbor, studying another Gram-negative bacterium, *Salmonella typhimurium.* Subsequently, the latter worker found a small amount of turnover (Cooper *et al.* 1993). The measurement of turnover is not simple. First, it is hard to get a labeled precursor of peptidoglycan into the cell; it is even harder to get one that will be uniquely used for wall synthesis. To overcome these problems, double mutants in lysine biosynthesis were developed that prevented DAP from being formed or being used to form lysine. Those blocks cause the pools from exogenous non-labeled DAP to become so large that they prevent a sharp buildup of pulses of incorporation into wall after the administration of the tracer. Moreover, because the pool is large, a chase technique will not rapidly chase the label from the cell's metabolic pools and so radioactive walls continue to form. Parallel studies with *S. typhimurium* are quite different; DAP enters many times faster, presumably entering via a permease whose primary purpose is to pump in cysteine (Cooper and Hsieh 1988). This feature of rapid uptake greatly improves the sharpness of incorporation into the wall, but so far can be criticized because rapid uptake defeats the ability to chase the pool label back outside the cell quickly. Distinguishing between these two results is essential to understanding how the Gram-negative wall is enlarged. For the moment, the issue of whether wall turnover is general in Gram-negative organisms has not been not resolved definitively.

The Elasticity of the Sacculus of *Escherichia coli*

It is now well established that the peptidoglycan sacculus is the stress-resistant, shape-determining structure covering the prokaryotic cell. Its covalent structure makes the sacculus the largest macromolecule in nature and imparts to the bacterial wall the strength to resist tension. According to the surface stress theory, the ability of a peptidoglycan fabric to stretch is fundamental for the mechanism for the enlargement of the sacculus during growth, both for the side walls and the formation of poles. (It also gives flexibility to the organism.) The elasticity, as measured by Young's modulus, is the degree that a body is stretched (strained) by a given force (stress). At high enough stress, a body's elastic limit is reached and it cannot be stretched further without rupture of integral covalent bonds. The elastic enlargement of a cell's wall fabric depends on conformational changes that involve rotations and easily made reversible rearrangements of hydrogen bonds. Enlargement exceeding the elastic limit must involve rupture (hydrolysis in an aqueous environment) of covalent bonds.

Sacculi are too small for engineering materials testing methods to measure either their elasticity or their elastic limit. Although the sacculus has not yet been affixed at two ends and mechanically stretched, the relationship between sacculus net charge and surface area has been determined. Steven Woeste and

I studied sacculi of *E. coli* from cultures in balanced exponential growth (Koch and Woeste 1992). Although the swelling of fragments of the wall of Gram-positive organisms has been studied, no previous studies were made with the walls of Gram-negative organisms. The sacculi were prepared free of almost all nonpeptidoglycan materials and the mean surface area of the sacculi under different conditions was measured by low-angle laser-light scattering. The charge on the sacculi was altered with either acid or base, or by succinylating or acetylating the amino groups of the peptidoglycan. Additionally, the effect of saccular charge was reduced by raising the ionic strength of the medium, and the effect of secondary structure was reduced by breaking internal hydrogen bonds with high concentrations of urea.

The Acid/Base Chemistry of Peptidoglycan

At this point we need to make an aside to consider the acid/base chemistry of the components of peptidoglycan. In the tetrapeptide, the ionizable groups are one α-carboxyl group of the glutamic acid ($pK = 4.6$), one carboxyl group of the terminal alanine ($pK = 4.5$), one ϵ-carboxyl group of the diaminopimelic acid ($pK = 3.75$), and one ϵ-amino group of the diaminopimelic acid ($pK = 10.2$). Consequently, the net charge of the tetrapeptide ranges from -3 to $+1$ as the pH is changed. In the basic crosslinked peptidoglycan tetra-tetra dimer structure containing eight amino acids, the average of the charges of all ionizable groups can range from -2.5 to $+0.5$ per muramic acid residue. The pK of one of the diaminopimelic acids is changed because it is involved in peptide linkage. This of course eliminates the titratable amino group and raises the pK of the carboxyl group because it is no longer part of a zwitter ion. (The carboxyl group of the D-alanine of the other chain is also eliminated.) To calculate the charge range of the preparations of sacculi, additional factors must be included: (i) the other minor types of crosslinkings, (ii) the contribution of the 10% of amino sugars that are deacetylated ($pK = 5.7$), and (iii) the lysine residues (amino $pK = 10.2$) left after trypsin hydrolysis of the lipoprotein. Because there is one lipoprotein molecule per nine disaccharides, there is one lysine ϵ-amino group remaining attached for every nine diaminopimelic acid molecules after trypsin hydrolysis. These factors give a charge range from -2.6 to $+1.1$ per muramic acid, consistent with our measurements.

Preparation of Sacculi and their Derivatization

Because of the possibility that autolysis may occur during the time necessary to inactivate autolysins as the cells are treated with sodium dodecyl sulfate (SDS) and heat, Woeste and I developed a method equivalent to flash pasteurization. We felt that quick heating was necessary because as the cell becomes heated the rate of enzymatic hydrolysis increases rapidly and is halted only when the enzymes become thermally inactivated. We developed an apparatus so that cultures while still actively growing were pumped to a Tee, the other arm of which pumped 20% SDS at a slower rate. The mixture from the Tee led directly to a copper coil immersed in a rapidly heated water bath. In this way there was only a small fraction of a second between the time when the cells were in an aerated medium at 37 °C and the time when they were switched to 4% SDS at 95 °C. The lysate was concentrated, washed, and then treated with RNAase, DNAase, and trypsin at 37 °C, washed, and then stored at 5 °C. In some experiments the amino groups of the sacculi were altered with either acetic or succinic anhydride. Acetylation, by converting the single amine group to an *N*-acetyl group, prevents the nitrogen from becoming positively charged, and succinylation removes this amino function and replaces it with an ionizable carboxyl group. Now Woeste and I had the material, but we needed a method and a machine to study them. But first let me present an outline of the theory needed to interpret our light-scattering results.

Theory for the Calculation of Mean Surface Area of Sacculi

Low-angle laser-light scattering can be used to measure shapes and sizes of physical objects nondestructively; this includes the mean sacculus surface area from sacculi suspensions that appear almost transparent to the naked eye. This method was also chosen because the measurements are not affected by the fixation and drying artifacts common with electron microscopy. The theory and computer programs for calculating surface areas of ellipsoidal shells of revolution are briefly summarized in this section. To fully develop the theory from first principles could be the subject of a full university course.

Bacteria have dimensions comparable to the wavelengths of visible light. But light scattered from particles much smaller than the examining wavelength of light does not depend on the size and shape as shown by Lord Rayleigh a century ago. In that case, however, 10 parameters are needed to calculate the absolute intensity of the light scattered in a given direction. The Rayleigh formula for the case of very small particles is:

$$I = \frac{8\pi^2 r^6 n_0^4}{R^2 \lambda^4} \frac{[(n/n_0)^2 - 1]^2}{[(n/n_0)^2 + 2]^2} \, v I_0 \nu [1 + \cos^2(\theta)]. \qquad Rayleigh\,formula$$

In this equation I is the intensity of the scattered light, r is the radius of the spherical equivalent of the actual particle, n_0 is the index of refraction of the suspending medium, n is the index of refraction of the particle, v is the volume illuminated, I_0 is the intensity of the incident light, ν is the particle concentration, θ is the angle of observation, R is the distance of the observer from the sample, and λ is the wavelength of light *in vacuo*. The factor $[1 + \cos^2(\theta)]$ is the only one depending on the angle of observation θ. This factor contains two terms. Both are needed to take into account the two mutually perpendicular beams of plane polarized light that together comprise unpolarized light.

If the particle's radius is comparable to the wavelength of light in the suspending medium (i.e., where $\lambda' = \lambda/n_0$), then the shape, size, and orientation become very important. This is the case for many kinds of bacteria and, of course, for their isolated sacculi. If the particle's mass is too small to cause significant phase shifts of the light passing through it, then the *Rayleigh formula* applies when amended by multiplication with appropriate quantities called *P*-functions. There are available *P*-functions for a variety of shapes, sizes, and orientation of particles. For ellipsoidal shells of revolution, the appropriate *P*-function (Koch 1989a) depends on λ' and θ, defined above. Additional parameters include: β, an angle that describes the orientation of the ellipsoid; a, the semiminor axis; A, the axial ratio; and B, the ratio of the semiminor axis at the inside of the shell relative to that at the outside. This last factor is a measure of the thickness of the sacculus wall. Only the variables of this particular *P*-function have been mentioned; the function itself is quite complicated and involves many trigonometric functions (see Koch 1989a).

In spite of the large number of variables, light scattering is useful for size and shape determination by comparing light scattered in different directions from the same suspension of particles. In the expressions for the ratio of light scattered in two different directions, almost all the parameters in the *Rayleigh formula* cancel out. The computer program starts from a derived formula that contains only the elements that do not cancel and calculates the relative angular light-scattering intensities of a randomly oriented suspension of ellipsoidal shells of revolution. However, it also takes into account the elongation of the average cell over a 2-fold range during the cell cycle, computing the averages by repeatedly using the *P*-function for ellipsoidal shells with different values of the parameters corresponding to different orientations of the sacculi and different lengths for cells in different phases of the cell cycle. The program, as written, requires the insertion of the value of the axial ratio of the sacculi of newly divided cells, A, and the factor describing the thickness of the sacculi, B. These are readily obtained by measurement and from the literature and are $A = 2.45$ and $B = 0.995$; this value of B corresponds to the wall thickness being 0.005 of the radius of the sacculus. The computer program then makes the best match (least squares fit) of the observed light-scattering data and the calculated light-scattering intensities to get the best measure of the single parameter, a, the width of the ellipsoidal shell of revolution. The program then computes the mean area and the mean volume of the ellipsoidal shells of revolution. These values should apply if the sacculi are representative of a population of cells in balanced growth.

Low-Angle Laser-Light-Scattering Apparatus Assembly

This work was done at a time when there were almost no funds available to us, but we did have some equipment left over from a more affluent time. This included a helium-neon laser which emitted red unpolarized light at a wavelength of 632.8 nm, near the long wavelength limit of detectability of the human eye. The value of the laser light source is that it emits a beam of light that is very well collimated. The laser was placed in a cradle attached to a large board to which a large turntable, constructed from an old phonograph, was also attached. Sample suspensions were placed in a square quartz cuvette, which was positioned on a small, nonrotating metal platform centered on the turntable. (This was possible because, fortunately, the phonograph had been of the record-changing kind.) A small metal cradle attached to the turntable held an assembly of nested brass tubes 30 cm long. Light passed through only the central tube, which had an inside diameter of 0.97 mm and ended in a small silicon detector. Because the detector was very small it had little background noise. The important point for our use was that silicon detectors have an output that is very accurately proportional to the light intensity. A second point was that one cost only $5.00. The scattering angles were measured in 2° increments from 4–12° from the laser beam's direction of travel. Because light is refracted on exiting the cuvette, Snell's law was used to calculate the actual angle of light scattered within the cuvette.

Angular Light-Scattering Data and Measurement of Sacculus Area

The computer program (Koch 1989) is based on the acceptable assumption that the sacculi approximate the shape of prolate ellipsoidal shells of revolution. The observed angular dependence for light scattered by sacculi obtained from cells grown in rich medium (brain, heart infusion, BHI) or in a minimal medium with L-alanine as the carbon source is shown by the data in Fig. 10.8. The best fitting computer generated scattering curves are shown as the solid lines. The fitting program normalizes the curve so that it has a value of 2 (i.e., the value of [1 + $\cos^2(\theta)$] applicable for unpolarized light at $\theta = 0°$), by fitting a single scale parameter C. (C contains all of the constant factors of the Rayleigh formula given above, as well as some of those contained within the P-function). It can be seen that the light-scattering data match the fitted curves for average sacculus radii of 0.47 μm in the BHI medium and 0.29 μm in the L-alanine medium.

The size of growing Gram-negative enteric organisms *E. coli* and *S. typhimurium* depends on their growth rate and the osmolarity of the medium. Using identical conditions and using the same strain of *E. coli*, the doubling times reported by Zaritsky *et al.* (1979) were 20 min for BHI-grown cells, and 135 min for L-alanine minimal medium grown cells, compared with our data of 24 min and 146 min, respectively. Assuming that the cells were cylinders with hemispherical poles, the calculated surface areas from their measurements would be 11.25 μm^2 and 3.79 μm^2, with a ratio of 2.97. The best computer-fitted curve to the light-scattering data (Fig. 10.9) gave values of 7.9 μm^2 and 3.1 μm^2, with a ratio of 2.5.

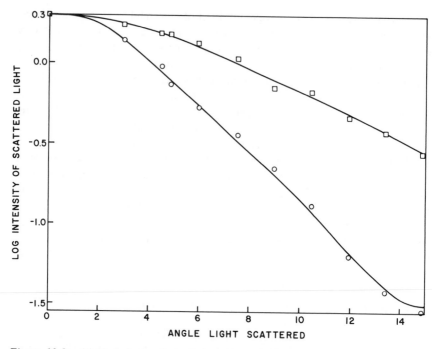

Figure 10.8. Light-scattering intensity versus scattering angle for BHI and L-alanine grown sacculi. Symbols: □ experimental scattering intensities for minimal medium L-alanine sacculi; experimental scattering intensities for brain heart infusion (BHI) sacculi, o. The solid lines are the best fits calculated form the light-scattering intensities under the assumption that the sacculi are a population of ellipsoidal shells of revolution. Measurements were made at 2, 4, 5, 6, 8, 12, and 14°. The angles at which the light was scattered after passage from the cuvette were corrected using Snell's law to give the angles at which the light was actually scattered while still in the cuvette. Scattering intensities at the corrected angles were compared to theoretical light-scattering intensities for the same angles. In the fitting process it was assumed from other data that the axial ratio of the cells at birth was $A = 2.45$ and that the wall thickness was 0.005 of the radius, fixing B at 0.995. Both the calculated and experimental light-scattering values were normalized so that the scattered intensity at $\theta = 0°$ was 2. The best-fitting, computer-calculated radius and area were 0.47 μm and 11.25 μ², respectively, for the BHI-grown sacculi and 0.29 μm and 3.79 μm², respectively, for the L-alanine-grown sacculi.

Mean Surface Area as a Function of Charge, Ionic Strength, pH, and Acylation of the Sacculi

From the results shown in Fig. 10.9 and Table 10.1, it can be seen that the sacculi of *E. coli* could be expanded and contracted reversibly over a 4-fold range in area. This may not be so surprising because molecular models of the tail-to-tail crosslinked dimer peptide connecting peptidoglycan chains can be

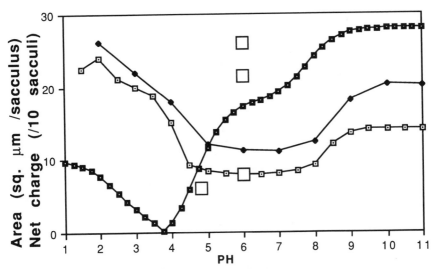

Figure 10.9. The mean sacculus surface area as a function of pH. As the experimental data were recorded, acidic titrations started at pH 6.0 and ended at pH 1.5 and the samples were back titrated; similarly, basic titrations started at pH 6.0 and ended at pH 12.5 and then were back titrated. The light-scattering measurements under different conditions were converted into areas of ellipsoidal shells of revolution and are reported as μm^2. The mean sacculus surface area is shown as a function of pH in the absence (small open squares) and presence (filled diamonds) of 10.1 M urea. The large empty squares ranging from top to bottom are: succinylated sacculi, acetylated sacculi, untreated sacculi at the same pH, and untreated sacculi in 1 M KCl. The filled small squares are for the theoretical net charge, which was computed from the average charge per sacculus calculated from the composition of the sacculi and the pK values of the ionizable groups. Reprinted from Koch and Woeste (1992).

expanded or contracted to this degree. The mean sacculus surface area in high ionic strength media (1 M KCl) at neutral pH was comparable to the mean surface area of intact cells grown under the same conditions used by Zaritsky *et al.* (1979).

The mean sacculus surface areas of BHI grown cells tested under various conditions are presented in Table 10.2. They are expressed as the average area of the sacculi. The smallest mean surface area was observed under isoionic pH conditions at high ionic strength. Under these conditions the sacculi are still ellipsoids of revolution as determined by comparison with the electron microscopic observations. Succinylation of sacculus amino groups converts them to residues containing carboxyl groups yielding a charge of −2.6 per muramic acid residue at pH 7.0. From the pK's of the titratable group, there is essentially no positive charge at pH 12.5, and no negative charge at pH 1.5. At pH extremes, and after succinylation, the mean sacculus surface area ranged from 20 to 26

Table 10.2. Mean sacculus surface area after various treatments[a]

BHI grown sacculi[b]		BHI grown whole cells[c]	
	μm^2		μm^2
pH 4.8	1 M KCl 6.1	Agar filtered OsO_4 .... 6.9	
pH 1.5	H_2O 22.5	Agar filtered Glut 8.1	
pH 1.5	Urea 24.5	Phase contrast 8.6	
pH 12.5	Urea 20.4		
pH 6.0	Acetylated 21.5		
pH 6.0	Succinylated 25.9		

[a]The mean sacculi surface area measurements from the light-scattering measurements of Koch and Woeste (1992) and the mean estimates of the surface area of bacteria from the electron and phase microscopic data of Woldringh (Zaritsky et al. 1979). All refer to experiments with strain B/r H266 grown in brain heart infusion (BHI) media at 37° C with aeration.

[b]Growing cells were harvested, lysed with SDS, purified with trypsin, washed extensively, and analyzed by the low-angle light-scattering method developed in my laboratory (Koch and Woeste 1992).

[c]Data from Conrad Woldringh for agar filtered cells fixed either with osmium tetroxide or glutaraldehyde and by phase contrast microscopy in the living state (Woldringh and Nanninga 1985).

μm^2, 3- to 4-fold greater than the isoionic (minimum) area. The important generalization is that roughly the same expansion of the peptidoglycan fabric can be obtained in a number of different ways, suggesting (consistent with the molecular models) that the maximum expansion of the fabric without breaking covalent bonds is 20 to 26 μm^2 of the BHI grown cells.

pH Dependency of Size Changes of the Sacculus

Figure 10.9 shows the net charge and mean sacculus surface area from pH 1.5 to pH 12.5. It should be noted that the net charge calculated from the pK's and amounts of different titratable groups in the wall preparation is zero at a pH slightly below 4. This is consistent with the isoelectric point of peptidoglycan preparations, and implies that placing a positive charge on the isoelectric sacculus with acid causes the dramatic expansion. On the contrary, as the pH is increased, the first two negative charges are not effective in expanding the wall. The third negative charge, however, causes an expansion, though not quite as extreme as that caused by acid. The implication is neither the carboxyl groups on the glutamic acid residues or the free carboxyl groups of D-alanines or the ε-carboxyl of diaminopimelic acid residues whose adjacent amino groups are engaged in peptide bond formation have much effect on the size of the sacculus. On the other hand, the removal of a proton from the ε-amino group of the zwitter ion to leave only the ε-carboxyl ionized of the diaminopimelic acid leads to expansion. Therefore, the expanded form depends on a net charge on the ε-regions of the DAP residue independently of whether that charge is positive or negative.

Estimation of the Normal Stretch of the Peptidoglycan Fabric During Balanced Growth

An essential feature of the surface stress theory as applied to Gram-negative and positive rod-shaped bacteria is that a newly inserted and linked element of peptidoglycan, when subject to stress, should expand. Although stretching is expected, the degree is not predicted by any theory, although the chemical models and the light-scattering experiments of the previous section suggest that a factor of 3- to 4-fold is possible, if not impeded by the inner or outer membranes. This section is devoted to the question concerning the state of expansion of wall fabric in the growing cell beyond its relaxed configuration. The analogy could be made to a stocking stored in a drawer versus one stretched on a small or on a big leg; the stocking accommodates to all these situations. Several ways have been used to measure the *in vivo* expansion (see below). They give a much smaller number than that obtained by *in vitro* experiments with preparations of sacculi discussed above (see Fig. 10.10). This figure shows the unit structure of peptidoglycan network, the tessera, in a version expanded when stretched in all directions.

Stopflow Turbidity Measurements

Many studies have been reported in which the osmotic pressure of the medium in which bacteria are suspended is shifted and the change in optical density or some other measure of cell size is followed. Most of these experiments can be criticized because the experiments take so long to execute that it is possible that the bacteria adapt fast enough to the new external osmotic pressure so that the measurements are confounded by the change in internal osmotic pressure. Many of the published studies (see Koch 1984b for an extensive reference list) are difficult to interpret because of the limitations of the analytical technique employed.

One study cannot be faulted on the grounds that the bacteria may have adapted to the altered osmotic environment, because measurements were made with sufficient rapidity to allow the initial events to be observed. In this study (Koch 1984b) a stop-flow device was used in conjunction with a narrow-beam spectrophotometer. The narrow-beam instrument allowed clear interpretation of the turbidity measurements in terms of the contraction of the bacteria. Cells growing under balanced conditions were pumped to a Tee and mixed with the same growth medium supplemented with a substance to increase its osmotic pressure, such as NaCl and various nonutilizable sugars. It was found that pentoses, such as ribose, xylose, and arabinose, were useful for such studies because they would not penetrate the cytoplasmic membrane and had a low enough molecular weight so that they changed the index of refraction to a smaller degree than did nonutilizable hexoses or sucrose. NaCl contributes only a small amount to the index of

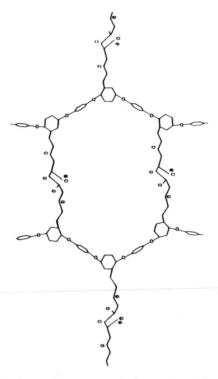

Figure 10.10. Stretched view of a surface enclosing portion of the sacculus, the tessera. Two oligosaccharide strands are shown crosslinked through their peptides at two sites eight saccharide units apart to form the unit of structure called a tessera. How the structure would appear upon being stretched in all directions in the plane is shown.

refraction, but the fact that it is highly ionized raises questions concerning ionic strength effects.

On the assumptions that the normal turgor pressure was 5 atmospheres and that when the external osmotic pressure was raised to match the cytoplasmic osmotic pressure, the cell would shrink but would not plasmolyze, and on the further assumption that the wall was isotropic, one could calculate the stretching in area. It was found that the mean area decreased to 80% of its former value with a shift of 5 atmospheres. This finding corresponded to the wall of the cells initially being stretched by 11.8%; i.e., $1/\sqrt{0.8}$.

Shrinkage of Growing Filaments when the Membranes are Ruptured

A fully independent approach that directly measured the degree of stretch of the peptidoglycan fabric was developed based on measuring the length of growing cell filaments (Koch *et al.* 1987). Normal *E. coli* cells are not long enough to be measured accurately in the phase microscope, but filaments can be long enough to allow their individual lengths to be accurately measured. Usually the light microscope is thought to be accurate to only a few tenths of a micron. Consequently, cells that are only 2–3 μm long would be measured with poor accuracy, whereas filaments 7–20 μm long could be measured with sufficient accuracy. The second hurdle was to immobilize a filament sufficiently well under the microscope to allow following its size while it was being perturbed experimentally. So the adopted protocol used a temperature-sensitive strain TOE13 of *E. coli* strain K12 that is *fts*A. When grown at the permissive temperature it divides normally, but at 42 °C it grows as a filament. After two to four doublings at the higher temperature the cells were introduced into a rectangular capillary that had been glued with epoxy to a microscope slide. Capillaries were used that had been previously treated with a polylysine-containing solution. This polycation binds to the negative charges of the glass and has positive charges left over to bind with the negative charges on the outside of the bacterial filament. Then a filament was chosen for study in which the entire filament was in view and in the plane of focus of the phase microscope. Even though bound, the filamentous cell continued to elongate normally.

For the experiment, the fluid in the capillary was changed to one that would disrupt the cellular membranes. For our purpose, an agent was needed that would disrupt the inner and outer membranes enough to allow leakage and so reduce the turgor pressure to zero. Solutions of SDS of a range of concentrations in the growth medium were used to destroy the integrity of the cell membranes. The filaments were observed and photographed as they shrank. Before we carried out the experiments, we had to anticipate that it might take some time for the shrinkage to be expressed because of the binding by charge-charge interaction with the polylysine holding the outer cell membrane to the glass slide. Therefore, there might be some time for the weak bonds to open and to reform, allowing "creep" to the equilibrium position. The shrinkage was slow enough to be measured, but fast enough so that the length became stable in a few minutes. The average shrinkage measured in this way was 17%. If the circumference shrank by the same length, it was concluded that the area of the sacculus of the growing cell was 45% greater than in the relaxed state.

Significance of the in vivo *Expansion of the Sacculus Above the Relaxed State*

Because both approaches yield much smaller degrees of expansion (20% and 45%) above the presumed relaxed state *in vivo* than are observed by the physical and chemical manipulation of 300% of the purified sacculi preparations *in vitro*, there is a quandary. The possible errors in the two *in vivo* methods are quite different. The contraction of filaments might be underestimated because of the adhesion to the microscope slide, but that is not a problem with the stopped flow. The stopflow involves physical theory and corrections for the contribution to the index of refraction of the medium due to the substance added to increase the osmotic pressure of the suspending medium; an error could be introduced during this process. The filament contraction method is not affected by distinctions between the periplasm and the cytoplasm, but there is some ambiguity in the light-scattering case. (Light scattering measures the biomass contribution above the water content, and even if the cytoplasmic membrane plasmolyzed and the outer membrane did not alter shape or size, there would be little change in the light-scattering signal unless substances leaked out of the cell.)

Because we now know that when the cell shrinks the lipid bilayers are not capable of shrinking, the wall envelope must wrinkle. Therefore, with either the osmotic challenge or the detergent treatment, the murein layer may not be able to achieve its equilibrium, relaxed conformation and the "smoothed" surface area remains larger than the murein when freed of attachments to lipid membranes.

Generation of the Rod Shape and Development of the Pole

From this point in the chapter a synthesis of two theoretical models (the Variable-T and the Three-for-one) that concern different aspects of growth will be presented. Of course, our understanding is still far from complete, but the outlines of the Gram-negative growth process are emerging. Both elongation and cell division of thin-walled bacteria must involve adding new wall to the old in such a way that only precisely regulated and targeted cleavages are allowed. For safe growth a list of important questions can be posed: (i) What are the physical mechanisms that make both the elongation and constriction reliable? (ii) What molecular mechanisms are implemented for stable growth? (iii) What details of enzymatic reactions and choice of reaction pathways make growth stable? (iv) What forces maintain the cylinder shape? (v) What circumstances lead to the invagination? At present there are two relevant models: the Variable-T model (Koch 1982a, 1982c; Koch et al. 1982a, 1982b, 1983; Koch and Burdett 1984) and the Three-for-one model of Höltje (1993) for both elongation and division. These two models are not alternatives, but represent different aspects of the entire process.

The Problems Caused by Being Thin Walled

Growth and maintenance of a rod shape for a thin-walled, single layered organism present apparently insoluble paradoxes that are indeed confusing. Having rejected the idea that mechano-proteins and enzymes act to produce a cylindrical shape via a cytoskeleton; having rejected templating mechanisms, having rejected the inside-to-outside mechanism that functions in the Gram-positive rods; and having rejected several more a priori possibilities, there seem to be no options left. Worse yet, if only physical forces function, the fact that hoop stress is greater than axial stress would seem inevitably to lead to rounding up into a spherical shape. On the other hand, it is undeniable that the Gram-negative organism elongates over its entire cylindrical region and is so thin that wall units must be inserted into the stress-bearing wall and then quickly come to bear stress. Based on the observation that the peptidoglycan layer is thin, the idea of insertion emerged (Weidel and Pelzer 1964; Verwer 1979; Koch *et al.* 1981b; and Labischinksi *et al.* 1983). Of course, with the development of the surface stress theory and appreciation of the necessity of utilizing make-before-break strategy for safe growth, it was necessary to reverse the break-before-make order postulated by Weidel and Pelzer.

The Contrast with the Gram-Positive Strategy

It was shown in Chapter 6 that the physics of elastic solids will not provide a workable mechanism due to inevitable bulging, but those of plastic fluid membranes would work under proper constraints. To see the problem clearly, consider again the Gram-positive rod solution to the problem. A quite reasonable mechanism to implement this set of constraints was presented in Chapter 9. It was that a layer of new wall was added uniformly over the inside of the extant wall. Little orientation for the polymerization can be directed by the stresses in the solid wall immediately above it. Then as the wall elongates, its ability to expand has been expended in the axial direction and is no longer available for hoop direction expansion when the layer of wall moves farther outside to where it is no longer totally supported by more peripheral wall. This preemption concept is an extension of earlier ideas of Previc (1970). At this point, the wall is in part supported by helical bands of still-older wall that give it a mechanical strength and balance out the 2-fold greater stress in the hoop direction that would be felt by a cylindrical shell pressure vessel of uniform material and thickness. (Mechanical engineers frequently reinforce hoses by winding strong wires or straps helically around the hose. Usually, two coiled wires winding in different directions are used. The pitch of the helix is determined by the fact that the hoop stress is twice the axial stress; it then turns out that the optimal angle is $\tan^{-1} \sqrt{2} = 54.7°$.)

The Gram-Negative Strategy

How can this series of circumstances be modified so that fluid membrane theory applies to the Gram-negative rod? Clearly, the stronger hoop stress must be overcome. Just as clearly, the same bond oriented in the hoop direction must be more susceptible to autolysin action than one oriented in the axial direction. So the secret has to be that the crosslinking of new wall, which certainly is independent of the stresses in the functional wall, is either at random or in an orientation so that the growing wall behaves as if it were a plastic fluid membrane.

Evidence for partial orientation of the wall has been presented by Verwer *et al.* (1978, 1980). These workers showed that if the sacculi were partially loosened or degraded with ultrasound or with endopeptidase, the sacculi became frayed. Their electron micrographs showed a frayed or shredded appearance with the residual strands going around the cell circumferentially. This indicates that cleavages by these agents were preferentially axial. The orientation of strands was by no means regular. The lack of regularity is strong evidence that a template mechanism does not function to maintain the diameter of the rod, but the existence of partial order could be key to an explanation of how the growth process behaves in a way to simulate the Young-LaPlace equation. A relevant point has been presented by Labischinksi et al. (1983) based on the fact that the peptide chains are extensible and the glycan chains are capable of very little extension. So if the chains were always inserted with the glycan chains oriented in the circumferential direction, this would help in solving the elongation problem and help in combatting the bulging problem. As mentioned, however, it would not solve either problem totally. Pressure-volume work must be converted into surface tension-area by the tension dependence of the autolysins, but the shape of the area must be defined by the arrangement of the crosslinks that were made prior to the cleavage events that permitted enlargement.

The Variable-T Model for Gram-Negative Pole Formation

We start from the fact that the *Young-LaPlace* equation guarantees stably cylindrical elongation given the four conditions specified earlier. If $P = T/r$, then a cylindrical surface is generated, but in order for wall growth to lead to constriction, the relationship must become $P > T/r$, not all over, but only in the middle of the cell where constriction is to take place. As before, r is the radius of the cylindrical side wall portion. Because P must equal T/r in the region that is to remain cylindrical, then the way that constriction and then division could occur would be for the cell to do something to make T become smaller in the central portion of the cell wall than it is in the rest of the side wall; this would force constrictive change. Before we explore this strategy, let us administratively exclude another alternative, viz., a local change in P. P cannot change because the hydrostatic pressure inside the cell must be the same at every point inside the cell; this is true of any fluid body in a closed vessel.

As discussed in the section on autoradiography, the rate of wall formation is intense where division is about to occur. A local increase in the rate of insertion of peptidoglycan units through the membrane would not by itself increase the rate of formation of surface area because of the limitations set by the autolysins. Somehow autolysins must locally increase their activity to favor wall enlargement.

Different organisms exist that grow in various shapes; these could be achieved by the use of the Variable-T model (Koch et al. 1982; Koch 1982a, 1982c) if the cell could somehow alter T in particular regions of the cell. These papers just cited attempted to account for a number of phenomena dealt with in various parts of this book, such as apical growth (Chapter 11). Primarily, however, the model was devised to account for the fact that the constriction in Gram-negative rods forms as a very narrow V-shaped structure that continuously changes shape as the sister poles develop and finally separate. Such a change in shape indicates that wall growth continues over the entire pole as it forms and is quite different from the case for the Gram-positive coccus, *Enterococcus hirae.* This sequence of events was largely deduced from the morphological studies of Conrad Wold-ringh (see Fig. 10.5).

POTENTIAL CELLULAR MECHANISMS THAT COULD ALTER THE SURFACE TENSION PARAMETER

For the Variable-T model to lead to cell division, the amount of work needed to make a new unit of surface area must be smaller in the constricting regions than in the cylindrical regions. There are a number of possibilities that must be considered (a long list is given in Koch 1982c) and that list should be amended to include the observation of Bi and Lutkenhaus (1991) that the *ftsZ* protein accumulates at the constriction site. These possibilities all presume the ability of the cell in some way to alter wall metabolism in local regions. One possibility is that the proton pumps that function in the chemiosmotic energy generation process are either localized or function to a different degree in the annular region where cell division is to occur. Such localization would affect the local pH, which in turn would lower the free energy of the transpeptidation reaction because of the pK's of the reacting amino acids in the tail-to-tail formation of the peptide bond and thus the biochemistry of the polymerization process. There is ample precedent for localized currents in microorganisms other than bacteria. The alga *Fucus* and the water mold *Achylia* are the prime examples (see Harold 1990).

Another possibility is that several glycan chains are linked together before the array, or raft, of nascent chains is inserted to enlarge the surface area. This will decrease T because the insertion event into the stress-bearing wall increments the wall by a larger area (Koch and Burdett 1984). This would lower the amount of pressure-volume work to force the autolysin to cleave stress-bearing peptidoglycan. This raft formation idea was incorporated into the idea of the "smart" autolysin model (Koch 1990c) to allow safe enlargement of both the side wall and the pole and modified in many important details in the new "Three-for-one" model.

Although a number of mechanisms can be imagined that would change the value of T, unfortunately, none has yet an experimental basis. I hope that it will soon be possible to choose between the array of hypothetical ones or to find and demonstrate an alternative process. Elucidation of this key problem is important not only for general microbiology, but also because a new generation of antibiotics may result that are selective against Gram-negative bacteria.

CALCULATION OF THE SHAPES OF THE DEVELOPING POLE

Shapes can be calculated by formulae developed from the *Young-LaPlace* equation by the calculation of variation. The mathematics is quite sophisticated (see Koch 1982c, 1983, 1984d). The equation that can be used for cases in which T does vary is:

$$(1 + S^2)/(1 + S_0^2) = \{(1 + S_0\,\Delta/r_0)/([T_0/T] + [P\,S_0\Delta(1 + S_0^2)^{0.5}/T])\}^2 \,.$$

Variable-T

In this equation S_0, r_0, and T_0 are, respectively, the slope, radius, and value of T at a point on the surface. The same symbols without the subscripts are the values of these variables a distance Δ along the axis of symmetry. The computational strategy is to chose the values of S_0, r_0, and T_0, formulate a rule to calculate T, use the *Variable-T* equation to calculate S, then calculate r from:

$$r = r_0 + S\Delta.$$

In the next step the new values are substituted to make a new set of S_0, r_0, and T_0 and the calculation repeated. The computer program can keep T constant, make it change discontinuously, change progressively, or respond to current slope or radius. Fig. 10.11 shows the shapes generated for a progressive change in T in the constriction zone wall relative to that of the cylindrical wall.

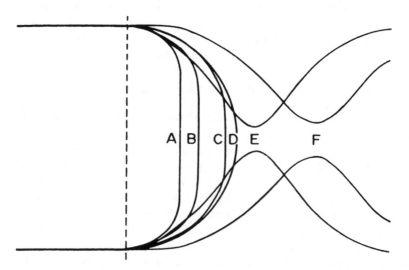

Figure 10.11. Pole shapes calculated by the Variable-T model arranged in a way to simulate the formation of the Gram-negative pole. Reprinted from Koch (1984d).

New Models for Growth of the Gram-Negative Wall

Several new models have been proposed and named: the allosteric, the holoenzyme, and the three-for-one. The last is the culmination of the previous two and has the virtue of ensuring safe growth of the Gram-negative cell and being consistent with many details of mechanical constraints in the wall's chemical structure, enzymology, and the growth physiology of *E. coli.* Some aspects of the two earlier models will be given (Fig. 10.12) because they emphasize important problems.

ALLOSTERIC MODEL AND "SMART" AUTOLYSIN

To apply the surface stress theory to the monolayered wall of Gram-negative cells, it was reasonable to suggest that the nascent glycan chains were created outside the cell membrane but inside the murein layer and each would form multiple peptide bonds to different parts of the existing fabric essentially at random; i.e., whenever a donor nascent peptide happened to interact with both the transpeptidase and an acceptor from the existing fabric. Then selective cleavage of stress-bearing bonds would pull the nascent chain into the fabric and enlarge it. But there is a problem: What if the cleavage enzyme cleaved existing stress-bearing wall in such a way as to yield an unfilled gap? Consequently, it was suggested (Koch 1982c, 1990c) that the autolysins that performed these cleavages would have to have an allosteric site in addition to their catalytic one to control cleavages so that rupture of the cell fabric would not be a consequence. The regulation of such enzymes would have to be very sophisticated because the regulatory site would have to recognize a nearby tail-to-tail linkage that was in an unstressed conformation in order to arm or activate the catalytic site. It is essential not to activate it to cleave a similar stress-bearing crossbridge that is not so "protected" by nascent linked but unstretched wall. (I have thus defined "protected bridges" as those bonds that are under stress and are adjacent to completely crosslinked although unstressed peptidoglycan dimer chains.) Then autolysis will result in the transference of stress when the cleavages occur. To be able to reliably make such distinctions, an enzyme would have to be very "smart"; in fact, it would have to possess properties beyond those of any known enzyme.

An important property of this model, which in the long run may lead to its revival, is that it would decouple the synthesis of wall from the existing stresses. New wall would be added at random and not in relation to the stress in the existing wall. This is essential for the maintenance of the cylindrical shape during growth of the wall to occur via the *Young-LaPlace* law for the plastic mode appropriate to soap bubble physics and not via the dictates of the hoop and axial stresses demanded by the physics of elastic solids (Koch 1993d).

Subsequent to its proposal a serious problem was recognized—the large distance between the old and new peptide strands. In Fig. 10.13, a portion of a

A. Allosteric Model

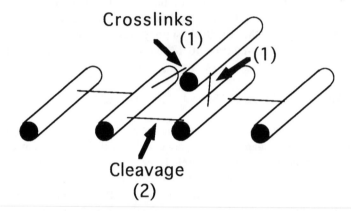

B. Holoenzyme Model

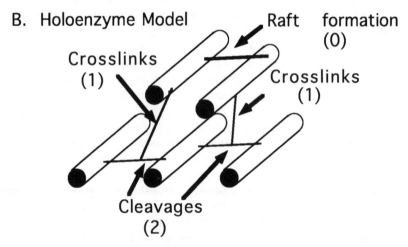

Figure 10.12. The Allosteric and Holoenzyme models. Two early models are shown that preceded the "Three-for-one" model. The allosteric model assumes that new oligopeptides are linked to stress-bearing wall and that a very "smart" enzyme is selective and only cleaves the stress-bearing wall if the region is protected by having a crosslinked but unstressed oligopeptidoglycan chain present. The holoenzyme model presumes that a raft of two nascent chains is linked together, then this assembly is linked to stress-bearing wall, and only then are cleavages carried out, liberating the central chain and replacing it with two new chains. This was originally suggested to account for turnover of the wall.

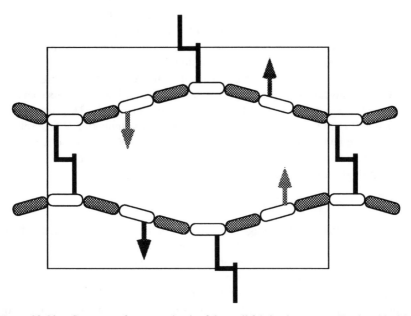

Figure 10.13. Structure of a stressed unit of the wall fabric: the tessera. Depicted inside of the boxed area are two oligopeptidoglycan chains connected by crosslinking peptides in two places to each other. The model is based on the supposition that it takes four disaccharides to allow the glycan chain to twist through a complete rotation. The glycan chain twists very little from saccharide to saccharide because of bulky lactyl groups. If the lactyl groups and the peptides were removed from the muramic acids, then the product would be chitin. Chitin makes a complete twist every four sugar residues instead of every eight. Because of the reduced amount of rotation of peptidoglycan from the unit outlined by the box, there are two peptides that extend outward in the plane to be part of the adjacent units. In addition, a pair of peptides points upward, which we can take as the direction of the outside of the cell. The other pair would face inward and form bonds to more central murein. Neither pair would be part of the fundamental ring unit of the tessera. The unit is shown in its most extended conformation. This means that few of the peptide bonds are in their lowest energy conformation; i.e., the conformation expected if the unit was not part of a fabric. The structure as depicted is inferred by the experimental results for the conformation of ring units of isolated sacculi at extremes of pH or in the chemically modified sacculi. In an intact growing cell, a less expanded structure exists; it is expanded over the most compact state by 45%. If the cell membrane is ruptured, then the unit would contract further as the covalent bonds adjusted to their lowest energy, subject still to the constraint that the wall is a covalent fabric. This relaxed isoionic form would have about one-fourth of the area of the unit depicted.

monolayer of wall is shown diagrammatically in planar section; i.e., all residues shown are very close to one plane except for the "up-peptides" and the "down-peptides." The allosteric model required that the new linkages to the stress-bearing strand would occur by addition of a nascent strand becoming linked to a series of down-peptides. Then, when stress-bearing bonds in the plane of the indicated strand and also those in another strand were cleaved, the down-peptides of the stress-bearing chains were imagined to rotate enough to bring the new oligopeptidoglycan chain into the stress-bearing plane. However, this cannot be the case because of the lack of free rotation in glycan chains. They have a very limited ability to twist (Labischinksi et al. 1983). The limitation to free rotation is due to the bulky lactyl group of the NAM. This problem is not fully resolved and must remain open until a better grasp of the energetics of conformations of the peptidoglycan polymer is obtained.

HOLOENZYME MODEL

Because of the evidence suggesting turnover of the wall (see above), a different model was suggested (Koch 1990c) in which the wall is enlarged by pre-forming a raft of two chains of pre-crosslinked oligopeptidoglycan and linked by an insertion process to the stress-bearing wall. The model required that a coordinated group of enzymes (i.e., a holoenzyme aggregate) would function in an orderly fashion to form the raft, link it to the stress-bearing wall, and cleave out an old chain in a lock-step kind of way. In this model for every two chains inserted, one old chain would be removed. This raft concept had previously been developed as a mechanism to vary the energetics of wall enlargement of the developing constriction as compared with the side wall in the Gram-negative organism (Koch 1985b; see also Nanninga 1991).

HÖLTJE'S THREE-FOR-ONE MODEL

The third model to be considered is a great improvement because it not only is consistent with the limited flexibility of the glycan chain but also provides a chemical explanation for the carefully controlled cleavages by the autolysins that actuate wall enlargement. This model is presented in Fig. 10.14 (one extra bond is shown, and indicated by an ellipse, this will be discussed below). To understand the model, it is necessary to appreciate that a third disaccharide peptide monomer can be added to the tail-to-tail crossbridged dimer by a new covalent linkage to form a trimer. This is possible because only one of the two diaminopimelic acids of the dimer has been involved in a tail-to-tail linkage and the ε-amino group of the other is available to be an acceptor for the donor carboxyl group of the penultimate D-alanine (i.e., the fourth amino acid of the chain) of a third disaccharide pentapeptide chain. As in the usual crosslink formation, the linkage is formed at the expense of cleavage of the terminal D-ala-D-ala peptide bond. (This linkage would not be possible if the D-alanine residue of the new chain

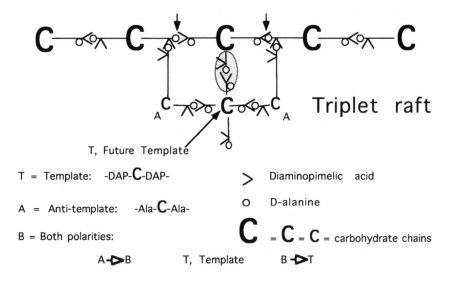

Figure 10.14. The "Three-for-one" model of Höltje. A cross-section through a portion of developing wall is diagrammed, the chains are to be imagined to continue in and out of the plane of the paper. See text for explanation and details. The key point is that certain chains are designated "template." These serve to fix the length of new chain forming and determine the orientation of the crosslinks to the two side chains of the raft. The side chains are linked by forming trimers to the stress-bearing wall dimeric cross-bridges. Then the original template strand is excised by multiple cleavages and the wall becomes enlarged. Note that all peptidoglycan molecules were secreted as pentapeptides. The terminal D-ala residues are all removed during the formation of crosslinks or by carboxypeptidase. No terminal D-ala is shown, but up to four D-ala-D-ala bonds could be present on all the peptides shown in the indicated cross-section of the triplet raft.

Minor modifications to Höltje's model are also indicated in the figure and discussed in the text. They are first that the central chain of the raft is designated "future template" and that it is covalently linked to the template of the stress-bearing wall. At a later time, the future template serves to template the linkage with two side chains to form the three-chained raft and at a still later time the bonds between the template and the future template are cleaved. These changes could account for the donor-acceptor patterns observed; see text.

has been removed by DL carboxypeptidase.) The trimer linkages allow a nascent chain to be appended to a chain of stress-bearing wall as a non-stress-bearing addition (i.e., as a true appendage, like the short fibers sticking out of a piece of velvet) to the stress-bearing fabric.

Höltje's insight was that the trimer components found in traces in the wall were the secret to wall growth of the Gram-negative rod. Although the observed concentration of the trimers is found to be quite low (and in fact, they could not be observed until high-pressure liquid chromatography (HPLC) methods were developed for the analysis of bacterial wall), they could be very important because trimer formation and subsequent cleavage allow the insertion of new wall without the steric problem of the first two models. Incidentally, both earlier models could be resuscitated by incorporating trimer formation and resolution. (In fact, some evidence is accumulating that a regulator site is functioning because an antibiotic, bulgesin (Templin *et al.* 1992), has a structure unrelated to the active site of the enzyme it inhibits and therefore binds to a quite different site.)

The safety constraints, of course, are always critical, and in the "Three-for-one" model the strategy to prevent cell rupture is an extension of the holoenzyme concept of the successive coordinate action of a series of enzymes. The selectivity by the cleavage enzyme of "deciding" which bonds to cleave and which to leave intact is not as difficult as in the earlier model. Although the two tail-to-tail bonds in the trimer are chemically identical, the endopeptidase could be specific for trimers and then only cleave the stress-bearing one. Additional cues could be present, such as the removal of the fourth amino acid (D-ala) from a stress-bearing dimer before trimer formation.

The model postulates that there is the formation and insertion of a raft of three chains for every chain removed. For this reason, Höltje calls his model the "Three-for-one" model. The presence of the third chain (instead of the two chains of the Holoenzyme model) is important because this feature also places less demands on the omniscience and omnipotence of the enzymes carrying out the joining and the cleavage steps. In fact, it creates reliable chemical cues having to do with the chemical orientation of the stress-bearing dimers as they form the trimer architecture; these can orchestrate the selective cleavages of only "protected bonds." The key idea is that the two outside chains of the triplet unstressed raft will be linked (vertical bonds) to form trimers with the stress-bearing peptides emanating from both sides of the "template" strand. When the template strand is excised, the cell's turgor pressure pulls the three chains of the raft into the plane of wall stress where the one template strand was before. (See Fig. 10.14). This model requires obligatory turnover of peptidoglycan at a rate of one disaccharide and its peptide removed for every three disaccharide pentapeptides entering the wall polymer. At present, the turnover data are not critical enough to experimentally validate this numerical prediction.

Another idea embodied in the three-for-one model is that the chains in the raft are created to have precisely the same length as the template strand. The reason for the synthesis of chains of exactly the same length, Höltje postulated, is a way to account for the rod-shaped nature of the bacteria and the constancy of the diameter of the cell when growing under constant conditions. In this book I have strongly objected to the idea that rod-shaped organisms are kept that way by either cytoskeletal elements or by belt-like connections around the cell; however, many other workers adopt one or the other of these mechanisms. Höltje retains the latter hypothesis and I will explain his point of view and retain his terminology. His ideas are based on the assumption that the glycan chains are oriented in the circumferential direction. In fact, the electron microscope studies from Nanninga's laboratory that have already been cited suggest that the glycan strands are largely oriented in approximately the hoop direction. This process, he feels, could give width constancy because the glycan chains are capable of little expansion; he postulates that the templating process generates new glycan chains of the length of their template (possibly of a mode length of 14 saccharides). In order for the wall to form as a series of belts around the circumference of the cell the glycan chains must be maintained in a circumferential orientation and linked via peptides to other glycan chains to provide a continuity around the cell. He also imagines that the enzyme aggregate would bind to one end of a template strain, cause the formation of a segment of a raft (i.e., 3 chains wide times 8 to 20 saccharides long), link these by trimer formation around the unit in the template strand, and eject the old, used part of the template strand.

Note: I do not accept this part of his model because the glycan chain is capable of some expansion and twisting. Moreover, the stress-bearing strands are maintained in a zig-zag structure when stressed and are not colinear at all. Moreover, it is not likely that the glycan chains are always and exactly perpendicular to the cell axis or parallel to each other. More, the peptide crosslinks are much expanded in the stressed states, so that the peptide crossbridges to other glycan strands to provide spatial continuity around the cell are extended in the template, but not in the nascent wall. Thus, for several reasons, generation after generation the cell would become rounder and fatter because the stresses are greater in the hoop direction than in the axial direction of the cells. A second consequence is that the templating procedure increases the number of new templates with exactly the same length. Over generations this would lead to cells in which all the glycan chains are of exactly the same length, unless there is some random autolytic activity that continues to generate different length of chains. With such a templating process, the peptidoglycan would acquire a semicrystalline nature that should have been detected in the X-ray crystallographic studies, although order has been searched for and invariably has not been found. Finally, it is well known that the diameter of the cells does change under changing environmental conditions; this would be difficult to accommodate to this reading of Höltje's model. A consequence of any length templating is that all the glycan chains in the cylindrical part of the wall should eventually end up with exactly the same number of saccharide units, which should be a multiple of eight-sugar residues. In the linking process the templating action insists that three chains of the raft are exactly as long as the template chain in the stress-bearing wall. This is a testable feature. But so much for criticism of minor parts, let us return to the explanation of the model.

Five carbohydrate chains are indicated by large C's; these are presumed to extend in and out of the plane of the paper and to be a dozen sugar residues long. Each chain originally had peptides protruding in different directions as shown in Fig. 10.13. The natural angles in a saccharide chain are such that it could be naturally twisted so that the peptides are offset exactly 90°, but not 180°, from one saccharide to the next when incorporated into a stressed planar monolayer. The oligopeptide structure is sufficiently rigid that requiring successive disaccharides to rotate 180° is totally unlikely from an energetic point of view. The chains projecting above and below the plane of Fig. 10.14 are not fully indicated. In the stress plane, the chains are crosslinked by extended peptides oriented with respect to the terminal amino group of the *meso*-DAP-D-ala as indicated. These orientations are critical for the functioning of the model. Of course, if viewed three-dimensionally, the crosslinks would be in quite different planes both in front and behind the plane of the paper because of the eight sugar residues necessary to complete one revolution of the glycan chain.

Certain oligopeptidoglycan chains are called "templates" in this model, as designated in Fig. 10.14. The templates may serve an even more important role: they determine the polarity of the crossbridges. All the peptides on the template chains participate in crosslinking as "acceptors." In their crossbridges they contribute their amino groups from the tails of their DAP residues, which are then connected to the carboxyl groups of the alanine residues of the donor chains. These acceptors are doubly marked in the figure both by being labeled with an "A" and, in the cartoon of the peptide chain being marked by a "V" oriented in the direction of the cross-linking bond. The V symbol was chosen to suggest the zwitter grouping of amino and carboxyl groups at the termination of the diaminopimelic acid. The other component in the crossbridges ends in a D-ala. These later peptides have served as "donors" because they have donated the energy of their D-ala-D-ala bonds in forming the crosslink in the transpeptidation event. They are designated by "D" for donor and also are indicated by small circles in the diagrammatic structure.

The model requires that the synthetic transpeptidase/transglycosylase use particular stress-bearing strands as the template to form, in a stoichiometric and processive way, a new glycan raft. The formation of the raft structure would require enzymes clever at elongating the chains by templating the addition of disaccharide pentapeptides. This templating process would be most easily explained by enzymes that are both transglycosylases and transpeptidases, such as PBP2 and PBP3. It may be that the directionality of the bonds are determined by which chains are stressed or by other signals, such as whether or not the D-alanine that was originally in the fourth position has been previously removed by DL carboxypeptidase. The strand to the left or right of the strand marked "template" would not serve because of the wrong orientation of the dimers. They must be pointing in opposite directions from the template strand. In the depiction in Fig. 10.14 this means that every fourth disaccharide of the central strand of the raft is matched up with the template strand. The template strand fixes, it is assumed, the number of disaccharides in the middle chain of the raft and the adjacent two side chains. Then all three chains of the raft, when it is integrated into the stress-bearing wall, will be of the same length as the template.

Concomitant with excision of the template strand, the triplet raft will be pulled into the stress plane. This enlarges the wall by the net addition of two strands. An important aspect of the present version of the Höltje model is that the formation of the trimer and its subsequent hydrolysis back to form a different dimer has the property of leading to the net addition of one new template strand whose DAP residues are linked to other strands in the acceptor format and one new strand whose D-ala residues are involved in tail-to-tail linkages in the donor format. This can be best appreciated by tracing in Fig. 10.14 the bond orientations before excision and in the resultant stress-bearing structure afterwards. It can be seen that the chain to the left of the template strand converts from a donor in both directions into one that is a donor in one direction and an acceptor in the other strand; this can be designated a bipolar strand. The chain on the right template converts from a bipolar strand to one with acceptor linkages on both sides.

MINOR MODIFICATION TO THE THREE-FOR-ONE MODEL

I have made a slight modification to the published version of the Three-for-one model; this is shown in Fig. 10.14 by the bond enclosed by the circle. In the variation of Höltje's model depicted here, the template strand also forms a link with the nascent chains, designated "future template." The idea is that it forms with some of the peptides that are not oriented in the plane of the wall, one up and one down. Such a bond will be cleaved later. Of course, the future template strands have been added at a later time to the stress-bearing structure than was the template strand and are appended below the monolayer superstructure. The future template serves as a pattern subsequently for the other two chains that make up the triplet raft determining their length and polarity.

This further feature of the suggested changes in the templating process is that the future template still later forms tail-to-tail bonds with its two partners in the raft by contributing the DAP amino groups as acceptors and never providing D-ala carboxyl groups as donors. The side chains of each raft are subsequently linked to the crosslinks of the stress-bearing bonds of the template strand through the formation of trimers. The orientation of the bonds to the template strand is required to be as indicated in Fig. 10.14 so that the template strand can be completely excised by endopeptidase action. As in the parent model, the relevant autolysins must be only able to act on trimers; they also must choose to cleave only the stress-bearing tail-to-tail peptide links instead of the unstressed tail-to-tail peptides, even though they are identical from the chemical bond point of view. (There are some differences in the near vicinity, however, that could be used as chemical clues.) From the arguments presented in earlier chapters, this selectivity can be simply a matter of stress lowering the energy of activation for hydrolysis. But additionally, for truly safe growth these hydrolytic events can only occur if all possible trimers linking a given template strand have been formed and if only then all trimers from both sides of the template strand have been cleaved. Of course, as the bonds are cleaved, the lowering due to stress of the energy of activation of the remaining ones will become ameliorated.

At some stage in the process, the crosslinking bonds between the template and the future template have to be cleaved. The timing of the cleavages does not appear to be critical, but the actual relations could affect the interval between insertion of a future template strand and its subsequent excision, and thus affect the ADRR (acceptor-donor radioactivity ratio) measurements. These measurements are made by reacting the amino groups of wall sample with a marking reagent and then hydrolyzing and separating marked and unmarked DAP molecules. Only DAP residues engaged in tail-to-tail linkage will remain non-tracer-bearing and thus be classified as acceptors.

The variation of Höltje's model impinges on two currently utilized experimental methods. First, the formation and resolution of trimers greatly confounds interpretation of available ADRR measurements, because a pentapeptide can give up its terminal D-alanine as a donor. The chain can later serve as an acceptor for another chain and then be classified also as an acceptor. Of course, the pentapeptide can only serve as donor once. Second, because there is a period of time between the formation and attachment of the future template to the template, and then another period of time between the event and the formation of the two side chains in the raft, and yet another period of time before the raft is incorporated into the stress-bearing wall, the ADRR measurements can have quite different meanings depending on the duration of the tracer pulse. The key point is that if the three chains of the triplet raft were formed very fast, then a brief pulse would lead to radioactivity in both donor and acceptor linkages. In the variant model presented in Fig. 10.14 a brief pulse would find its way only into the two side chains. These measurements are further confused by the formation of the poles in which the rate of incorporation and the interconversions of donors to acceptors is much more rapid. Finally, rapid recycling can return a labeled residue that has been liberated from the wall to enter the cell again and be newly synthesized into the pentapeptide to act as a donor.

Conclusions About the Nature of the Peptidoglycan Fabric of the Gram-Negative Organism and the Growth Process

Peptidoglycan is formed into a strong, but elastic, fabric that resists the expansive effects of the cell's turgor pressure. The cornerstone of the surface stress theory as applied to Gram-negative microorganisms is that the nascent wall is capable of being considerably stretched. The surface stress theory postulates that for any of the models, nascent oligopeptidoglycan strands in their relaxed conformation become linked at multiple points to existing strands. The nascent strands are subsequently stretched into an extended conformation when cleavages of old stress-bearing wall occur, stressing the new wall and permitting growth and enlargement of the sacculus and an increase in the cell volume. Thus, a seven-stage cycle operates as follows: (i) biosynthesis of cytoplasm, (ii) increase in turgor pressure, (iii) increase in stress in the wall, (iv) increase in autolytic action on existing wall, (v) transfer of stress to the new wall, (vi) stretch of the nascent oligopeptidoglycan strands, and (vii) increase in area of the sacculus, increasing the cellular volume and lowering the turgor pressure. The data presented in Table 10.1 and Fig. 10.9 demonstrate the great elasticity of the Gram-negative sacculus.

These seven steps describe the cyclical process of wall growth, but that is not all there is to the wall growth process. Concurrent with these seven stages is the biosynthesis of the disaccharide pentapeptides, their transport to the outside of the cytoplasmic membrane, linkage to saccharide chains already on the outside of the cytoplasmic membrane, and crosslinkage (tail-to-tail) with other peptides. It is presumed that ordinarily none of these is limiting for wall growth. Formation of unstressed wall can be limiting, as during penicillin treatment, and then this can result in either rupture or bulging. Under normal conditions, however, adequate new wall units are firmly integrated into the stress-bearing wall and patiently wait for their turn to become part of the stress-bearing structure. When that time arises, the peptide crossbridges elongate and the new unit is pulled into the plane of the fabric, thus increasing the surface area both by insertion and by expansion, i.e., unfurling the peptide.

With the advent of the Three-for-one model there is a *raison d'être* for trimer formation and, possibly, for the existence and action of DL-carboxypeptidase. This enzyme removes the last D-alanine from the peptide chain, preventing that chain from being a donor in trimer formation. Then the template strand is capable only of being an acceptor, forcing insertion of a raft of three chains surrounding the old template strand. The new template strands are generated as shown in Fig. 10.14 and then the process can continue indefinitely, in a way that greatly reduces chance of wall breakdown.

The Unit of Gram-Negative Wall Architecture: The Tessera

The results amassed in this chapter have implications about the fundamental unit of the wall fabric. Although the chemistry and biochemical studies have of necessity focused on the linear course of the polymer of either the crosslinked peptides or glycan chains, it is the area of the cell protected by a structural unit of wall that is the real subject at hand. The smallest units of surface to compose a mosaic are its tesserae. The tessera is the two-dimensional equivalent of the unit cell of crystallography. It is the geometric entity that upon translation in space could fully form a single layered complete surface, which in our case is the curved two-dimensional surface that would cover and enclose the cell. The covalently bonded structure that corresponds to the edges of the tesserae of the units of wall structure is a two-dimensional network, i.e., a ring with both glycan and peptide elements. Each unit is part of other rings. From the connectivity of the fabric please conjure a physical model by picturing a fence of chicken wire that has hexagonal holes. In that case, a segment of wire defines the limits of the individual structural unit; whereas, in the bacterial case, it is the crosslinked peptide and a portion of glycan chain that do so. The smallest possible unit of crosslinked two-dimensional peptidoglycan is diagrammed in Fig. 10.13. A smaller one was incorrectly suggested earlier (Koch 1990a). The earlier one is not possible because the hexosamines of the glycan chains of the murein rotate relative to each other in such a way that a complete revolution requires about four disaccharides instead of two. With this correction, the smallest structural unit is the region enclosed in the rectangle of Fig. 10.13. It is the unit of peptidoglycan such that when many of them are connected they could totally cover a bacterium's surface with no overlap. The structure shown is comprised of two decasaccharides connected by two pairs of crosslinked peptides emanating from the first and ninth saccharide of both chains. The crossbridges are formed by the usual tail-to-tail linkage of D-alanine of one strand with the ϵ-amino group of diaminopimelic acid on the other. The third and seventh saccharides of both chains would be muramyl glycosides with their lactyl groups pointing above or below the plane of the figure. Obviously, other rings with different connections between the parts also occur in the stress-bearing wall.

The peptides originally linked at the third and seventh positions do not appear to be quickly degraded; there is only a remote possibility that they may serve in linking nascent oligopeptidoglycan or may have served to form links to oligo-peptidoglycan that has yet to be discarded. (This latter possibility probably may occur, and is part of the variation of the Three-for-one model presented in Fig. 10.14.) In any case, these peptides are not serving the function of covering the cell surface area. The fifth saccharide would extend in the plane of the cell surface but extend away and be part of two other tesserae. A perfect monomolecular network constructed of identical units in this fashion and under tension would be the ideal structure that could cover the cell and provide the most static strength. Considering just such units, each would have eight N-acetylmuramic acids, eight N-acetylglucosamines, and four tetrapeptides linked to each other in two pairs (although a very small proportion may still be pentapeptides and another fraction tripeptides for the template strands) engaged in crosslinks. There would be six N-acetylmuramic acid linked peptides that are not part of the particular surface unit, composed of three pairs: two N-acetylmuramic acid residues with peptides with the lactyl group pointing out of the plane of drawing, two N-acetylmuramic acid residues with peptides pointing below the plane, and two N-acetyl muramic acids residues extending in the plane, but pointing outwardly (see Fig. 10.13).

Although these *N*-acetyl muramic acid residues originally bore pentapeptides, they may be present as penta-, tetra-, tri-, di-, or monopeptides, or (very rarely) may have been completely removed. The tetra-, tri-, and dipeptides are readily observed, but for lack of analytical methods, it is not clear how large a percentage of the *N*-acetylmuramic acids have completely lost their amino acids (Cooper, unpublished). Crosslinking is usually calculated as 100 times the ratio of diaminopimelic acid residues molecules in ε-peptide bonds to total diaminopimelic acid residues. Just considering the ring of the tessera itself, the value is 50%. It is diminished to the extent that there are uncrosslinked peptide chains in the stress plane that still retain the diaminopimelic acid. It is artificially increased to the degree that DAP molecules have been removed from uncrosslinked chains. Finally, the peptides maybe involved in linkage to new murein centripetally or bound to more peripheral wall centrifugally, which would tend to raise the percentage of crosslinks.

Implications of the In Vivo Structure not being Maximally Extended

Living *E. coli* cells shrink 45% in surface area when the turgor pressure is eliminated by detergent treatment; they shrink 20% when the osmotic pressure is raised to a point assumed to erase the turgor pressure. The finding that the relaxed isolated sacculus can readily expand 300% in area with quite minimal stress forces seems contradictory to these findings. There are two interpretations of these results. The first is that the inner and outer membrane truly contribute a good deal to supporting the stress due to turgor and keep the murein from expanding to its fullest extent (see Chapter 13). The second (much less likely) interpretation is that the turgor pressure does not present as strong a stress as the electrostatic repulsion that we applied by causing the sacculi to bear a net positive or negative charge.

The second unexpected result of the studies reviewed in this chapter is the key role of the zwitter group of diaminopimelic acid. When the carboxyl and amino groups are this close to each other due to their connection to the same carbon atom, they cause the p*K* of each to be more extreme. The finding that saccular expansion is associated with either pH extreme shows that this part of the sacculus is key to its expansion whenever and however the zwitter character is destroyed to result in a group that has a single charge of either sign. This finding may be particularly important in considering the implications of elasticity or inelasticity on the ability to expand in comparing the shapes of the poles of *E. coli, B. subtilis,* and *E. hirae.*

11

Apical Growth of Streptomycetes and Fungi

KEY IDEAS

Apical growth is the tactic of both the hyphal growth of filamentous streptomycetes and mycelial fungi.

It is also the tactic of higher plants, but with tissues many cells thick.

The problem of apical growth is that the growing structure is supported only from its base, giving little support to the continuously renovating tip.

A special strategy is needed for the prokaryote because it has no cytoskeleton to perform mechanical work, but instead has a passive exoskeleton to resist the turgor pressure.

It is hypothesized that the tips of Streptomycetes *elongate by rapid turnover of the tip.*

Turnover maintains the integrity of the wall during growth because an intact, covalently linked portion of the tip is always present.

The "apical inside-to-outside" model depends on a high degree of elastic expansion of the murein composing the sacculus.

A radically different strategy is needed for lower eukaryotes, which have a cytoskeleton, but have no totally enclosing, strong, crosslinked fabric.

The hyphal wall of the eukaryotic fungi resembles fiber glass in that the wall is composed of fibers embedded in a plastic phase that gradually sets to become rigid.

In mycelial fungi, the tip wall is enlarged as the result of fusion of vesicles containing needed substrates and enzymes.

The "Moving VSC" model places a key role on an organelle called the "Spitzenkörper," idealized as the "vesicle supply center."

The current forms of the moving VSC model do not take into account that tip wall will be displaced radially.

The "Soft-spot" model assumes that the vesicles fuse only if they reach points in the wall that are sufficiently new to be still plastic.

Neither the Moving VSC nor the Soft-spot model is sufficient and a fusion of the both models and the incorporation of the time course of the viscoplasticity of new wall as it hardens will be necessary.

The Problems

An entirely different set of problems must be solved by rod-shaped cells that grow at only one end than those microorganisms that grow by elongation over their central cylindrical region. The latter organisms solve the problem of maintaining shape because the pair of existing poles forms a supporting framework for central enlargement (see Chapter 6). Both mycelial prokaryotes and eukaryotes have developed successful techniques for apical growth, though the strategies are more difficult than for cylindrical growth of prokaryotes. Mycelial organisms use these mechanisms to burrow through soil or rotting wood. By using the combination of apical growth and branching, these organisms can exploit resources effectively throughout a three-dimensional solid substratum.

It is relevant to draw the analogy to the architectural construction of a dome. Apical extension by these organisms is a development beyond the more straightforward elongation of the middle of the cylinder, just as Roman and Gothic domes required more sophisticated construction techniques than those needed for the more elementary earlier lintel and flat roof forms.

The mechanisms for hyphal growth that prokaryotes use must be different from those of the eukaryotes. Prokaryotes do not have cytoskeletal elements to serve as internal scaffolding, to provide stress-bearing braces, and to do the mechanical work of extension and contraction (see Chapter 2). The bacterial wall owes its unique strength to the crosslinked peptidoglycan (murein) of which it is composed. Instead, the surface stress theory presumes: (i) that cytoplasmic growth, which increases the turgor pressure, concomitantly increases the tension in the sacculus network that encloses the cells, and (ii) that the tension in turn increases the activity of the autolysins present in the wall. The loosening of the peptidoglycan results in the pulling of nascent, crosslinked (but unstressed) wall into the stress-bearing part of the wall fabric.

On the other hand, eukaryotes do not have the two- or three-dimensional covalently crosslinked fabric of peptidoglycan. Still, the tip must elongate and exert considerable force while growing through media containing solid obstacles. It is probable that the cytoskeleton serves a role in this process. The walls of fungi (Wessels 1990a,b) contain one-dimensional linear polymers, *viz.* chitin and cellulose fibrils. These substances serve the equivalent role of the glass fibers in the fiber glass. Various polymers—mannans, glucans—serve as the substances equivalent to the epoxy component of fiber glass. The analogy to fiber glass requires that the fibers be distributed in space and surrounded by a mastic phase, which then is allowed to crosslink to itself and to the fibers as it hardens. Only then can the new wall serve a stress-resistant structural role. After curing, the fibers bequeath stress resistance, and the combination of fibers and mastic resists compression and flexing. During hyphal elongation of eukaryotes, the tip is the least cured part of the wall, but it must exert the highest force on the environmental substratum.

The purpose of this chapter is to explore these problems and speculate on the disparate solutions employed for hyphal growth in the two kingdoms. The basic premise, as in the rest of this book, is that engineering tactics devised by humans to solve their problems are similar to those evolved by lower creatures to solve theirs.

The Elastic and Plastic Components of Growing Walls

When not growing, the cell's architecture would be best served by a strong, elastic material that would not deform or creep under continuing pressure and stresses. Yet in order to grow, the wall must be plastic enough to enlarge. The engineering of both elastic and plastic materials is clearly relevant to hyphal elongation by tip growth. Particularly for bacterial rod elongation, the interplay between plastic and elastic behavior is clearly demonstrated in many ways in this book. What is the interplay between the two physical forms of materials during apical growth of hyphal streptomyces? What is the equivalent strategy utilized by hyphal fungi?

Variable Stress Model for the Prokaryote and Eukaryote Kingdoms

The surface stress theory was originally proposed in a simple form to explain binary fission of Gram-positive cocci. Much more elaborate mechanisms were clearly needed for the growth of other types of microbes. Consequently, the theory was quickly enlarged in order to apply to hyphal growth of fungi (Koch 1982a) and to pole formation of Gram-negative bacteria (Koch and Burdett 1984). These two studies proposed that hyphal growth of fungi and the cell division of the thin-walled Gram-negative rod-shaped bacteria could be understood if the cell had a way to alter biochemical mechanisms in order to lower the parameter T (the analogue of surface tension) in a narrow zone localized at the tip of the hypha or in the middle of the Gram-negative bacterial cell.

This Variable-T model in a more elaborate form serves as the theoretical basis of the models for both the apical growth of prokaryotes and of eukaryotes presented below. The name "Variable-T" indicates a more general formalism than a mechanism-oriented model. It includes the "Apical inside-to-outside" model to be developed in the next section for streptomycetes, a mechanism in which the tip is rapidly turning over so that nascent wall is able to expand easily near the tip. It also includes the century-old idea of insertion into the existing wall, a process called in mycology "intussusception" (Reinhardt 1892), in a new form developed below.

The Sophistication of the Growth Strategy of *Streptomyces coelicolor*: The "Apical Inside-to-Outside" Model

A possible solution of the inherently difficult problem of performing hyphal elongation in streptomycetes without invoking a cytoskeleton or mechano-proteins and mechano-enzymes is presented here. This mechanism for tip growth is essentially the inverse of the one used by the Gram-positive rod-shaped fission bacteria for side wall growth. The contrast between the two is shown in Fig. 11.1.

It is known that the side wall of *B. subtilis* elongates by the inside-to-outside mechanism that permits safe elongation; it requires new wall incorporation on the inside, movement to the outside, and cleavage and release of fragments to the medium. However, the side wall of *S. coelicolor* probably turns over very slowly. Certainly, there is very little incorporation of tracer into the older cylindrical wall (Gray *et al.* 1990; Miguelez *et al.* 1988). Even the little tracer that does enter the side wall could come from some metabolism of the tracer compound *N*-acetyl-D-glucosamine into other labeled compounds, which might then enter the wall as other polysaccharides or as wall proteins. Miguelez *et al.* (1988) believe that they have excluded true turnover in the side walls, and thus the radioactivity would correspond to a slight wall thickening.

With respect to the poles, there is a profound difference between cylindrical and apical prokaryotic growth patterns. The evidence discussed in Chapter 9 is conclusive, that once formed, the pole in *Bacilli* turns over only very slowly. Two laboratories working with *S. coelicolor* showed that the tip rapidly becomes labeled. Neither set of measurements, however, was performed in a time scale or in a way to allow the measurement of the metabolic stability of the tip. Such measurements require new experiments that search for the almost immediate release of labeled turnover products.

The new suggestion is that the tactics of *S. coelicolor* are just the opposite of

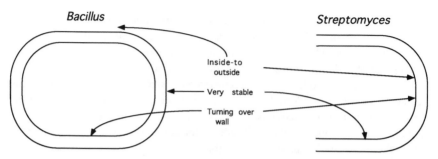

Bacillus   **Streptomyces**

Inside-to outside

Very stable

Turning over wall

Figure 11.1. Difference in the tactics of nonapical and apical growth of two prokaryotes. *Bacillus subtilis* has been shown to elongate safely by the inside-to-outside mode; i.e., with the continuous addition of peptidoglycan diffusely to the entire inner face of the side wall and diffuse dissolution on the outside. As the wall moves outward, it is stretched, mainly in the longitudinal direction, before it comes to bear stress in the hoop direction. This longitudinal enlargement and expenditure of the capability of elastic expansion before it becomes subject to hoop stress permits rod growth even though the hoop stress is twice the axial stress. Eventually, old wall is cleaved and no longer supports the turgor pressure. The pole is formed by creating a septum that is then split and bulges to its definitive shape as turgor pressure comes to be applied to the entire pole. The pole wall is almost inert and turns over extremely slowly.

On the other hand, the "Apical inside-to-outside" model proposed here assumes that another Gram-positive organism, but one that grows as a filament, *Streptomyces coelicolor*, does just the reverse. The side wall is very stable, and the tip turns over in an inside-to-outside mode. This affords continuous integrity to the tip. For elongation of the *S. coelicolor* hyphae to occur, the wall, as laid down, must be capable of a high degree of expansion when stressed by the autolysis of peripheral outer layers. This frees an underlying layer that stretches and bulges forward and outward to allow the tip to grow and press through the environment. Note that neither of these strategies utilizes cytoskeletal elements, which appear to be lacking in prokaryotes.

the tactics of *B. subtilis;* this provides a method to avoid both uncontrolled bulging and rupture at the growing tip without invoking cytoskeletal participation. As discussed in Chapter 9, murein is cleaved when it has moved through the side wall as an obligatory part of growth of the stress-bearing part of the side wall of *B. subtilis*. Theoretically, the wall must be cleaved, but usually it is turned over; i.e., the cleavages are sufficient to release fragments into the growth medium. In the "Apical inside-to-outside" model for *S. coelicolor* it is postulated that there is a rapid synthesis of wall at the tip region and there is obligatory turnover in this region only. There is very little degradation of wall, however, at any distance from the tip, except at sites where a branch is to form. The idea of inside-to-outside turnover of the pole is novel; I have found no precedents in the experimental or theoretical literature. Still, it must be considered as an essential part of the mechanism for safe apical growth of streptomycetes.

Turnover can describe a number of processes. Classically, it refers to the uptake of a labeled compound to form macromolecular molecules that are subsequently degraded and released as small molecules. When the synthetic and degradative rates are the same, then the organism is a constant size and is in a steady "dynamic" state.

The assumption that the tip wall of *S. coelicolor* engages in extensive turnover makes it possible for the tip to elongate with the continuing maintenance of intact stress-bearing wall; this is shown in Figs. 11.1 and 11.2. A co-requisite assumption is that the wall, once formed, is capable of elastically stretching a good deal. Although there are no data on the *Streptomyces* wall, there is evidence that other peptidoglycans can be stretched by a large factor (see Chapter 10). The model proposes that layers are added continuously in the tip region on the inside of existing wall but on the outside of the cytoplasmic membrane. At present, no attempt is made to explain how this addition is directed almost exclusively to the tip regions. The rate of addition decreases with distance very rapidly from the tip, as evidenced most clearly by the electron microscopic autoradiographic data from studies of pulse-labeled cells of Gray *et al.* (1990). For the apical inside-to-outside model three processes must go on concurrently: (i) the addition of new layers of unstretched peptidoglycan only in the tip region, (ii) the stretching forward and bulging outward of layers laid down earlier, and (iii) the autolysis (rupture) of still older layers that are now outermost. This third process transfers stress to underlying layers that then expand a good deal in response.

An integral part of the model is that the murein, as laid down, is unstressed and unstretched and is not affected by the cellular turgor pressure, because it is inside a stress-bearing shell. When the most apical part of the tip peptidoglycan is autolyzed sufficiently to lose its structural integrity, the underlying and previously unstressed peptidoglycan becomes stressed, resulting in bulging and radial expansion of what will be side wall. However, the covalent integrity of the fabric covering the tip is retained. As the newly laid-down tip wall stretches and bulges, the hypha grows forward. The forward movement results because the elongation due to the bulging is much greater than the lost wall by autolysis from the tip.

To visualize the apical inside-to-outside model consider a stack of cups. One may think of styrofoam or plastic cups without handles, but the analogy would be much better if they were of highly elastic rubber. The operation simulating growth is the repetitive addition of a new cup at the inside of the stack (left side in Fig. 11.2), the removal of the bottom of the oldest cup because it has been stretched and bulged too far, and the extensive stretching of the intermediate cups as they come to support the turgor pressure. The cup analogy shows how the apical inside-to-outside model could yield stable growth because there are always complete layers of intact unstressed and unstretched wall and more peripheral intact both stressed and stretched wall.

The cylindrical side wall does not progressively bulge in diameter nor inappropriately enlarge its diameter for several reasons. The major reason is that, at a

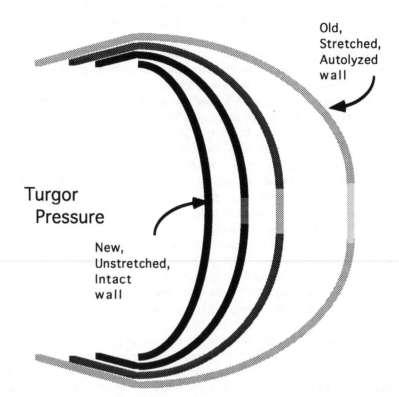

Old,
Stretched,
Autolyzed
wall

Turgor
Pressure

New,
Unstretched,
Intact
wall

Figure 11.2. The "Apical inside-to-outside" model for growth of streptomycetes. The figure depicts the addition of new wall as the most darkly colored line. The darkness of the line corresponds to the density, which in turn reflects a lack of having been stretched or autolyzed. Subsequent to its formation, a layer becomes stressed. The stress supported by a unit thickness of wall in both the meridional and hoop direction is least near the margins of the pole because there are many stress-bearing layers in this nearly cylindrical region, even though the total stress (particularly the hoop stress) is greater. Autolysis at the tip applies stress to the wall that had previously been non-stress-bearing, causing it to stretch outward and forward. The stresses of the stretched wall at the tip become very high because of the thinness of the wall. High stress favors autolysin action, which in turn leads to further deformation of the uncleaved portion of the layer, favoring its stretching, both in the forward and outward directions. The side wall is built up of these pre-stressed portions of wall that can no longer expand in the hoop direction as well as some wall that is subject to lesser stress because it was originally laid down under the existing cylindrical wall.

small distance behind the tip, the wall is made up of several conical layers, each sleeved inside the next and each already stretched from the bulging process and sharing the stress originating from turgor pressure. Because many layers are sharing the stress, the maximum stress on any layer would be small. Compare this with the case of the inside-to-outside growth process of *B. subtilis,* in which the stress is concentrated in a few central layers in the wall that have been stretched but not yet subject to autolysis. The inner layers inside have not been fully stretched and therefore are not bearing much stress, and the more peripheral layers have already been subject to partial cleavages and their stress partially relieved.

The proposed model works only if the peptidoglycan is capable of a good deal of elastic expansion. From the experimental data and the models for wall growth of various bacteria, the degrees to which different nascent wall can expand when stressed can be estimated. The expansion factor of surface area measured for *Escherichia coli* sacculi is 4 (Koch and Woeste 1992; Chapter 10); the factor for the *B. subtilis* pole *in vivo* is 1.5 (Koch and Burdett 1986ab; Chapter 9). On the other hand, *E. hirae* wall probably expands much less once externalized (although it probably can be distorted slightly), implying that the expansion factor is very close to unity (Koch 1992; Chapter 8). For the murein of *E. coli,* we showed that the area of the sacculus depends critically on its net charge (Koch and Woeste 1992). Therefore, a detailed knowledge of the chemical structure of the streptomycetes wall and the wall pH could support or falsify the model.

Morphometric Analysis of the Apices of Streptomyces coelicolor

Measurements of the dimensions of poles of *S. coelicolor* have been made by Gray *et al.* (1990). They fitted the electron microscopic data of the tip profile both to an ellipse and to a power series. With more terms, the power series gave a better fit than the ellipse, but a power series will not be considered here. It is inappropriate because it does not extrapolate to a constant radius at large distances behind the tip. The ellipsoidal model pieced to a cylindrical shell is more realistic. In the elliptical fitting, however, there is the subjective difficulty of choosing the point dividing the pole from the side wall, so actually there is an additional parameter value that must be assigned in the fitting process. In Fig. 11.3, the data from Gray *et al.* (1990) have been replotted; these data are from measurements of a single apex examined in the electron microscope. I have achieved a better fit (solid line) to an ellipsoid of revolution by assuming that the pole actually started at 0.2 μm farther from the tip of the pole than the value (0.978 μm) that they had assumed. The new elliptical fit was made because then the engineering formula for the hoop and meridional stress given in Chapter 6 can be used to calculate the stresses. Fig. 11.4 shows the stresses (long dashes for hoop and shorter dashes for axial or meridional stresses) on the additional (and incorrect)

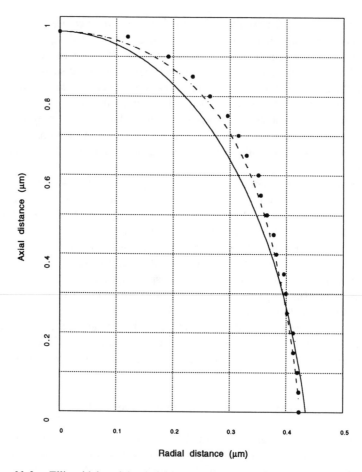

Figure 11.3. Ellipsoidal and hyphoidal fits to the measured shape of a pole of *S. coelicolor*. The data are taken from Gray *et al.* (1990) and derived from measurements from an electron micrograph of a single median section. These authors fitted their data to an ellipsoidal pole shape; the ellipsoidal fit presented here (solid line) is closer to the experimental data and was obtained by assuming that the elliptical portion of the tip extends farther than Gray *et al.* (1990) had assumed. The data show the better fit to the Hyphoid model (dashed line), which in this case has only one virtue. This is an important one; it is that it is a one-parameter, single-equation curve that predicts a nonpointed tip and a cylinder region at large distance from the tip.

assumption that the tip wall is made of material of uniform thickness. These calculations are relevant only because they show that no matter what the details of the fitting procedure, the "stress resultants" are smaller at the tip if the wall is of uniform thickness and composition and is not a composite material made of many layers with different degrees of stress. In order to have greater stress

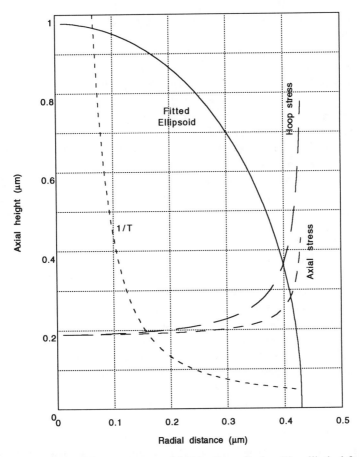

Figure 11.4. Axial and hoop stresses and 1/*T* for *S. coelicolor*. The elliptical fit to the data of Gray *et al.* (1990) shown in Fig. 11.3 is reproduced here by the solid line. From formulae given in Chapter 6 the stresses were computed from this curve by assuming the wall to be a nongrowing elastic solid of uniform thickness. The hoop (circumferential) stress is shown by the line of long dashes. The axial (meridional) stress is shown by the line of medium-length dashes. These would reflect the stresses in the actual rigidified portions of the wall, if the stress-bearing part had constant thickness. Two conclusions follow: the tip must have a thinner wall supporting the stress for the Apical inside-to-outside model to function, and the stresses near the edges of the pole will force a forward and outward movement. This figure also shows, by short dashes, the plot of 1/*T* that is appropriate to the degree that the wall can be treated as a plastic, fluid material. This is the parameter of the surface stress theory that is a measure of the wall's ability to grow.

and to invoke autolysin action directed by stress on the substrate murein, one must presume that the wall at the tip that actually bears the stress is much thinner than the side wall as the result of the continuing autolysis of the outermost layers of the tip wall.

These calculations given above show that the hoop stress is larger (2-fold greater) near the edge of the pole than is the meridional stress. This leads to the radial expansion of the wall while it is in the tip and then limits expansion at an elastic limit so that the radius of the cylindrical part of the hypha has a maximal value and cannot increase progressively.

One can use the morphological data in a second way by employing the framework of the Variable-T model described in Chapter 6 to calculate the parameter $1/T$. Koch (1982a) considered the tip wall to be plastic and showed that the value of the biological analogue of T was approximately equal to $2r/SP$, where S is the slope of the pole profile at any point of interest. It is thus easy to calculate from the image of a pole how T increases with distance from the apex. The more accurate form of the equation is $S = [(Pr/2T)^2 - 1]^{1/2}$. This latter form has been used to calculate $1/T$ over the entire pole in Fig. 11.3. This in turn can be thought of as the measure of the turnover of the tip for the prokaryote (or the kinetics of the hardening of the new wall for the fungal eukaryote). This demonstrates that $1/T$ decreases dramatically with distance, and it can be interpreted in the streptomycetes case as reflecting in some way the intensity of turnover.

A second fit is shown in Fig. 11.3, using the hyphoid equation discussed below. It is appropriate simply because the hyphoid is the simplest mathematical form (only one parameter is involved in fitting) that meets the general requirement of an unpointed tip and a constant radius at large distances from the tip.

The Strategy of Filamentous Lower Eukaryotes

Superficially, the hyphae of actinomyces and of fungi are similar. Obvious differences are that they belong to different kingdoms and that the radius of a eukaryotic hypha is generally an order of magnitude larger than that of a prokaryotic hypha. Because the force acting to thrust the hypha forward is in proportion to the cross-sectional area times the turgor pressure, the growing eukaryotic hypha can do more work. From the formulae given in Chapter 6, however, it follows from the difference in size that stress in the wall will be larger in a typical eukaryote than in a typical prokaryote if the turgor pressure and wall thickness is the same. If the tip were hemispherical, then T would equal Pr. It appears unlikely that the turgor pressure in the eukaryotic hypha is lower than it is in prokaryotes but likely that the stress is borne by a greater thickness of wall. That the larger cross-sectional surface of the wall suffices to exert a higher amount of hydrostatic force on the environment may explain why eukaryotic hyphae typically elongate an order of magnitude faster than those of mycelial prokaryotes, and why the wider ones elongate faster.

The crucial difference between the two types of hyphae is the existence of a cytoskeleton in the eukaryote and its absence in the prokaryote. This difference is exceedingly relevant because the cytoskeleton of the eukaryote is probably involved in movement of vesicles to the tip of the cell. The cytoskeleton also could be involved in various stress-bearing roles. It could force extension, prevent bulging, and guide in the movement of the vesicles under the action of kinesin and dynein.

When the vesicles reach the wall, it takes less work to put new material into a region that has unpolymerized wall than into other regions where the wall has partially, or completely, become set and "rigidified." New wall takes time to harden and thus still newer wall can readily fuse to it, a process that might not occur at all in regions that have had time to fully harden (or mature, or age, or crosslink). Frank Harold (personal communication) called this model the "Soft-spot hypothesis"; this is the name that will be used here because it is more specific and descriptive than Variable-T or the "Steady-state model." The former allows one to calculate shapes without specifying the details of the process and the latter is the name used by Wessels *et al.* (1983) and Wessels (1990a,b). The Wessels model was predicated on three kinds of observations: (i) the newly synthesized wall at the tip is chitin in a noncrystalline and nonfibrillar form, (ii) the new glucan is not alkali insoluble and has formed no branches or crosslinks, and (iii) stoppage of elongation causes the wall over the apical tip to assume a structure indistinguishable from that of the side walls. These observations are exactly what one expects from a plastic composite material as it is formed, ages, and hardens. We will return to the Soft-spot model after we consider one that is quite different.

The Moving VSC (Vesicle Supply Center) Models

Bartnicki-Garcia *et al.* (1989) presented a new model that appeared at first quite attractive. I conclude here that even in a revised form its mathematical forms are inadequate to account for the shape of apices. Nonetheless, the biology involved is appropriate for the apical growth and for the purposes of this book. This model depends on the formation of a variety of vesicles in the Golgi apparatus, their movement along cytoskeletal elements, and their collection near the tip of the hyphae. When seen in some species, this aggregate body is called the "Spitzenkörper." Although not seen in all fungi, these authors assume that a depot of some kind exists, but is too transient to be identified. So this entity, whether visible or not, is called a vesicle supply center (VSC). New vesicles are continuously recruited from the more distal parts of the cell. Vesicles, perhaps after being altered or matured, continuously leave the Spitzenkörper or VSC and move toward and fuse with the wall, permitting it to enlarge.

Some vesicles contain enzymes that have a role in wall formation (see Harold 1990); others contain the precursors of chitin. All must become attached to the cell membrane, be transported to the outside, and form the growing wall. The glucan is partially polymerized, presumably in the Golgi apparatus, and then passed through the membrane by vesicle fusion. There the glucan is gradually crosslinked to the chitin.

The VSC, besides serving as a depot for the temporary storage of vesicles with individual components needed for wall formation, somehow maintains a fixed distance behind the tip of the hyphae. That is why the model is called the "Moving VSC" model. The Bartnicki-Garcia et al. (1989) model postulates: (i) that it is the controlled movement of the VSC that sets the rate of fungal tip extension, (ii) that the vesicles move out from the VSC in all directions, (iii) that they proceed until they encounter the wall, (iv) that they fuse with the wall, and (v) that this results in the local enlargement of the wall. For simple geometric reasons, the tip receives vesicles at a greater rate because it is close by and therefore enlarges more rapidly than the cylinder region. If the Spitzenkörper were a point source, if it emitted vesicles at a constant velocity, N (s^{-1}), if the vesicles radiated uniformly in all directions in a plane, if they proceeded in straight lines, if they fused and increased the wall area locally, and if the VSC moved with a velocity V while remaining at a distance d behind the tip, then, as the authors showed, the shape of hyphae would be mathematically described by:

$$z = r \cot(r/d), \qquad\qquad \textit{Hyphoid}$$

where

$$d = N/V.$$

In this formulation the center of the cylindrical coordinate system is the center of the VSC, r is the radial distance $(r = x = y)$, and z is the height in the axial, i.e., the cylindrical, direction. (I have changed the nomenclature of their coordinate system to agree with the mathematical development for other cases considered in this book.)

This equation describes a mathematical curve called the hyphoid. Therefore, this term will also be used for this model of tip growth. It has two very desirable features: one is that a single equation describes the cylindrical shape of the side wall and the nonpointed shape of the pole, and the other is that it has only a single parameter, d. Many previous experimental studies and the theoretical models for prokaryotes have made a distinction between pole formation and the elongation of side walls and therefore depend on an extra parameter simply because there are two sections to the structure. Earlier models thus presumed a discrete break between pole and the side wall. Justification for this division has been given for the Gram-positive and Gram-negative rod-shaped bacteria. For the growing mycelium of either eukaryotes or prokaryotes, these arguments are not appropriate.

There are three incorrect assumptions in the derivation of this form of the model. First, the published mathematical treatment actually dealt with the two-dimensional analogue of a hypha. The authors felt that the formula would still apply to growth of a three-dimensional shell and gave qualitative arguments. Secondly, both their Monte Carlo simulation by computer and the mathematical equation given above were derived from a model that actually calculated the increase of the cell substance rather than calculating the increase in the surface area surrounding the new cytoplasm. These two problems were later overcome by a more refined analysis by Gierz, communicated to me personally. His results, however, led to a similar, but by no means identically shaped curve. He derived an ordinary differential equation:

$$d\rho/d\beta = (C/(1+\cos \beta)^2 - \rho^2)^{1/2}, \qquad \text{``3-D'' model}$$

for the three-dimensional, "3-D" model, which can be solved numerically for this cylindrically symmetrical case. C is an arbitrary constant and, in a cylindrical coordinate system, ρ is the distance dimension from the origin (placed at the VSC) and β is the angle from the axis of the cylinder.

It should be emphasized that even though the Hyphoid model is unrealistic as presented in Bartnicki-Garcia *et al.* (1989, 1990) and Bartnicki-Garcia (1990), it is still useful because it is the simplest mathematical framework for describing a tube of constant diameter that ends in a nonpointed tip. In the Hyphoid model, the diameter of the hypha at a large distance from the tip is predicted to be 2π times d, the distance from the center of the Spitzenkörper to the tip. It also predicts that the diameter when measured at the level of the Spitzenkörper should be one-half the maximal cell diameter at long distances from the tip. These properties of the curve lead to an easy test for its applicability. Measure the diameter of the cylinder at a large distance from the tip. Divide by 2π to obtain an estimate of d. Measure back from the tip this distance, and then measure the diameter at this point. For the hyphoid equation to apply accurately, this distance should be one-half the diameter of the cylindrical region. For the biology to apply, the distance d behind the apex should be at the microscopically observed center of the Spitzenkörper, if it is visible.

For the more relevant "3-D" model, only two changes are needed. First, instead of dividing the diameter by $2\pi = 6.283$, one should divide the maximum diameter by 8.0 to find d, the distance to be measured back from the tip to locate the center of the Spitzenkörper. Thus, for the "3-D" model the VSC is proportionately closer to the tip. Second, the diameter at this point should be $\pi/4 = 0.7854$ (instead of one-half) of the maximum diameter. Thus, the "3-D" curve proportions are quite different from those of the hyphoid.

In inspecting the curve fittings of either the hyphoid or the "3-D" model to electron micrographs of various fungi, two questions must be considered. Is the degree of fit of the equation to the cell shape satisfactory? Is the point at a distance d from the tip, in fact, at the center of the Spitzenkörper? Because there is only one adjustable parameter, the fitting is easily and unambiguously done when the photograph extends far enough into the cylindrical region. The maximum diameter can be measured and set equal to $2\pi d$ or $8d$ for the two variant models. For *Polystictus versicolor* (Bartnicki-Garcia *et al.* 1989, Fig. 5) the fit for the hyphoid model appears very good, but what I would judge as the center of the Spitzenkörper is 1.5 times d behind the tip instead of 1.0 times d; the fit is poorer for the "3-D" model. For *Armillaria mellea* (their Fig. 7) the hyphoid model fit is good, but the Spitzenkörper is not quite centered on the origin of the coordinate system. In addition to these, I considered three organisms whose electron micrographs or cell dimensions, from the work of others, were presented in Harold's review (1990). *Sclerotium rolfsii* is blunter than predicted and the center of the Spitzenkörper is a factor 2-fold closer to the tip than predicted by the Hyphoid model, but the fit is a little bit better than in the "3-D" model, whereas *Saprolegnia ferax* gives a poor fit indeed. The measurements on *Neurospora crassa* show it to be blunter than predicted by either model. Its radius at the presumed center of the coordinate system is 30% greater than predicted by the Hyphoid model. Similar statements can be made from other photographs presented in the literature, but the criticisms given in this paragraph are possibly minor and do not consider biological variation. Moreover, some of these discrepancies could be attributed to the selection of slightly off-axis electron micrographs.

Both the Hyphoid and the "3-D" model were based on the idea that a vesicle would leave the VSC and travel outward in a random direction, but proceed in straight line. With only slight elaboration of the model this assumption could apparently be modified so that the vesicles leave the Spitzenkörper and travel in an irregular diffusion path until they collide with the wall. At this point I have added, or refined, the idea that the vesicles come into the VSC on cytoskeletal elements, are armed there, and are liberated to diffuse at random not attached to the cytoskeleton. This is the basic idea of the Moving VSC model: the closeness of the VSC to the tip would increase the chances of a vesicle encountering the tip instead of the cylindrical part of the cylinder. It is evident that random motion (see Berg 1983) would lead to a narrower cylindrical region than predicted by either model.

I have not succeeded yet in elaborating the theory in detail for diffusive movement for the "3-D" case. This is because of other conceptual difficulties that appear to render these refinements moot. Because random diffusion could be important, I have saved my attempts for resuscitation in the future when a better biological base is available.

The Inadequacy of the "Hyphoid" and "3-D" Implementation of the Moving VSC Models

It was only when I tried to simulate models that included diffusion that I found I could not develop a definite shape. I could not, in the sense that the shape was initially indeterminate because in my formulation I had added extra flexibility to the system by permitting the outward (radial) movement of some of the yet-unhardened wall. The model used by Bartnicki-Garcia *et al.* (1989) and its "3-D" version give stable steady-state shapes only because both are constrained by the unstated assumption that the new material must remain at the distance from the cell axis where the vesicle had originally hit the wall. However, I had modified the model shown in Fig. 11.5 to approach that shown in Fig. 11.6, under the tacit assumption that Wessels' Steady-state hypothesis, Harold's Soft-spot

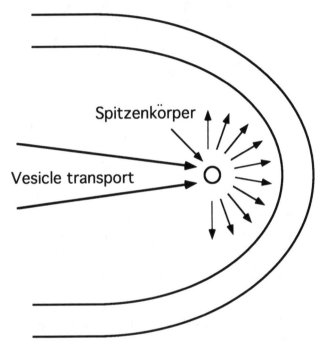

Figure 11.5. Eukaryotic strategy for apical growth according to the Bartnicki-Garcia *et al.* (1989) Moving VSC model. Vesicles are made in the body of the organism and move, via cytoskeletal elements, to the "Spitzenkörper." This organelle, called the vesicle supply center (VSC), is an aggregation of many vesicles, and it is assumed that it exists even in those fungi where it cannot be visualized. The vesicles are activated there and then dissociate from the VSC and move out in all directions. Reaching the wall, they fuse and enable the process of wall expansion. A key point of the Bartnicki-Garcia *et al.* (1989) model is that the VSC moves forward at a constant velocity. The model further assumes that simply because the Spitzenkörper is nearer to the tip wall than to the side wall that more vesicles will fuse in the polar region. A mathematical theory leading to the "hyphoid" equation was derived by Bartnicki-Garcia *et al.* (1989) and used to predict the shape of some eukaryotic hyphae quite accurately. In spite of this success, there are a number of biological and mathematical problems with both the Hyphoid and the "3-D" variant of the moving VSC model, although the 3-D model corrects for two of the problems of the earlier model.

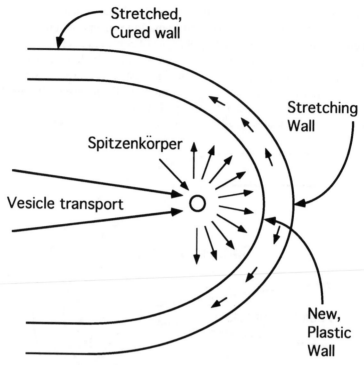

Figure 11.6. A possible synthesis of the Moving VSC and the "Soft-spot model." Figure 11.5 has been augmented to add some of the factors that may determine apical tip shape. These include the plasticity of the new wall and its ability to yield due to the turgor pressure. It is not a model but an outline of a model. Additional factors that must be measured by *in vivo* and *in vitro* studies concern whether vesicles follow random or straight paths from the VSC, the kinetics of deposition of vesicles, the kinetics of the change of degree of viscoplasticity, and the influence of turgor pressure in causing the vesicles to fuse with the wall.

hypothesis, or my own Variable-T model was functioning. It seems necessary to hypothesize that new wall added near the tip is sooner or later moved peripherally and eventually comes to be no longer part of the tip, but instead part of the cylindrical region. Given this extra flexibility of peripheral movement, one can assume that a new increment of area in an annulus around the tip not only changes the slope of the tip in the immediate vicinity but also in part migrates outward. These considerations point out that in a full model other factors can be important in determining the shape of the tip, and that the velocity of the VSC is not a sufficient condition to define the shape. How fast the wall hardens, the relative amounts of movement in the radial and in the axial direction as a function of distance from the tip, how the viscoelasticity changes with time after vesicle fusion, and the effect of turgor pressure on growth are other factors. Without additional constraints, the system is "underconstrained" (Kuznetsov 1991).

The Role of Surface Stress

In the Moving VSC model in either version there is a logical deficit because these models are not concerned with what controls the rate of tip growth. In the absence of vacuoles, surely enlargement is not autonomous, but is controlled by the success of the organism in accumulating resources and processing them to make cytoplasm; wall growth is only a response to such success. This is, of course, the keystone of the surface stress theory preached in this book and in fact is the theory's second postulate. Bartnicki-Garcia *et al.* (1989) pointed out that their theory would operate just as well whether the Spitzenkörper were pushed by the cellular mass or pulled by the tip, but they did not address the question of what could make either movement possible.

The surface stress theory holds that the growth of cytoplasm in the body of the cell tends to increase cellular pressure and this in turn must force both the tip and, in this case, the Spitzenkörper to move and lead to the elongation of the cylindrical region. Because the cell's turgor pressure is felt only at a wall separating the cell from the outside, the controlling force (or at least the intermediate control point) has to be the tension developed in the wall at the tip. Tension, at least momentarily, must then increase if cellular wall enlargement lags behind volume increase, and vice versa. If tension in the wall facilitates the fusion of vesicles or their action, then elongation of the hypha will be linked to cellular growth *per se*. Thus, we must imagine that the partially hardened wall is a viscoplastic material that not only is deformable by stress but is also rendered more amenable to incorporation of new wall material from recently arrived vesicles. It must further be considered that the very tip is most plastic because it was most recently formed by accretion of new vesicles.

Fig. 11.7 shows diagrammatically a non-biological demonstration of the key elements of the Soft-spot model. This exercise makes two points relative to apical growth of fungi: first, that fusion takes place more readily with the younger wall; and second, that fusion takes place more readily with a wall that is under tension (in the biological case, due to turgor pressure). These assumptions can be readily demonstrated with commercial "20-min" or with "Two-ton" epoxy preparations. Films about 1 mm thick and 3 mm wide are made on a rubber sheet (actually I use a piece of rubber tubing with the second epoxy). The figure shows that after only 5 min the rapidly polymerizing epoxy, no longer admits freshly mixed epoxy, but when stretched even at 10 min would allow the fusion of the droplet of newly prepared epoxy. For the second epoxy, the same results are obtained, but the time scale is prolonged. Such a film of 35-min old material is found to accept a newly mixed droplet of epoxy less well than a film of 30-min old material does. When the rubber support is stretched at the 40th min, it

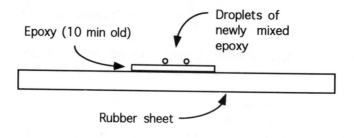

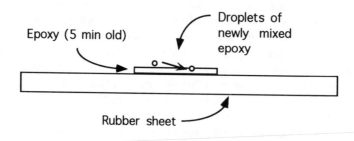

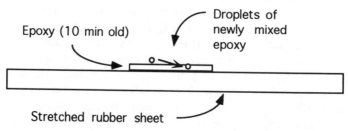

Figure 11.7. An analogue of the "soft-spot" model. The soft-spot model is at least 100 years old and has been modified and discussed by many workers; it holds simply that the newest portion of the cell wall is soft and can take up wall precursors in a form of material that gradually hardens. Depending on the kinetics and spatial location of the several processes, the shape of the apical tip is determined. Here is presented an experimental analogue to show whether this type of model can explain the growth needs of the cell. Freshly mixed epoxy is applied to a rubber sheet or tubing. After an appropriate time, droplets of freshly mixed epoxy are added, and it is observed whether they remain upon the surface or fuse into it. The top frame is a control in which enough time has elapsed so that the band of epoxy has hardened too much and droplets do not fuse. The middle frame shows that, at an earlier time when it was more plastic, it would accept a droplet of fresh epoxy. The bottom frame shows that increased tension in the viscoplastic material favors fusion. This experiment was repeated twice once with "20-minute" epoxy and the second time with "Two-ton" epoxy, each time with the same results, but for the latter material the times were different.

is found that a freshly mixed droplet is still accepted and is integrated into the layer.

Consequently, the modern version of the Soft-spot model is that successful fusions of vesicles to the wall occur near the tip, and that creates a gradient of the average age of the wall away from the axis. This favors further vesicle fusion (of competent vesicles) to the tip. This would occur particularly under conditions where the turgor pressure is high, leading to tension in the substratum. This mechanism can account for Reinhardt's 1892 observation and the more modern versions (reviewed by Wessels in 1990a,b).

From these considerations, it can be anticipated that a much more elaborate model that includes the mechanical engineering of viscoplastic materials will be presented to account for the growth of fungal apices; it will have many parameters to be added to the V and d of the Hyphoid and 3-D model. Other parameters of obvious importance will have to do with the rate of curing of wall, the rate of vesicle fusion, and the action of stresses in favoring fusion of the vesicles.

Outlook for the Future

Much remains to be done in order to understand the hyphal mode of growth. For prokaryotes, the prime challenges are to measure turnover of wall components at the very tip and to test critically for the presence of contractile proteins. This first requires brief pulses of tracer, effective chases, and sampling of the chemical species that should be released by turnover. Critical use of high-resolution electron microscope autoradiography might also be valuable. The second may be accomplished when the sequence of the entire genome of any prokaryote is completed. (It is nearly completed for *E. coli,* and still there is no published evidence of working contractile protein.) For the eukaryotes, it now seems that the Bartnicki-Garcia *et al.* (1989) Hyphoidal model and the more solidly based "3-D" model not only do not fit the range of tip profiles in the literature but are constrained in a way that nature is not. They have not been replaced with more appropriate models, however. We need to understand how (or whether) biomass growth drives fungal hyphal elongation and to explore and extend the mechanisms presented here. Maybe the problem will be solved during the second century of our fascination with the process of hyphal growth.

12

Twisting and Rotation During the Growth of Gram-Positive Rods

KEY IDEAS

Bacillus subtilis *grows in such a way that one end rotates relative to the other.*
Rotation of the ends does not appear to be due to helical insertion of murein.
During the inside-to-outside growth process the peptidoglycan experiences different
    stresses in different directions in different levels of the wall.
Because total stress is twice as great in the hoop direction as in the axial direction,
    splitting tends to occur in the direction of the longitudinal axis. The older, more
    peripheral wall is stressed in the direction of the elongating cell axis, and this
    favors circumferential cracks.
These two processes combine to form a system of helical cracks, grooves, or
    crevasses, resulting in the rotation of one end of the cell relative to the other during
    growth.
Variations on this mechanism may account for curved rods and vibrios.

Introduction

This chapter covers the topic of processes that break cylindrical symmetry; that
is, the processes that curve or twist the organism so that it is not a radially
symmetric entity around a straight central axis. Among these symmetry-breaking

processes are the rotation of one end of the organism relative to the other end; this occurs during growth of the rod-shaped Gram-positive bacterium *Bacillus subtilis*. The growth process is examined below, along with the possibility that it may be the paradigm for other morphological cases. The qualitatively different case is that of the spirochetes in which the cell is wound around one or more central axial fibrils or filaments. One may think of these as flagella that are retained internally. This arrangement prevents them from interacting with a host immune system. The net effect is that the organism is pulled into a spiral and can move by a flexing or rotating motion, thus pushing forward in a screw-like motion. This is an effective method for movement in a highly viscous environment. Such organisms will not be considered further here, because there is nothing more to add concerning the details of the connection of the filament from end to end, how the tension builds up to cause the spiral, or how flexing and rotation are achieved.

Relative Rotation of the Poles During the Growth of Gram-Positive Rods

Three key facts have emerged through the study in many laboratories concerning the growth and division of *Bacillus subtilis*. These are: (i) that the mechanism for cylindrical elongation is an inside-to-out one analogous to the formation and molting of the outgrown skins of snakes or insects, (ii) that one end of the cylinder rotates relative to the other as growth takes place, and (iii) that the outside surface of the wall is very irregular and exposes a much larger surface than that facing the cytoplasmic membrane. These facts, together with an analysis of stresses in different planes within the wall, lead to a biophysical explanation for the source of torsion in certain prokaryote systems, leading to twisting. Though helical cracks have not been observed, it is postulated that they form during growth due to stress distribution in the side wall, and that this leads to the rotation of one end of a rod-shaped organism relative to the other. This, in slightly different situations, causes other manifestations such as curved rods, spirals, double-helical patterns of growth, etc. The hypothesis that torsional stresses develop as the result of inside-to-outside growth is an alternative to Mendelson's conjecture (1976) that ". . . the cell surface components in rod-shaped cells are organized in a helical pattern. The addition of new cell surface, during growth, follows this helical organization."

The evidence on the first point is now clear; it is that the side wall of the Gram-positive rod is made by an inside-to-outside growth mechanism (see Chapter 9). The key point of the inside-to-outside growth process is that a layer becomes stretched as the cell grows, and it comes to support the tension once borne by more peripheral wall before it was autolyzed. When this wall, for the first time, is cleaved by autolysins and cracks form, they underlie cracks present in the more peripheral layers. These cracks, grooves, or crevasses temporarily lower the tension within a layer.

The second key fact is that one end of the growing cylinder rotates relative to the other. This fact was adduced by Mendelson (1976). He observed that the growth of chains of cells led to supercoiling of cell filaments when the ends of the chains remained attached. He studied germinating spores from a division-inhibited mutant that also maintained its attachment to the spore coat. Growth produced double-stranded helices of cells similar in structure to a covalent closed supercoiled DNA (Fig. 9.4). The same observation can also be made from published pictures of other authors for a variety of conditions and organisms; thus, rotation is a general phenomenon. All that is required is a chain of growing cells that loops out of a growing mass and returns to its starting point. The rotation is not a result of the constraint of closed circular topology, because I observed directly under phase microscopy such rotation of unconstrained growing filaments of wild type *B. subtilis* in depression slides (Koch 1990e). Fig. 12.1 shows the readily carried out experiment of growing a filamentous organism in a depression slide. A portion of the filament attached to the glass cover slip can be observed as it grows to lift itself off the surface, rotate, and fall back to the glass surface as if rotational torsion were developing in the filament as it grew.

The third key fact relative to Gram-positive rod growth is the finding that the outer surface of the wall of a nominally smooth-surfaced strain of *B. subtilis* is not as smooth as the inner face and exposes a much larger surface area (Fan *et al.* 1975; Anderson *et al.* 1978; Sonnenfeld *et al.* 1985). It was found by the latter authors (see Fig. 12.2) that a much larger number of sites (more than 50-fold greater; Sonnenfeld, personal communication) is available for the adsorption

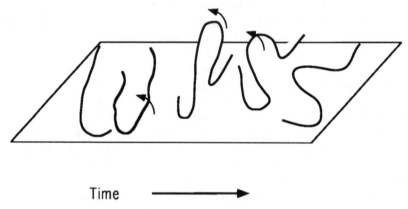

Time ⟶

Figure 12.1. Diagrammatic representation of the movements during growth of a filamentous mutant of *B. subtilis* growing on a glass surface. On the left-hand side, the filament is initially lying flat on the surface. As growth takes place it not only elongates, it gyrates. Although most of the filament most of the time adheres to the glass, there are episodes where a region of the filament rises off the glass and, as shown for the second, third, and fourth idealized time frame, rotates through the medium, and comes to rest again on the glass surface having completed a 180° rotation.

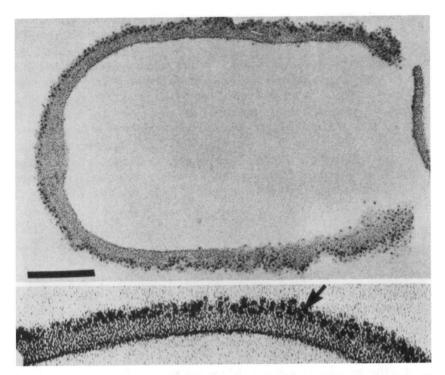

Figure 12.2. Distribution of cationized ferritin granules used to stain a section of wall of *Bacillus subtilis*. It can be seen that the positively charged ferritin probes adhere, in much larger numbers, on the outside. This and the morphology of the surfaces suggest that the inner wall is smooth, but that the outer wall is very rough. No doubt this is caused by the action of the autolysins, but because of the distribution of stresses is not simply at random, it could be formed in a way to make a series of helical grooves on the outer face leading. Reprinted from Sonnenfeld *et al.* (1985).

of cationized ferritin on the outside than on the inner face of the wall, indicating a highly irregular outer surface. The more extensive external surface is probably the consequence of cleavage of more highly stressed bonds, resulting in isolated ("frayed") regions of less stressed wall that persist for a long time. Eventually, these are autolyzed enough to be lost from the cell surface. In localities where there are imperfections, e.g., where bonds had failed to form or had previously been ruptured, the local stress on neighboring bonds is larger; there further autolysis will be mechanically favored. This is just another application of the Griffith principle (1920) of the domino theory for the propagation of cracks.

The rough nature of the outer surface is also apparent from the turnover kinetics of the side wall. It is observed that a pulse of tracer compound enters the wall, moves through the wall with none being lost, and then that label from the side wall is lost over several generations in a gradual (exponential) way, the half-time nearly equal to the doubling time of the bacteria (see Koch and Doyle 1985; Doyle *et al.* 1988 for a review of turnover). If the outer wall were smooth and of fixed thickness, one might expect an abrupt and almost complete loss of a pulse of label within a quite brief period by an inside-to-outside growth process. The observed half-lives during this phase are nearly equal to the doubling time in many experiments in the literature and this relationship is consistent with turnover being due to the stress generated by the growth process (Koch and Doyle 1985). It is argued that from the exponential loss during this phase most of the autolysis is due to a system of eroding peripheral cracks. The formation of a system of cracks is consistent with the physics of solids, and it is observed in nature under many circumstances, such as geological erosion or aging of solids.

Biophysics of Wall Development

Turgor pressure inside the bacterial cell creates stresses on the enclosing peptido-glycan sacculus. The precise distributions of the stresses depend on the shape of the cell as well as the turgor pressure. Man-made pressure vessels are fabricated and then subjected to pressure. The Gram-positive cell, however, builds the wall structure while the cell is under pressure. It does so in a two-stage process: the constructive insertion of wall material through the cytoplasmic membrane and subsequent polymerization, and the destructive activity of the autolysins on the periphery of the wall. This means that the wall polymer is in different states of strain and intactness in layers formed at different times. In the following sections I outline the interplay of these physical and biochemical ideas leading to helical cracks.

Stress Distribution in the Thick Walls of Gram Positive Rod-Shaped Bacteria

The cylindrical wall of a typical Gram-positive bacterium is thin (25 nm) compared to its radius (400 nm). The fact that mechanical engineering analysis of such a shaped pressure vessel with a uniform wall demonstrates that the circumferential (or hoop stress) is dominant has been alluded to many times in this book. The stresses in a thin-walled pressurized cylinder are twice as large in the hoop direction as they are in the axial direction (see Chapter 6). The hoop stress is greater than the stress at any other angle, and therefore, in this ideal homogeneous wall, cracks would tend to form parallel to the long axis of the cell.

Because the cell is a growing system, however, the cracks will not be parallel to the long axis because of the accumulation of partially stress-bearing, partially fragmented, older wall outside the most stressed intact layer (Fig. 12.3). The contribution of patches of semi-intact wall located more peripherally tend to swerve the cleavage plane toward the perpendicular to the axis of the cell. The reason for this can be readily grasped from a series of four mechanical analogies that at the end approximate the geometry of the growing *B. subtilis* cell.

First, imagine that many sheets of rubber are glued together. Imagine that the topmost layer, however, has been stretched in one direction. The next layer has been stretched to a lesser degree in the same direction, and so on, so that the bottommost layer is not stretched at all. This is shown on the left side of Fig. 12.3. Imagine that, after the glue has set, the whole assembly is then progressively stretched in the same direction. The topmost layer will fail first and will crack or tear at right angles to the direction of stress, but the structure as a whole will remain intact. With further stretching the second layer will crack. This crack will preferentially underlie the pre-existing crack on the outermost layer, because of the partial support that the fragmented outer layer still provides. With further stretching, a new crack will develop, splitting both layers and again with a preferential direction at right angles to the direction of stretch. Further stretching will lead to a crack in the third layer, again underlying existing cracks, but it will also lead to the formation of a new crack penetrating from the top down through the top layer and then the next layer. This is shown on the right-hand side of Fig. 12.3. At the end a force sufficient only to rupture a single sheet when applied continuously will finally suffice to rip the structure in two.

In the case of growth of the Gram-positive rod, the inside-to-outside growth process and the elongation have a dynamic character. For the second analogy

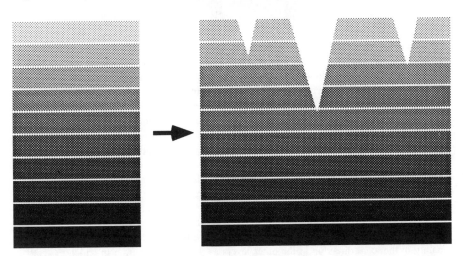

Figure 12.3. Illustration of a hypothetical constructs leading to groove formation. See description in the text. Successively stretched strips of rubber are glued together and then the whole assembly is stretched until various portions of the rubber break.

(see Fig. 12.4), let us additionally assume, that as stretching continues, new unstressed rubber sheets are periodically glued onto the bottom of the stack, the outermost layer is continuously dissolved or cut off (corresponding to the autolysin action in wild type cells), and the ensemble is continuously elongated. In such a dynamic steady-state situation, the stack will become longer in the direction that corresponds to the axis of the cell. Progressively longer unstretched layers

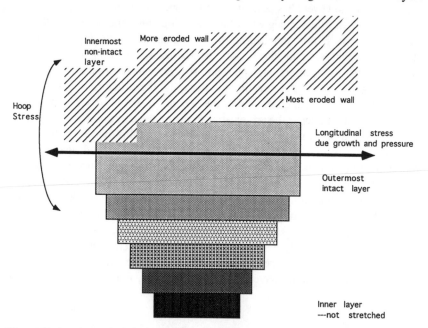

Figure 12.4. An exploded view of a small portion of the side wall of a growing Gram-positive rod. The axis of the bacterial rod is indicated by the double pointed arrow. The stresses are shown with double-headed arrows. The nascent wall bears no stress either as extruded through the cytoplasmic membrane or immediately after being crosslinked to form the most innermost part of the wall fabric. As the cell grows, it is underlain with new wall, pushed outward, and stretched. It is stretched only slightly in the circumferential direction, but a great deal in the longitudinal direction. If only one homogeneous layer bears all the stress due to turgor pressure, then for physical reasons, the stress is twice as strong in the hoop as it is in the axial direction, but that is not at all the case here. After cleavage events start to occur (indicated by diagonal gaps in the shortened upper segments), the wall becomes fractured. The fractures would be parallel to the axis of elongation if the wall were homogeneous; however, because there are wall fragments on top of the intact layer they are on a diagonal. Partially cleaved wall, however, continues to bear a portion of the stress through connections to the underlying wall. The stress is mainly in the axial direction in the outer layers as this is the direction in which the cell is elongating. These interactions modify the stress distribution and, consequently, the fracture pattern in the lower, hitherto intact layer. Eventually wall fragments are liberated into the medium.

of rubber must be glued onto the bottom as the layer immediately above it becomes stretched. In this case, cracks will initiate, penetrate through a part of the stack, and finally appear to go no deeper because the rate of penetration equals the rate at which the layers are added and other layers dissolved. Pairs of such cracks will move apart from each other with growth, and new cracks will be initiated in between.

For the growth of the cylinder of the Gram-positive rod, the hoop stress in a cylindrical pressure vessel would tend to give the cracks an axial component, and this must be combined with the dynamic growth process just outlined. Therefore, the third version of our rubber model analogue would simulate the effect of turgor pressure in a vessel with cylindrical symmetry by the addition of another stress to some of the middle layers in the stack of rubber sheets, but at right angles to the existing stress. As in the previous case, an existing crack would favor further cracking, or more accurately, the ruptured outer layers would still give some support and protect regions in the lower layers from tearing. The effect is the deepening of existing crevasses, with the cracks progressively assuming some oblique angle. Initially, some chance imperfection determines the angle, but eventually it is the growth processes that will determine and maintain the angle.

With a final modification of this mechanical analogy, we will have all the important elements of the inside-to-outside growth process of the cylindrical elongation part of the Gram-positive rod's growth. This modification is generated by wrapping the previous version to form a cylinder, capping it, and subjecting it to an internal hydrostatic pressure. Of course, it is necessary to continue the formation of the new wall and the turnover process of the old. In the dynamic state of inside-to-outside growth this would result in helical cracks on the cylindrical surface. For a special set of conditions these helical grooves would lengthen indefinitely as growth proceeds (Fig. 12.5). But even though the material lining the edges of the grooves and the material between the grooves are continuously thinned by turnover, as in the second analogy, grooves in the steady state would persist with a constant cross-sectional shape and constant pitch, but elongating in a spiral fashion as the cell cylinder grows. As cracks are propagated inwardly from outer layers, the cell would be able to elongate further. Subsequently, when the layer has been elongated 2-fold (one doubling time after the incorporation of a layer) the autolysin turnover processes in wild type *B. subtilis* will have continued to widen the cracks and have cleaved enough bonds so that some of the wall material would begin to be shed from the cell. After another doubling time, 50% of the layer will have become autolyzed (because the half-time for turnover is also equal to a doubling time), and the original layer will have been distributed in patches to cover nearly four times the original surface area. This illustrative case is explicitly described in Fig. 12.5.

The pitch of the helical grooves is increased by the stresses due to elongation in the outer layers and is decreased by the dominant hoop stress in the inner layers. If the balance between these processes is just right, as in the case discussed above, coherent helices will form that elongate coordinately with cellular elongation. Then no new cracks will form and the existing cracks, given their dynamic character, will elongate with constant pitch. Otherwise, either the spacing between cracks will increase and new cracks will appear, or old cracks will make tight spirals and finally stop while new cracks open up in between such sites.

It can be seen that, in this model, the three key facts about the physiology of Gram-positive rods fit together with consideration of stress distribution and activation of autolysins due to stress on their peptidoglycan substrate during growth to generate a system of helical cracks on the cell wall. In the next section it is shown that this growth process leads to rotation of one end relative to the other.

Rotation During Growth

The continuing propagation of cracks in a helical pattern of fixed handedness as proposed generates rotation of one end of the cell relative to the other in the same way that insertion of material in a helical pattern would do, as in the proposal of Mendelson (1976). His model will not be considered further because there appears to be no mechanism that could accomplish helical insertion of new material between segments of old wall throughout the entire thickness of the wall (see Koch 1988b). The physical forces creating the grooves cause twisting of one end of the cell relative to the other in the same way that unraveling a cardboard tube (such as used for mailing or the tube at the center of a roll of

---

Figure 12.5. Two diagrammatic views of a crack in the external surface of a Gram-positive rod. The cracks are formed by autolysin action; the crack direction is more probable at right angles to the direction of maximum local stress. Probably the bulk of the autolytic events occurs on the external surface because of the accumulation of autolysin on the surface. If the physical properties of the wall were uniform, then the cracks would form in the axial direction, but the cracks tending to go in the hoop direction are found in the outermost region because the cells are elongating, and consequently grooves form because of the support by patches of old wall. This interplay results in grooves that wind around the cell in a helical fashion. A shows the cross-section to the grooves. B shows a view from the surface. All such grooves generally have the same handedness on a given cell and its descendants. There may be shorter cracks with a different orientation. The cracks grow longer as the cell elongates so that the density of grooves stays more or less constant. At cell division and separation, two cells are formed that have the same handedness as the maternal cell. New cracks may form and some old ones may terminate. The figure is unrealistic in that the grooves probably could be very numerous and much more irregular. The grooves would be very narrow at their bottom. The continued deepening of the crack combined with the inside-to-outside pattern of growth causes one end of the cell to rotate relative to the other.

A

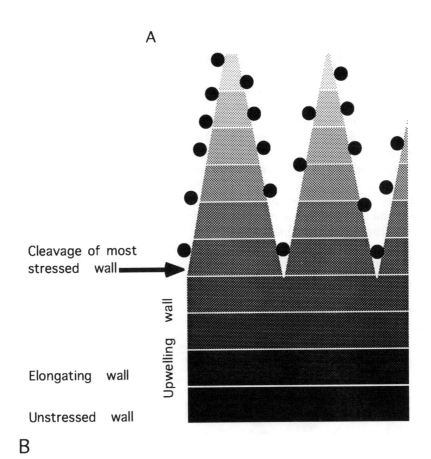

Cleavage of most
stressed   wall ➔

Upwelling   wall

Elongating   wall

Unstressed   wall

B

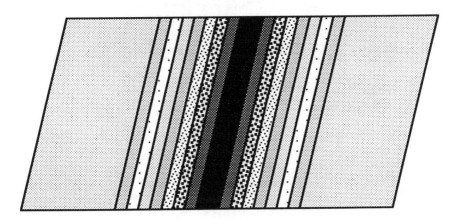

335

toilet paper) along the helical grooves while keeping the diameter constant leads both to rotation of one end of the tube relative to the other (see Fig. 12.6) and to elongation. Because there is tension in both the axial and hoop directions at the bottom of an existing crack, with rupture the sides of the cracks move apart both axially and circumferentially. Deepening of the crack will rotate cell wall material at one end of the structure relative to the other. The argument is the same no matter how dense the grooves are. Thus, the helically wound system of remaining peptidoglycan ribbons of cell wall will twist as the cracks propagate inward and autolysins function to digest the outer surface.

For this helical groove mechanism, the source of energy for the rotation comes directly from the turgor pressure. This pressure is generated by the osmotic pressure differential between the inside and the outside of the cell. This elicits stress and the development of a hydrostatic turgor pressure inside the wall. As the cell contents enlarge, pressure-volume work is done, and in this case, because it occurs via crack extension, rotation of one end of the cell relative to the other takes place. Thus, the rotation of the ends is powered by the biochemical energy

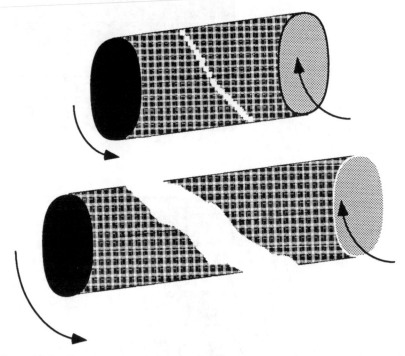

Figure 12.6. Rotation caused by helical cracking. The upper part shows diagrammatically the incipient cracks in the outermost intact cylindrical layer. The lower part shows the cracks after they have formed in this layer. Generally, narrower cracks underlie wider cracks in more outer layers. The consequence to the cell of the crack enlargement as the cell elongates is that the two ends will rotate in opposite directions from each other.

that the cell mobilized initially in the act of accumulating solutes against a concentration gradient.

The morphological effect of the relative rotation of the ends varies in different cases. There is simple rotation in the case of *B. subtilis* filaments. On the other hand, the rotation may distort a cylinder into a comma, vibrio, or spirilla shape. It appears that filaments of *E. coli* do not rotate as they grow. Similar shapes form when a spring is over-or underwound. Not only is a spiral habit of growth the normal morphology for many prokaryotes, but it can be induced in many instances by treatment of rod-shaped organisms with growth inhibitors. For the cases of the Gram-negative organisms, the process described above may function as indicated except that the outer membrane may serve the supporting role ascribed to the outer layers of the Gram-positive rod.

There has been only one explicit suggestion in the literature (Koch 1988a) of some other mechanism whereby peptidoglycan could be laid down in a roughly helical pattern. This model is that in the main stress-bearing layer cleavage would occur and partially orient the wall, so that new wall would tend to orient in parallel. This could be a mechanism for the insertional model. But it would be no better able to account for the observed influence of strain differences, ionic environment, temperature, and inhibitors on the handedness of macrofibers (reviewed by Briehl and Mendelson 1987; Mendelson and Thwaites 1988). Seemingly, these environmental and genetic changes do not affect the primary structure of peptidoglycan and, in fact, even in a fixed genetic background permit the handedness to be quickly changed. On the other hand, the critical experiments of Briehl and Mendelson (1987) demonstrate that under particular environmental conditions the handedness of the microfibers depends on the genotype and physiological environment and not on templating due to pre-existing wall.

A possible merit or demerit of the model is that it does not predict the handedness of the rotation other than to predict that the handedness will remain, ordinarily, the same as the handedness of ancestral cells. Of course, if the wall were removed to form spheroplasts or protoplasts (as in the experiment of Briehl and Mendelson 1987), or if autolysins (or externally added lytic enzymes) had a chance to act in the absence of growth to smooth the outer surface, then the handedness might change when the cells are allowed to grow, because in the model presented here the handedness of the rotation of the ends of a cell is not predetermined. The handedness might be determined, however, in relationship to the predominate direction of rotation of flagella, for example. Possibly, the relative proportions of the autolytic cleavages due to the two types of autolysins in some way control the handedness. Other possibilities may be imagined, and Mendelson and Thwaites (1988) have listed some, but the point is that the handedness is determined by the genotype and environment and is not explained by either conjecture without additional assumptions.

Systematic helical insertion has not been shown, and the recent work from Archibald's laboratory (Merad *et al.* 1989) demonstrates that at least the teichoic acid component of wall is laid down on the inside of the wall as a uniform cylinder and not in helical bands. That argues against helical insertion. On the other hand, negative staining techniques have not shown helical grooves. That argues against helical crack formation. But the resolution of the electron microscope technique is probably too low to have shown surface cracks. The increased external area relative to the internal area is clearly demonstrated by the binding of cationized ferritin (Sonnenfeld et al. 1985); this result is consistent with the formation of helical grooves and random dissolution of external peptidoglycans, but is by no means diagnostic. There have been several studies from Nanninga's laboratory (Verwer et al. 1978, 1980) with the electron microscope that show that there is a pattern in the peptidoglycan that can be visualized in both Gram-positive and Gram-negative organisms after treatment with enzymes or ultrasound. These papers have been used to support the claim that the oligosaccharide chains run transversely around the cell. Close examination shows that they might better support the idea that the patterning is a tightly wound, but probably not very precise, helix. This would support both conjectures but not distinguish between them. Perhaps these microscopic techniques could be exploited further to lead to a critical experimental test of the existence of a helical pattern in the walls of bacteria.

Other Cases of Noncylindrical Symmetry

Spirals and helices occur in many guises in microbiology. How they arise is a question not addressed critically heretofore for filaments formed by prokaryotes. Plants form spirals in a variety of situations. Higher plants are much more complex than bacteria, however, and have many more possible mechanisms for the generation of helices. The model proposed here is important because prokaryotes have neither the contractile proteins (such as actin) nor the cytoskeletons for a biophysical mechanism that in eukaryotic systems probably serves for the development of helices. A key point developed here is that rotation of one end relative to the other during growth can result from differential stresses and, at the different levels within the stress-bearing wall, in a way that is independent of interaction with other cells or surfaces.

   In my previous attempts to analyze the biophysical aspects of wall growth of Gram-positive rod (Koch 1988a,b), I pointed out that the direction of stresses in different layers varied, but I did not consider the combined effect of the tensions in different layers working together. This interaction means that the wall has quite different properties and that analogies to composites, fiber glass, and reinforced concrete are appropriate to some degree. The interplay of tensions would lead to helical grooves developing in the outer surface. As the sides of cracks widen, the cell elongates and its two ends rotate with respect to each other. This is a self-propagating process based on the Griffith principle, and the angle of the helix is determined by a balance between peptidoglycan mechanical properties, synthesis rate, and autolysin kinetics.

13

The Structural and Physiological Roles of the Layers of the Envelope of Gram-Negative Bacteria

KEY IDEAS

The envelope effectively contains six layers plus the periplasmic contents.
There are 10^5 lipoprotein molecules per cell holding the outer membrane to the murein layer.

Wall physiology has been studied largely by perturbation with "osmotica" and now with "periosmotica."

Osmotica, such as sucrose, penetrate the OM but not the CM.

Periosmotica are water soluble but too large to pass through the porins.

The phase microscope is useful because of the dark or reversed phase appearance of shrunken cells.

High molecular weight additives can be used to balance out the index of refraction of the cell to allow the visualization of internal structure.

Glycerol causes transient plasmolysis and distorts the wall structure.

Cellular adaptation to osmotic challenge involves many processes and can be fast.

The first response to osmotic up-shock is water extraction and collapse and wrinking of the wall.

Cryofixation and freeze substitution have allowed us to discern many subsequent responses to cellular collapse.

With mild up-shock, polar plasmolysis spaces and spaces in the neighborhood of a developing constriction site are common.

With the formation of endocytotic (and exocytotic) vesicles, laminar and side wall spaces may develop.

With strong up-shocks, the cytoplasm may be so severely shrunken that its surface is irregular, because it either becomes rough, becomes disk shaped (Scheie structures), or tubular structures develop (Bayer structures).

The basis for all these changes is that bilayers cannot contract because they form a nearly incompressible two-dimensional fluid.

The periplasmic space, in the absence of challenge, accounts for less than 10% of the cell's water content.

Analysis of the geometric restriction due to bilayer incompressibility generates shapes for plasmolysis spaces consistent with electron microscope visualization.

The Envelope of the Gram-Negative Bacterium

Although the envelope of an organism like *Escherichia coli* is usually thought to consist of three layers, it is functionally composed of five or six layers that together serve to accommodate the organism's needs in obtaining medium constituents, preventing toxic material from entering, maintaining cell shape, and resisting the osmotic forces that otherwise would tear the cell apart (Beveridge 1981; Cota-Robles 1963; Donachie *et al*. 1984; Graham *et al*. 1991; Koch 1988b; Leduc *et al*. 1989). Conventionally, the envelope is divided into three structural components: outer membrane (OM), murein (M), and cytoplasmic membrane (CM). The six physiological layers (Fig. 13.1), however, are the hydrophilic lipopolysaccharide portion of the outer leaflet of the OM, the hydrophobic inner leaflet, the 100,000 molecules of Braun's lipoprotein connecting the OM to the M layer, the monolayered stress-bearing peptidoglycan M layer, and the two hydrophobic leaflets of the CM. The two leaflets of the CM are different because certain proteins interact with one, the other, or both layers. There are intrinsic and extrinsic proteins in the CM; there are intrinsic proteins in the OM, such as

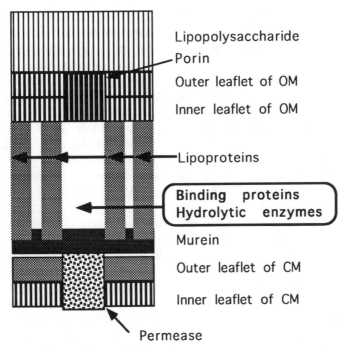

Lipopolysaccharide
Porin
Outer leaflet of OM
Inner leaflet of OM

Lipoproteins

Binding   proteins
Hydrolytic   enzymes

Murein

Outer leaflet of CM

Inner leaflet of CM

Permease

Figure 13.1. Anatomy of the Gram-negative cell wall. The layers are not drawn to size, but rather in a way to show their connectivity.

porins. In addition there is a periplasmic space contained between the OM and the CM. The proteins in the periplasmic space are very important; these are needed for enzymatic cleavage of those nutrients entering from the environment that cannot be taken up into the cell directly. Also present in the periplasmic space are binding proteins that ferry the nutrients from the periplasmic space to the transport sites (permeases) on the CM so that nutrients can enter the cell by carrier mediated transport (see Chapters 4 and 5). Although many of the activities that take place in the periplasmic space are understood, the size of the periplasmic space is debatable (Ferguson 1990, 1992; Hiemstra *et al.* 1987; Hobot *et al.* 1984; Wielink and Duine 1990; Schwarz and Koch 1995). Also debatable is the role of the membrane-derived oligosaccharides (MDOs) that are secreted into the periplasm when the cells are grown in low osmotic strength medium (Kennedy 1982).

Passage Through the Outer Membrane is Limited by Penetration of Solutes Through the Porins

Seemingly nothing could penetrate the OM of the Gram-negative bacterium: hydrophobic molecules should not go through the hydrophilic outer leaflet layer, and hydrophilic molecules should not go through the hydrophobic inner leaflet. This isolationism is useful to a cell that must avoid the challenge of bile salts and bile acids while in the intestine and must avoid many antibiotic and toxic chemicals present in all stages of its global life cycle, including those encountered during the trip from one gastrointestinal tract to another. Of course, the organism would starve to death unless it had devised a special transport mechanism to allow passage through the OM for a select class of compounds. The gatekeepers mediating permitted entry are the porins. The porins (Fig. 13.1) are trimers of a protein, each of which forms hydrophilic channel passing entirely through the OM. It is this trimeric arrangement that permits cellular control of porin function by blocking passage through porin monomer channels in particular situations (Nikaido 1993). These pores, when open, exclude water-soluble molecules bigger than about 600–700 Mol. Wt., while they let smaller molecules diffuse through freely. Thus, saccharides, amino acids, and nucleotides can pass through, as well as some small oligomers of these substances, such as short oligopeptides. Many antibiotics penetrate slowly or not at all.

The study of the physiology of the periplasmic space is fundamental to an understanding of the strategy of the Gram-negative organism. The osmotic properties of this space and of the cytoplasm have been intensely investigated. The porins permit small solutes to penetrate the OM readily, but many substances cannot penetrate the CM. Examples of this class are sucrose, certain small saccharides, and many inorganic salts. We will use the term "osmotica" to include the small molecules that do not readily pass through the CM. Recently, a class of molecules that are too big to pass the porins but can still have an osmotic effect, has been studied (Koch and Schwarz 1995). We have designated these substances periosmotica (Fig. 13.4) because they remove water from the periplasmic region that surrounds the cell proper. In our experiments we have used three molecular probes: Alpha cyclodextrin (ACD), Mol. Wt. 972; a polyvinyl pyrrolidone (PVP) preparation with an average Mol. Wt. of 10,000, designated PVPL; and another with a nominal Mol. Wt. of 40,000, designated PVPH. ACD is one of three Schardinger dextrins; they are created by a bacterial carbohydrase that acts by splitting starch without adding water, using instead a hydroxyl group at the end of the polysaccharide chain as the electrophilic donor. ACD has six glucose residues. PVP is a synthetic polymer that was used as a blood extender in World War II, before it was found to be carcinogenic. Although the effect of these substances is to extract water out of the periplasmic space and the cytoplasm, the action of these substances is limited because of the small osmolar concentration that can be obtained with them. The solubility of ACD is limited by its tendency to form hydrogen bonds with itself and 7.8 mosm is the maximal final concentration used in our studies. PVP is a water-soluble compound and when the smallest available size is used (10,000 Av. Mol. Wt.), a stock solution of 0.5 g/ml leads to a final concentration of 25 mosm. With a 40,000 Av. Mol. Wt. preparation, a 0.5 g/ml solution could be maintained only at elevated temperatures, and then the final concentration was only 6.25 mosm. These maximal concentrations are not sufficient to dehydrate the cytoplasm, as, for example, 500 mosm NaCl does very dramatically. On the other hand, they shrink the periplasmic space and block the formation of plasmolysis spaces, in contrast to osmotica. With this blockade, these agents result in the cell's contraction by wrinkling of the membrane and not in forming plasmolysis spaces when additionally challenged with osmotica.

Contrast with the Paradigm of the Higher Plant Cell

For longer than a century plant scientists have studied "water relations" in plants because an understanding of wilting and turgor are essential for agriculture and horticulture. The ability of a plant to resist drying out is of utmost importance. When plant tissue is treated with a sucrose or salt solution of sufficient concentration, plasmolysis takes place and the plasma membrane, corresponding to the cytoplasmic membrane of bacteria, is forced to separate from the wall and adapt to surrounding the smaller volume of cytoplasm resulting from response to the extraction of cellular water. If one varies the osmotic strength of the applied solution, one can estimate the osmotic pressure of the cytoplasm. Because plant cells are large enough and because in many cases it is reasonable to assume that the cell walls are quite rigid, it is easy to find the lowest concentration that just causes plasmolysis. As we have seen in Chapter 5 and will further see below, turgor pressure is not trivial to measure for bacteria because the wall is more flexible and as a general rule the entire cell collapses and shrinks; moreover, formation of plasmolysis spaces occurs only under some conditions, times, and places.

Use of the Phase Microscope

Modulation of the Index of Refraction Difference Between Bacterial Cells and Medium

Let us return now to the prokaryote, which by definition does not have its genetic information stored within a protective nuclear membrane. It was, and is, important to find how the chromosome is organized. An important tool for this and for the studies described below must be considered: this is the old technique of Mason and Powelson (1956). They observed by phase microscopy that if bacterial cells are suspended in high concentrations of gelatin, the index of refraction of the medium could be increased to match that of the cytoplasm. This works because such big molecules cannot penetrate even the outer membrane of the cell and are so big that their contribution to the osmotic pressure of the medium is so very small that they do not cause shrinkage. Consequently, water movement out of the cell is negligible and the cytoplasm is not appreciably dehydrated. With the index of refraction of the medium made nearly equal to that of the cell, the substructure of the cells can be discerned; in fact, it was under such conditions that bacterial nucleoids were first made visible in living cells (Mason and Powelson 1956). The nucleoids, the regions containing the DNA, are made visible with this technique because this part of the cell has a smaller index of refraction due to the paucity of proteins in the region containing the coiled DNA. The nucleoid usually occupies a central position in the cell. PVP (with a high average Mol. Wt.) has also been used for this purpose (Schaechter *et al.* 1962) (see Fig. 13.2), as has bovine serum albumin.

On the other hand, aspects of plasmolysis space formation and cellular shrinkage can be more clearly observed when the index of refraction of the medium is low. NaCl does not penetrate into the cytoplasm, as mentioned, and it does not raise the index of refraction of the suspending medium significantly. Therefore, it does not decrease the difference in index of refraction with the cytoplasm as does sucrose. A modestly high concentration of NaCl will extract water through the cytoplasmic membrane, and this causes the volume surrounded by the cytoplasmic membrane to shrink and its index of refraction to increase (Alemohammad and Knowles 1974; Knowles 1971; Koch 1984b). The dehydration of the cytoplasm can easily be detected because it leads to very dark-appearing cells at optimal focus and to very large phase halos as the focus is adjusted above and below the plane of a cell. The dehydration can be sufficiently severe that the index of refraction of the cell becomes much larger than that of the suspending medium, creating a phase shift so large that it converts portions of the phase-dark image into a phase-bright region that appears bluish (Koch 1961, 1984b)

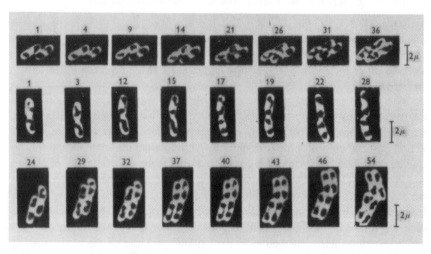

Figure 13.2. Visualization of the nucleoid by phase microscopy in the presence of PVP to balance the index of refraction of the medium. The photographs were copied to increase the contrast. Reproduced from Schaechter *et al.* (1962).

(Fig. 13.3). This latter effect is similar to the appearance of a highly refractile oil droplet, a rod-shaped cell when viewed on end, or the overlap when one cell is on top of another. The increase of index of refraction difference between cell and medium is independent of whether the cell develops localized plasmolysis spaces or just shrinks. Sucrose, fructose, and xylose create plasmolysis spaces and also cause collapse of the cells. With increasing molecular size, a higher weight concentration is needed for plasmolysis, there is a corresponding decrease

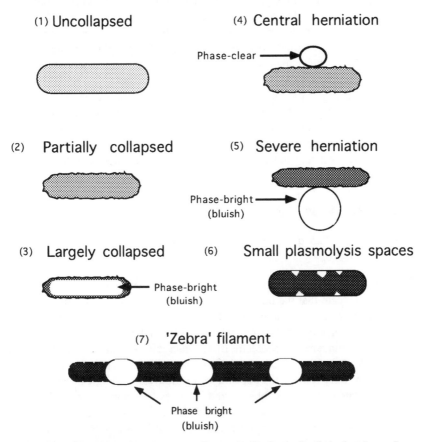

Figure 13.3. Diagrammatic appearance of osmotically shocked cells in the phase microscope. The unchallenged cell when viewed in the phase microscope appears dark when properly focused and is surrounded by a phase halo (not shown). Upon challenge the appearance as depicted in (1) can change to (2) or (6). More severe challenge can lead to such severe collapse that the center reverses phase and appears phase bright (3). Sometimes central herniation occurs (4) that may create such a large bleb that it appears phase bright (5). At the bottom, the appearance of a filament that has formed phase-bright regions interposed between shrunken phase-dark regions is shown.

in the index of refraction difference with the shrunken cytoplasm, and consequently either the shrinking or the formation of plasmolysis spaces is less obvious.

Use of PVP to Measure the Osmotic Pressure of the Periplasmic Space

Two PVP preparations were used in our studies to pull water out of the periplasm. At the same weight concentration the one with a nominal Mol. Wt. of 10,000 should have four times the osmotic effect of the other with Mol. Wt. 40,000; this would only be true if the particles are rigid. At the same weight concentration they have the same ability to equalize the index of refraction to match that of the cell cytoplasm. The lower molecular weight fraction prevented the formation of plasmolysis spaces (defined as enlargement of the periplasm resulting from osmotic challenge that can be seen in the light or electron microscope) produced with relatively low levels of osmotica, as expected (Figs. 13.3 and 13.4). To our initial surprise, so did the higher molecular weight preparation. The explanation has already been given in Chapter 5: large flexible molecules can behave as many independent particles in terms of their osmotic effect. So both contributed enough osmotic force to pull water out of the periplasmic space, while neither treatment was sufficient to cause general shrinkage or collapse at the maximal standard concentration used.

Electron Microscopy of the Wall

Cryofixed Freeze-Substituted Preparations Observed in the Electron Microscope

Although the electron microscope is an extremely powerful tool, it has only been as good as the care that the microscopist has been able to use to avoid artifacts. Many difficulties have been appreciated and dealt with progressively since the 1930s when the instrument became practical. Within the last decade the solutions to a number of fundamental problems for the microbiologist have been developed. These problems are not with the instrument, but with the techniques of preparing a sample that truly reflects the appearance of the living biological sample. Currently, the state-of-the-art technique first involves freezing a very small sample extremely fast so that the water freezes in a glassy state. All is lost if ice crystals form because the forces of crystallization are sufficiently powerful to greatly distort the image. Different mechanical arrangements have been used. They have their individual idiosyncrasies, but one can observe when and where they failed by looking for ice crystal formation.

The Heresy of the Use of Glycerol as a Cryoprotectant

Cryoprotectants, such as glycerol, have been used in the past. In fact, glycerol is still being used as a cryoprotectant to prevent ice crystal formation. Cryoprotec-

A. Action of osmotica in extracting water from the cytoplasm

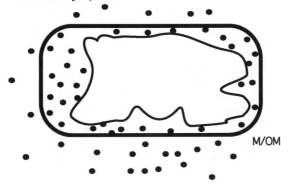

M/OM

B. Action of periosmotica in extracting water from the entire cell

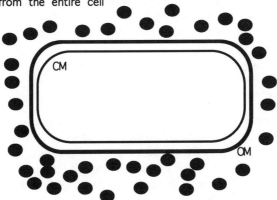

CM

OM

Figure 13.4.   Osmotic action of osmotica and periosmotica. A typical osmotic agent is sucrose; it acts because it cannot enter the cell proper. For sucrose and *E. coli* the match is perfect because there is no permease for sucrose and no invertase so that even if some sucrose enters, it cannot be consumed. An osmoticum needs to be small and water soluble so that an adequate concentration can be produced. Other osmotica of use are hexoses and pentoses. Note that they have to be big enough to not penetrate the CM naturally, so glycerol is not an osmoticum. NaCl is very useful because the cell extrudes sodium. NaCl has the advantage of being very small and ionizing into two particles. These agents favor extraction of water from the cytoplasm. Periosmotica are agents that are too big to pass through the porins and remain outside the cell. Inulin has been used (Mol. Wt. 5,600) as a tracer because of this property, but an adequate osmolar concentration cannot be achieved. However, the agents mentioned in the text can, and they therefore extract water from the periplasmic space as well. Small osmolar concentrations do this, which indicates that the osmotic pressure of the periplasm is only slightly greater than that of the medium.

tants are miscible with water and prevent the formation of ice crystals; however, the transient shrinkage creates artifacts that are especially dangerous in the study of the wall layers of bacteria. Reports of polar spaces in the presumed absence of plasmolyzing agents come from many studies in the literature. Although it is true that glycerol would penetrate the cell and therefore could decrease crystal formation both inside the cell and outside, the transient process of facilitated diffusion through the CM must create mechanical stresses that would affect the wall structure. Even though glycerol penetration is fast, stopflow experiments have shown (Alemohammad and Knowles 1974) that on the 100-ms time scale, the turbidity of the cells greatly increases, indicating either shrinkage or immediate plasmolysis, due to the temporary action of glycerol as an osmoticum. Then in the time scale of less than 7 s, the glycerol concentration equilibrates inside the cells causing a reversal of the shrinkage and/or plasmolysis process. But with either shrinkage or plasmolysis, realignments, slippage, and other morphological rearrangements would be expected to have taken place and the original structure probably would not be precisely regenerated. So the observation of polar spaces and spaces around the developing constriction site may simply represent the achievement of new stable structures as discussed below.

Osmotic Challenge

Cellular Adaptation

A range of phenomena is elicited by osmotic challenge because the direct effects of shifting from one osmotic environment to another involve very potent forces. Bacteria have necessarily developed an array of responses to such fates. The best-studied organism is *E. coli* and its best-studied response is taking up potassium from the medium when osmotically challenged to increase its internal osmotic pressure. In addition, these cells also take up or form proline, trehalose, and glycine betaine to counter an increase in the osmotic pressure of the environment. Other organisms accumulate still other molecules to protect themselves against either a high osmotic environment or dehydrating conditions. The champion survivors of dehydration are some fungi and lichens. When the goal of study is the osmotic effects *per se* and not the adaptive response, one needs to have recourse to ways of inhibiting the uptake and synthesis processes of accommodation. Sodium cyanide or azide has been frequently used for this purpose. Probably, however, it is better to make the measurements so rapidly that any metabolic recovery is negligible. Stopflow devices have been employed to follow turbidity changes (Koch 1984c; Knowles 1971).

Shrinkage and Plasmolysis Space Formation Studies with the Phase Microscope

To observe the morphological effects either in the phase microscope or in the electron microscope, fixation with glutaraldehyde or formaldehyde has been used to stop repair and other changes. But then the distorting effects of these chemicals become a significant issue. For light microscope studies the problem is more easily overcome. Our procedure was to pipet a very small volume of culture onto a microscope slide. Then one or more additional solutions prepared in the growth medium were pipeted as separate droplets onto the microscope slide to a total volume of 20 μl. All reagents contained 20 mM NaN_3 so that the final concentration was at least 10 mM; this was done to prevent the cells from adapting to the osmotic change, a precaution taken in case the observation period was prolonged beyond a minute. The droplets were then mixed with a #0 cover slide that was then placed on top of the mixture, and the slide viewed within a few seconds under oil with a 100X phase 3 objective.

Most observations were made of cells unattached to the glass surfaces, because additional forces might come into play as the bacteria adhered to the glass. For documentation purposes, however, the glass surfaces were treated with polylysine to cause the bacteria to adhere. The positively charged polylysine attaches to both the negative charges of the glass surface and those of the OM to immobilize the cells. Taking photographs greatly slowed the actual procedure, because it was necessary to locate a cell that had become attached and would remain motionless long enough for the necessary exposure. It is to be emphasized that the quality of the photographs of Fig. 13.5 is much poorer than the appearance when viewed through the eyepieces. Another difficulty in preparing photomicrographs was that in the presence of sucrose the cells do not attach as readily to the polylysine on the glass; as a consequence, most photographs were of those few cells that had grown to short filaments and were therefore more quickly and firmly bound. So the evidence for the assertions made with the technique depend more on the visual appearance of the free-floating cells observed as rapidly as possible (i.e., in about 15 s) than on the photographic images.

Cellular Collapse and Types of Gram-Negative Plasmolysis Spaces

The key observation is that shrinkage always occurs on sufficient osmotic challenge; only sometimes, in some locations and in some situations, is it accompanied or replaced by the formation of plasmolysis spaces. Thus, localized plasmolysis is a limited alternative response to osmotic shock. Stronger shocks favor greater general shrinkage and more plasmolysis spaces (see diagrammatic presentation in Fig. 13.3 and microphotographs in Fig. 13.5). Fig. 13.3 shows some of the patterns as most of the cell and its envelope shrink. Plasmolysis can take place in any portion of the cell (Fig. 13.6). Most commonly, the spaces are polar, sometimes they are partial and invade the cylindrical part of the cell, sometimes they are complete and divide the cell cytoplasm in two. Quite frequently in our hands (Koch and Schwarz 1994), but apparently not in those of Mulder and Woldringh (1993), the complete spaces form where obviously there has been or will be a cell division event. Even when no constriction was evident, probably physiological separation has already taken place (Clark 1968). These complete

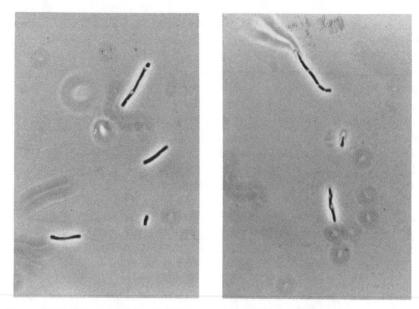

Figure 13.5. Phase microscopic appearance of lpp$^+$ and lpp$^-$ cells.

Panel I. Phase microphotographs of lipoprotein-positive strain of *Escherichia coli.* Sucrose-treated cell possessing polar spaces on both ends. These are difficult to see, but the poles bulge slightly outward from the more shrunken cylinder of the cell.

Panel II. Phase microphotographs of a lipoprotein-negative strain of *Escherichia coli.* A short filament in the presence of PVP that reveals cell internal structure. Structure is visible because the osmolarity is not sufficient to cause shrinkage or plasmolysis but the PVPH has increased the index of refraction and thus lowered the contrast and decreased the halo.

spaces may be examples of the periseptal annuli (PSA) seen by MacAlister *et al.* (1983) and Anba *et al.* (1984).

Besides the modulation of the degree of phase contrast that is a clear marker of the shrinkage process, two additional phenomena can be recognized. In the short filaments that are sometimes found, the collapse of the cell wall around the cytoplasm occurs, but between cellular units within the filaments complete plasmolysis spaces form at regular intervals that have the large diameters characteristic of the untreated cell and bulge out from the shrunken cellular regions. We designated such shapes as "zebras" because they showed an alternation of light and dark regions in which the dark regions are much narrower in width than the untreated cells.

The second unusual structure, seen much less frequently, is one in which a large bleb forms at a central location on osmotic challenge, presumably limited by OM (the location of the M layer has not been established). These blebs remain for at least one hour without bursting or being resorbed. Blebs have been seen

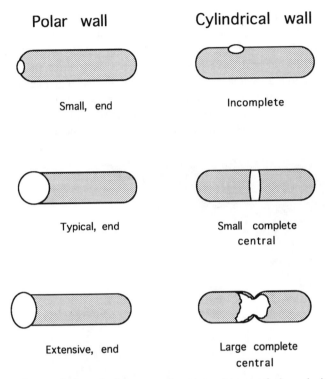

Polar wall   Cylindrical wall

Small, end   Incomplete

Typical, end   Small complete
                                 central

Extensive, end   Large complete
                                 central

Figure 13.6. Types of plasmolysis spaces. A variety of types of plasmolysis can be seen under the phase microscope. They can be much clearer and quickly seen in unfixed specimens in which the cells are floating freely in the challenge medium.

under many conditions before, but the hernias produced in this way usually contain enough substance so that they are wider than the unshrunken cell and contain enough material to appear phase-bright in the phase microscope. Presumably, the blebs contain the periplasmic constituents that have left the shrunken cylinder part for the central regions, but cannot escape through the outer membrane.

Shrinkage and Formation of Plasmolysis Spaces Viewed in the EM

Clearly, the best modern technique for the EM studies of bacteria is cryofixation, but unless a carefully designed and speedily executed protocol is followed, there may be a significant opportunity for bacterial adjustment to an experimental challenge before the organisms are prevented from mounting a response by the cryofixation. For this reason we minimized the time during which adaptation could take place and used azide as well. Still, our experiments would have been better if the chance for adaptation had been further decreased. Response to osmotic challenge when viewed alone in the phase microscope or in the electron microscope (EM) gives an incomplete picture; the former shows earlier and progressive changes and the latter the structural details. Figs. 13.7 and 13.8 show unchallenged and sucrose-challenged cells. Note three states of the wall: first, the smooth appearance of the wall in the unchallenged cells; second, the wrinkled nature of the entire wall due to shrinkage from osmotic challenge where plasmolysis space formation has not taken place; and third, the smoothed straightened form of the wall, particularly the CM, where plasmolysis spaces

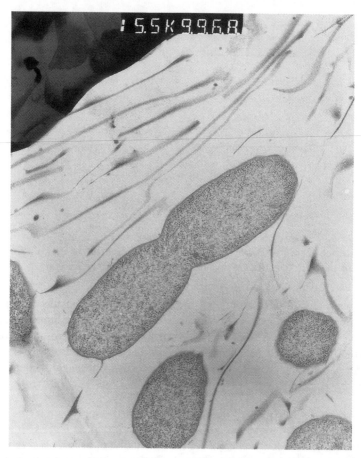

Figure 13.7. EM views of unplasmolyzed *E. coli*. The photograph is of a lpp⁻ (lipoprotein negative strain) but the wild type serves as well. The section was prepared by cryofixation and freeze substitution. The photograph was part of the series for the work of Schwarz and Koch (1995). The smooth nature of the wall and close apposition of the several layers are evident.

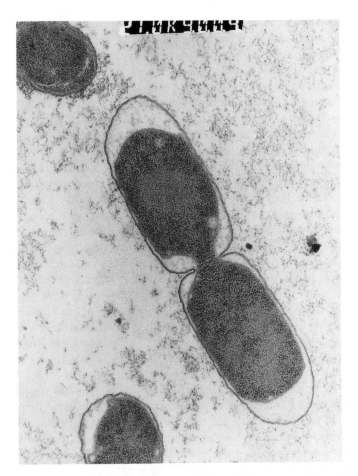

Figure 13.8. EM views of sucrose-challenged *E. coli*. The same strain and technique were used as in Fig. 13.7, but the cells were challenged 2 minutes before freezing to −180 °C with 18% sucrose in the same growth medium. Several new features are evident. These include the laminar side spaces, polar spaces, and plasmolysis spaces where the cell is to divide. Also note that the CM in regions of plasmolysis is again smooth, whereas the M/OM is wrinkled, but not as wrinkled as the cells or regions of cells that have not plasmolyzed, but have lost cellular water due to the osmotic shock.

have formed. Also note the appearance of the several types of plasmolysis spaces that can also be identified in the phase microscope.

There are two phenomena that can only be observed in the EM. One of these is the development of membrane-bound vesicles; these are generally endocytotic and very near the cytoplasmic membrane. More rarely, the endocytotic vesicles are large and, even more rarely, vesicles form that are exocytotic.

The second phenomenon that can only now be observed with the newer cryo-fixation techniques followed by freeze substitution and examination with the electron microscope is that of the formation of laminar plasmolysis spaces. Upon challenge the CM separates a short distance from the M layer over a large area in the side wall, and is no longer wrinkled and has been able to return to a planar form by the production of endocytotic vesicles.

Collectively, these observations can be explained by the fact that phospholipid membranes are capable of only very little expansion or contraction. Consequently, the usual results of shrinkage of the cytoplasm is that all the layers of the cell envelope must become wrinkled in concert because they adhere to each other and must now cover a smaller volume of cytoplasm. If detachment of the cytoplasmic membrane occurs, then plasmolysis spaces (or vesicles or bays) can form and enlarge (and they may move). The most likely sites for spaces to form are at a pole because inversion of the cup-shaped cytoplasmic membrane conserves the area, but allows the CM to surround a smaller volume of cytoplasm. This process can take place without the formation of vesicles. The thin lamellar spaces that form adjacent to the side walls are of quite uniform thickness and necessarily require formation of endocytotic vesicles. The evidence for these conclusions is: (i) in the absence of an osmotic shock, cryofixation and freeze substitution show the layers of the wall to be quite smooth and parallel to each other (with no plasmolysis spaces observable anywhere); (ii) with osmotic challenge the cell wall becomes highly wrinkled; and (iii) the CM of the plasmolysis spaces is again smooth and smoother than the corresponding moderately wrinkled outer layers above the lamellar spaces. Although endocytotic vesicles are present in the microphotographs of earlier workers, their essential role in the formation of plasmolysis spaces was not appreciated. Thus, bacterial membranes, like those of plant plasma membranes, have the ability to form vesicles, equivalent to the well-known process of endocytosis in animal cells when subjected to compressive forces due to the osmotic extrusion of cellular water.

The Physiology of the Periplasmic Space

Conventional View

The first key stage in the development of our understanding of the significance of the Gram-negative cell envelope as an important organelle was the discovery by Neu and Heppel that certain enzymes are released by an osmotic down-shock and others are not. This was followed by the realization that these special proteins resided between the OM and CM. Gradually, these were identified and their functions established. Stock *et al.* (1977) established that the volume of the periplasmic space could be measured by using radioactive water, sucrose, and inulin (Mol. Wt. 5,600). These could be used to calculate the volume of the periplasmic space because water could penetrate and equilibrate with the entire cell, sucrose only with the periplasmic space, and inulin could not enter the cell at all. By adding an osmotic challenge (with sucrose), they estimated the osmotic pressure of the periplasm; their interpretation that the osmotic pressure of the periplasm is similar to that of the cytoplasm will be questioned below. In the third key stage of the study of the periplasm, Kennedy (1982) found that the cell would form MDOs (membrane-derived oligosaccharides, see Chapter 5) when grown in a low osmotic strength medium. These would contribute to the osmotic pressure of the periplasm.

Revised View of the Size of the Periplasmic Space

Values in the literature vary from 5% to 70% for the fraction of the water volume of the cell devoted to the periplasmic space. I believe that a lower estimate of about 5–7% is correct, largely because I also believe that the osmotic forces push the CM against the M layer, the M layer is connected directly to the OM layer by way of the lipoproteins, and with these two constraints there just is not much space available. To resolve the issue, however, we need to mesh the measurements obtained from isotope dilution, those from the earlier electron microscopic data in the literature, and our observations and those of others with cyrofixation and freeze substitution techniques. To do this I calculated the cell's total volume from the widths of cells and an estimate of the average length, and computed the average cell volume in our experiments. I also computed the periplasmic space (defined as the volume of water belonging to the periplasm) from our microphotographs, and then divided the estimated volume of the cell by this (see Table 13.1) and we found the wall volume to be 11%. But there are a number of corrections that must be appended to this estimate. If the content of dry matter were the same in the cytoplasm and in the entire EM observable wall envelope, this value would be a valid measure of the periplasmic space as a percentage of the cell's water. The wall when averaged over its layers, however, appears from the microphotographs to be a little less dense than the cytoplasm. The water content of the cytoplasm has been estimated at 70–80%, and I have chosen the value of 73% and have also chosen 80% for the periplasm. Then the 11% estimate of total periplasmic volume leads to a value of 12.5% for the periplasmic space, which is still much less than some values in the literature. But this estimate needs to be further corrected for the fact that the lipid layers have a very low water content. This lowers the estimate by 20%. Second, the outer membrane has a covering of hydrophilic polysaccharides, associated with a high concentration of water. But this water is not part of the cell's water as measured by the isotope dilution method because much of it would be accessible to the inulin molecule used as a probe. This factor might give a second deduction of 20%.

On the other hand, there is another possibility: there might be a large periplasmic space at the poles of unchallenged cells resembling polar plasmolysis spaces, such as seen in Fig. 13.5 or 13.8. Could it be that such spaces exist and have been missed with certain EM techniques and found by the biochemists with their measurements utilizing tracers molecules of various sizes?

Could polar spaces have been neglected in certain studies with the EM [e.g., see review in Beveridge and Graham (1991)] that usually focused on the dimensions of side wall, so that these spaces might have not been considered in the analysis? It certainly was not the case in our studies (Schwarz and Koch 1995) because we scanned thousands of cells and never observed even a small polar space. (There were, however, extremely few lamellar spaces in unchallenged cells, but these were very small and thin.) Certainly, the volume of natural separation of the cytoplasmic membrane from the murein layer for the cultures in exponential balanced growth and cryofixed and freeze substituted as studied here would be totally insufficient to account for any value greater than 8%. Therefore, the probability remains that with some culture conditions and experimental techniques, enlarged periplasmic regions account for these higher values, but not normally.

An alternative explanation of the high estimates of the periplasmic space comes from considerations of the technique used in the isotope dilution studies. Bacteria growing under aerated conditions are harvested and concentrated into nearly a paste. Under these conditions they must become anaerobic and be subject to noxious products of their neighboring cells. It can well be imagined that this leads to physiological changes, temporary breakdown of the OM, allowing the osmotic pressure inside the cells to decrease. Then when diluted for the measurement of isotopic mixing the cells may be plasmolyzed by what ordinarily would be a normal osmotic milieu. Kellenberger (personal communication) has noted the effects of even brief centrifugation on the appearance of the cells. In any case, the cryofixation freeze substitution must be taken as the gold standard for the estimation of the normal value of the periplasmic space.

Speculations on the Primary Effects of Osmotica

When a cell is suddenly exposed to an osmotic shock, water rushes out of the cell to equalize the water activity inside with that outside. This happens within a fraction of a second (Alemohammad and Knowles 1974; Koch 1984b). Therefore, a smaller volume must be contained within the cytoplasmic membrane. As a consequence, the CM must wrinkle; it cannot contract because phospholipid membranes are not able to shrink more than 2–5% (Kell and Glaser 1993). If the CM comes loose from the rest of the envelope, then the M and OM would not have to wrinkle. It is likely, however, that in most regions of most cells the entire wall envelope becomes corrugated.

It is considered very improbable that the cytoplasmic membrane becomes immediately detached on challenge and moves inward while the remainder does not. Rather, it is more likely that the sandwich of CM-M-OM layers remains contiguous and wrinkles, as depicted in Fig. 13.1. One may conclude this because the M and OM, which remain together even in regions of plasmolysis, are somewhat wrinkled and not as smooth as the walls of unchallenged cells (Fig. 13.7). Moreover, it was generally observed in our work that the outer wall wrinkling over plasmolysis spaces is not quite as severe as the wrinkling in other parts of the same cell where shrinkage had occurred but no plasmolysis spaces formed. The wrinking in such regions was in turn qualitatively similar to that of the wall of nonplasmolyzed cell sections present on the same grid. The implication is that the separation is secondary and is powered by the forces tending to relax the M and OM combined layers, which had partially, but not fully, straightened out. On the counter assumption that plasmolysis is the immediate response to osmotic challenge, then there is no obvious way at all that the murein and outer layer could have been disturbed from their original position and become wrinkled.

The electron microscope is not suitable for studying short-term processes. We can only report the appearance of cells at 45 s after osmotic shock, and they were no different than those seen in Fig. 13.8. With phase microscopic studies (Schwarz and Koch 1995) we could look at the cells on a shorter time scale (but not less than 15 s). Therefore, both techniques are much too slow to study the primary event. The stopflow measurements from Knowles' laboratory (Alemohammad and Knowles 1974) are fast enough to show that the response is very rapid, but do not distinguish between shrinkage with concomitant wrinkling of all layers and the movement of the cytoplasmic layer from M and OM layers that remain nearly stationary. The distinction cannot be made because the turbidity measurements are mainly measurements of the anhydrous mass of substances contained within the cytoplasmic membrane (Koch 1961, 1984c).

Changes That Must Take Place as the Wall Wrinkles

If a pure phospholipid planar bilayer is caused to bend into an undulating structure, only three kinds of changes are possible because stretching or compression; i.e., shrinkage in the plane of the surface, is very limited. One possibility is that diffusion of phospholipid molecules will occur, leading in each bilayer to rarefaction in a region in one bilayer and a compaction in the corresponding position in the other leaflet. This then permits the bending and the balanced distribution of phospholipids in the curved bilayer. The diffusion process within each layer may consist of movements of single molecules, groups of molecules, or slippage of tails of regions of one leaflet relative to those of the other. Because the wrinkles are not regular, it must be supposed that centers for the positive and negative curvatures occur by a random process or are triggered by local events.

The second process is flipping of phospholipids from one leaflet to the other. Ordinarily, this is a slow process (Henis 1993), but it can be accelerated with enzymes or physical forces. Certainly, compression and flexing of a planar bilayer due to osmotic forces should be adequate to speed flipping. The readiness with which phospholipid vesicles become inverted in an ultrasonic field is also evidence for the mechanical generation of flipping movement of phospholipids from one layer to the other. Also in evidence is the formation of endocytotic vesicles from the CM, discussed above, to permit smoothing of the CM covering the cytoplasm. Because lateral diffusion of phospholipids, however, is fast compared to transversion events (i.e., flipping the molecule through the membrane), the lateral movement process is more likely as an initial event.

The third possibility is that each leaflet is composed of a variety of types of phospholipids and extrinsic proteins with varying abilities to accommodate to positive and negative curvature depending on the ratio of the size of the charged region to the area covered by the hydrophobic tails (Tanford 1980; Norris 1992). Some types would be sorted into regions of positive and others into regions of negative curvature. Sorting takes more time than lateral diffusion to fill a void created by positive curvature or the diffusion of molecules leaving a region of high surface concentration. Thus, unless there are additional factors, movement of phospholipid molecules generally within a layer seems more likely than this third possibility of clustering of phospholipid types.

Both inner and outer membranes have intrinsic proteins that go through both leaflets. These would have to move and adjust to accommodate to the wrinkling process. Again, these should be slower processes because of the large mass of these molecules relative to the size of the phospholipids. Still, all such changes may be faster than our time scale of measurement with either the phase or the electron microscope. It should be noted that because of the angular dependence of the light scattering (Koch and Ehrenfeld 1968), it would be possible to distinguish these alternatives if a stopflow apparatus were constructed that could measure the low- and high-angle scattering simultaneously. A stopflow apparatus could also be constructed to mix and dispense fine droplets into liquid propane for electron microscopy.

The creation of the undulations also requires rearrangements of attachments between the layers of the CM and M layers and between the M and OM. The cytoplasmic membrane during normal growth must be in close physical relationship with the murein because of its role in peptidoglycan metabolism through the action of bactoprenol, penicillin-binding proteins (PBPs), and certain membrane-bound autolysins (e.g., the membrane-bound lytic transglycosylase), etc. These attachments are noncovalent and possibly this lack of strong bonding is the reason that plasmolysis spaces almost always form between the cytoplasmic layer and the murein layer. Obviously, such separation would disrupt any integrated relationship of the CM with the OM. This would include protein transport processes of either the outer membrane proteins and export to the cell's exterior, as well as those for more sophisticated functions, such as a mechanism by which the cell could find its exact middle for the site of cell division (Koch and Höltje 1995, see Chapter 7), or for forming the elusive Bayer's junctions (1991).

Lipoprotein links the murein layer with the outer membrane. The surprising result is that even without this protein these two layers stay together except for rare cases. There is another situation in which the role of the lipoprotein is relevant. It has been shown that the lipoprotein's fatty acid tails insert quickly into the OM in a strong hydrophobic interaction, whereas it takes about a generation for the other end of the lipoprotein to form covalent linkages to the M layer (Hiemstra 1987; Heimstra *et al.* 1987). This means that newly formed poles do not have covalent linkages of the OM to the M layer, even though they will develop them as they become more mature. The absence of more general separation in the *lpp*− mutants together with the surprising fact that such mutants live, although they are more sensitive to detergents, suggests that there may be additional covalent contacts with the OM that can form. Possibly these kinds of linkages form only as a backup mechanism.

Development and Geometry of Plasmolysis Spaces

After shrinkage, separation of the CM from the M and OM layers allows the latter two to return to near the original position. This separation *per se,* however, does not allow the cytoplasmic membrane to re-extend itself because it surrounds the smaller volume of dehydrated cytoplasm. Consequently, the CM must remain wrinkled until subsequent endocytosis takes place allowing it to straighten out; alternatively, and very rarely exocytosis (defined as the invagination of the membrane with the pinching off of a vesicle into the periplasm) and separation may occur.

The curvature of potential plasmolysis sites determines the degree to which the formation of endocytotic vesicles is mandated. Consequently, certain locations in the cell could more easily develop plasmolysis spaces because the formation of vesicles is not needed or is less important. Thus, directly from the physical properties of the phospholipid bilayer, periplasmic spaces are more likely to form at the poles and where constriction sites are developing (see Koch 1995e). The simplest case not requiring endocytotic vesicle formation from the CM bilayer is when a plasmolysis space forms at the cell pole as an inverse image of the cell membrane because the membrane area need not change. At the perimeter of such a space, there now is a sharp bend in the bilayer and what was the convex surface is now the concave one, but otherwise the cytoplasmic membrane is as it was before in the unperturbed cell. Consequently, there must be some migration of phospholipids at the periphery of the plasmolysis space, but the energetic cost should be small. The argument for the low energy requirement is consistent with the prevalence of formation of polar spaces (Schwarz and Koch 1995; Mulder and Woldringh 1993).

At constriction sites where the wall has already grown inward but division is not yet complete, it is possible to form a cylindrically symmetric space as shown in Fig. 13.9. This type of space is seen frequently, such as in Fig. 13.8.

The finding of a simple energetic reason for the prevalence of these two types of spaces renders less attractive the concepts suggested by Rothfield's group (MacAlister *et al.* 1983) of their relevance to the cell division process. The concept of the formation of periseptal annuli, their movement along the cell cylinder, and the arresting of this movement to mark the site of the next division invoked by the PSA model is therefore moot. Also the concept of the Leading edge model (MacAlister *et al.* 1987; Wientjes and Nanninga 1989) as an extremely narrow zone at the very deepest part of the constriction being the unique site of intense murein insertion, originally deduced from sections such as shown in Fig. 13.8 is greatly weakened by the alternative explanation of Fig. 13.9. Additionally, the support for the leading edge model from autoradiography (Wientjes and Nanninga 1989) has been weakened by the recent studies of the effect of the random diffusion of the electrons from tritium decay in autoradiographic studies (Koch and Woldringh 1994).

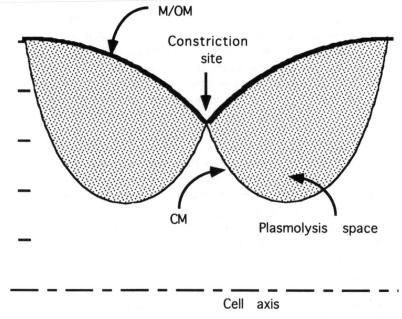

Figure 13.9. Simulated plasmolysis space developing at a site of constriction without the formation of endocytotic vesicles. The shape of the plasmolysis space was calculated by a program (Koch 1995e) that assumed that the area of CM had to remain constant and that the surface was not to be highly wrinkled.

14

Gliding Motility, Protonmotive Force, and Flagellar Rotation

KEY IDEAS

The isolated sacculus of Escherichia coli *can be expanded from its minimum surface area by lowering the pH.*

The chemiosmotic process can lower the pH by several units in the very near vicinity of the outside of the cytoplasmic membrane.

These facts suggest a model for gliding motility of Gram-negative microorganisms based on the assumption that the volume enclosed by the sacculus of the gliding cells is normally enlarged by chemiosmotic extrusion of protons.

It is postulated that the gliding cell is able to control special shunts that allow protons to return freely to neutralize the excess negative charges remaining inside the cell in local regions and at controlled intervals.

When these shunts are activated, a series of physical processes occurs, resulting in the shrinkage of the regions where the special shunts are functional, and slight expansion where they are not.

This expansion wedges the cell in the substratum by favoring hydrogen bonds between the cell surface and the environmental surface or slime layers.

When the shunts are subsequently turned off, the volume changes are reversed, and the cell inches forward.

This could be the prokaryote's substitute for mechano-enzymes.

It could be also the basis of the rotation of the flagellar motor.

Models for Gliding Motility in the Literature

Various bacteria move by gliding motility. The prokaryotes that do this come from a wide variety of morphological and taxonomic classes. Four characteristics seem to unify the process (see Burchard 1981; Pate 1988); these are: (i) the organisms are all Gram-negative, (ii) gliding motility only occurs when the organisms are on a suitable surface, (iii) slime may be an important component of that surface, and (iv) the process is powered by chemiosmosis. There appear to be no other common elements. When these bacteria are viewed microscopically, the means of motion are not evident because flagella are not present and sinusoidal movements are not detected. These observations have increased, rather than decreased, the number of proposed mechanisms. Earlier models, reviewed by Burchard (1981), have invoked forces such as osmotic pressure, surface tension, slime secretion, and contractile cell-surface waves. Other models involve the retraction of fimbriae (MacRae and McCurdy 1976), make-and-break of envelope interactions with the substrate (Dickson et al. 1980), and "electrokinetic" phenomena (Mitchell 1956). Dworkin and associates (Dworkin et al. 1983; Keller et al. 1983) introduced a model of surfactant release, Pete and Chang (1979) have suggested rotary motors, and Lapidus and Berg (1982) have imagined that gliding cells have continuous tracks under the outer membrane that revolve around the cell. Recently, a variant has been proposed by Ridgway and Lewin (1988) in which limited motion of many mechanically independent domains in the cell surface is functionally coordinated.

I have proposed one more mechanism (Koch 1990f), based on observations of the elasticity of the peptidoglycan of a Gram-negative cell (Koch and Woeste 1992; see Chapter 10) and the change in size of the isolated sacculus when the net charge is altered. The model is mainly applied to organisms like *Myxococcus* and *Cytophaga*. But variation can be applied to other cases of gliding motility, and the mechanism may power the more common flagellar motility typical of *E. coli*.

The Involvement of Protonmotive Force

The extrusion of protons by the chemiosmotic mechanism is known to be the source of energy for gliding motility as it is for flagellar motility. See Pate (1988) and Khan (1988) for reviews. Proton extrusion coupled to photosynthetic and nonphotosynthetic electron transport lowers the pH a very short distance from the outside of the membrane compared with the pH in the medium (Koch 1986a). It also raises the pH relative to the bulk of the cytoplasm of the cell in the first several nanometers inside the membrane. The pH changes are calculated to be several pH units different from the nearby bulk phases. The effect on the outside is relevant here because the low pH protonates the peptidoglycan. This neutralizing of carboxyl groups results in a net positive charge due to charged amino group leading to electrostatic repulsion. This in turn favors wall surface expansion. Unless the surface wrinkles, this will cause an increase in the cell volume. Consequently, the cell imbibes water and the cell wall surrounds a larger volume than it would in the absence of a functioning chemiosmotic system. This assumption itself derives from the assumption that the peptidoglycan layer of Gram-negative organisms is thin (on the order of 2 nm thick) and from the assumption that the turgor pressure keeps the cytoplasmic membrane pushed against the sacculus due to osmotic forces. Because of the limited ability of bilayers to expand or contract, the total surface of the cytoplasmic membrane (CM) or the outer membrane (OM) cannot change very much, but because the bilayers are fluid one end of the cell could enlarge while the other end contracted in surface area.

The key assumption of the model presented and developed in Koch (1990f) and represented here is that gliding organisms are capable of opening a class of special return paths that may be channels or carriers to allow protons to reenter into the cytoplasm. These are imagined to be gratuitous paths that neither form ATP nor serve in symport or antiport processes functioning to accumulate substances against an electrochemical gradient. In the act of returning via this kind of postulated shunt (or "short" or "channel" or physiological "proton-conductor pathway"), the charge separation created by the chemiosmotic process is neutralized, and power is dissipated in a futile cycle. As a result of the reneutralization of the cell constituents, the pH in the environment immediately adjacent to the cytoplasmic membrane generated by chemiosmosis is raised on the inside and lowered on the outside. These pH shifts are subsequently reversed when the shunt is opened and energy is expended.

The local pH at both inner and outer surfaces would approach the bulk phase of the periplasmic space or the cytoplasm, respectively, except for an additional phenomenon. This is that the pH is lower on both surfaces due to the fixed negative charges of the surface of the membrane resulting from the presence of phospholipids, etc. (see Koch 1986a).

It is to be emphasized that for the transduction of chemiosmotic energy into mechanical energy by the cycle of opening and closing shunts it is thermodynamically necessary that the protons readily return by low-resistance pathways to the inside of the cell, and in doing so do little or no electrochemical work. The switching on of these leaks is assumed to occur both periodically (possibly many times a second) and asymmetrically; i.e., for the case of gliding motility of rod-shaped organisms only at one end of the cell. Of course, in order to change direction, the cell must switch ends at which the pulsating leak takes place.

This short-circulating process when the shunt is activated (in the "on" state) reverses the pH change due to a chemiosmosis within a particular region of the peptidoglycan sacculus. The consequence is that the net positive charge on the wall polymer temporarily decreases. This allows electrostatic attraction of ionized carboxyl groups and protonated amino groups in the peptidoglycan and causes the peptidoglycan sacculus to contract in a localized region (see Figs. 14.1 and 14.4). When the net positive charge on the peptidoglycan decreases, the extended peptidoglycan conformation will tend to contract, and bonds that have free rotation will tend to approach more closely their most probable and more compact

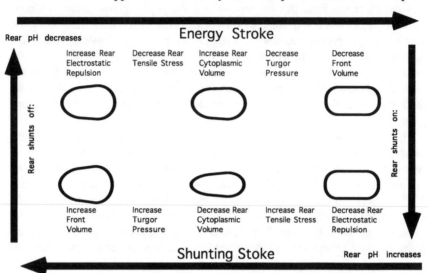

Figure 14.1. The cyclic process leading to gliding motility. The effect of opening and closing the shunts on the two ends of the cell leads to a series of processes contracting and expanding the two ends of the cell alternately and out of phase. Listed are the cyclic steps of the process of swelling and contracting alternately and sequentially. The mechanism depends on formation of hydrogen bonds with the substratum at the end that is swelling and the dissociation of hydrogen bonds at the end that is shrinking. Forward movement is generated in the same way used by a cross-country skier in that elongation of the organism occurs when it is affixed at the rear. The sacculus of the peptidoglycan is under stress because of osmotic pressure differences. In the regions of the cell where the local positive charge has been decreased by the opening of the special shorting channels (or possibly by activation of a proton conductor-like substance in the cytoplasmic membrane), the peptidoglycan will contract because the local pH in the wall rises and carboxyl groups on the peptidoglycan become deprotonated. Then entropic effects and the attraction of the ammonium groups to the residual carboxyl groups lead to a smaller surface area. Local shrinking forces water out and thus raises the cellular osmotic pressure, which then causes the other end of the cell to swell. The energy cost for movement by this model is paid by the energy-dependent extrusion of positive charge and the energy-independent return of that charge without doing electrochemical work. Instead, the energy is expended in doing physical work. It pays for the pressure-volume work of reducing the cell's total volume by squeezing water out of the cell. Because of the shifting of the strength of the local binding to the substratum in the manner of the motion of a cross-country skier, cycles of this kind lead to the net movement of the cells.

conformation. Therefore, the secondary consequence of the lessening of charge repulsion is an increase in the force constant (reciprocal of Young's modulus) in the sacculus, leading to a decrease in surface area. This decrease causes local egress of water, possibly wrinkling the CM and OM, and decreasing the volume of this part of the cell. It is assumed that water can leave the cell quickly, but other constituents do not. The loss of water causes some increase in the solute concentration of the cell, slightly increasing the internal osmotic pressure until a balance with the turgor pressure is reachieved. This should be a rapid process as would its equilibration over the entire cell.

For motility of rod-shaped organisms to occur, this contraction process must take place at the trailing end of the moving cell. The contraction at the trailing edge results in changes at the forward end of the cell mediated via the increased osmotic pressure in the cell. Because of the osmotic pressure increase and because of the elastic properties of the peptidoglycan, the osmotic pressure is partially ameliorated by swelling at the anterior end of the cell. As a result of the posterior end shrinking and the forward end of the cell bulging, this vanguard region wedges itself into the slime-covered surface to which the cell is adhering and leads to increased formation of hydrogen bonds linking the cell to the slime.

The physical properties of a functional surface (which I shall call the "substratum") probably could range from being quite solid to being a highly viscous thixotropic gel or semigel. [With minimal slime, or no slime at all, the extra force moves surface components of the cell closer to the electronegative atoms in the surface to favor hydrogen bonding to electronegative atoms in the surface of the substratum. This surface would give better traction if it can form many weak (hydrogen) bonds with the external electrophilic atoms of the bacterial envelope.]

The enlargement of the cell's advancing end's total volume not only causes this portion to expand, to form more hydrogen bonds, but also to extend its leading edge slightly. Then a short time later, when the protonmotive force at the rear of the cell is reestablished, the net positive charge in that region causes the sacculus area to increase, the stress in the wall to decrease, water to leave, and the volume in this portion of the cell to decrease. Because the osmotic pressure differential with the outside environment decreases, the forward region shrinks and that part of the cell shortens. During this shortening, at least part of the cell must move relative to the substratum. The part to move will be that part of the cell that has just been previously shrunk and loosened to some degree from its hydrogen bonding to the substratum. The cell will be pulled toward the forward region previously bloated and wedged into the substratum that has a higher concentration of hydrogen bonds with the substratum. Cycles of this kind will cause the cell to inch along.

Thus, transport by gliding motility is akin to the process whereby a cross-country skier moves. He varies the traction between the ski and the snow surface. By shifting his weight, the skier increases the pressure and traction on one ski, while decreasing that on the other ski. This ski is then easily slid forward without the other ski slipping backwards. Similarly, in gliding motility the cycle of traction intensity is initiated by expansion in the forward end pushing it forward and wedging it into the slime. Evidently, the shape change of the motile object needs to be coordinated properly to achieve efficient unidirectional motion. The process would be favored if the cell were surrounded by semisolid hydrogen-bondable slime. It should be enough for the cell to worm its way down along a shallow trough or over a hard surface if there is any possibility of hydrogen bonding to the surface.

As described, movement would occur along the axis of the cylindrical portion of the rod- or filament-shaped cell. There are some cases where the movement is crosswise to this axis; for these cases, the roughly cylindrical cross-section would have to oscillate between a more or less hen's egg shape and one that was more elliptical or circular. The model can also be extended to explain flexing in some species and circulation of surface components observed in certain other cases by postulating different arrangements of shunting sites with timing of the action of the shunts.

pH Near an Energized Membrane

The Debye-Hückel distribution is the well-known case of countercharge distribution around an ion. It has a parallel in a slightly different case of the distribution of counterions in the neighborhood of a membrane bearing fixed charges (the Gouy-Chapman-Stern theory), and this in turn has a parallel to the third case of relevance here of a membrane actively engaged in charge separation (Koch 1986a). The Debye-Hückel theory for ionic solutions develops the dependence of potential on the counterion concentration in the bulk phase and the dependence of potential on distance from a charge ion. The distance dependency is a function of the Debye constant. That constant in turn depends on the charge of the ions of the electrolyte and their concentration, but typically it corresponds to a $1/e$ exponential decay distance of a few nanometers. The Debye constant is given by:

$$\kappa = (2C_{\infty}/\epsilon_r\epsilon_0 kT)^{1/2}, \qquad \textit{Debye constant}$$

where C_{∞} is the concentration of a mono-monovalent salt like NaCl in the bulk phase, ϵ_r is the dielectric constant of the solution, ϵ_0 is the permitivity of free space, k is the Boltzmann constant, and T is the absolute temperature.

The second case of the Gouy-Chapman-Stern theory for membrane surfaces is mainly an extension to a linear geometrical situation. The curve of potential versus distance from the planar membrane also follows the same exponential law. The potential is given by:

$$\psi(x) = \psi(0)e^{-\kappa x}, \qquad \textit{Potential with distance}$$

where $\psi(0)$ is the potential at the membrane surface and $\psi(x)$ is the potential at a distance x away from the membrane. The third case is a membrane, which in addition to effects due to fixed charges, also has an internal energy-transducing process that causes charge separation. Such a charge-separation can occur by illuminating photosynthetic membranes, by electron transport in a membrane-bound redox system, or by the action of a membrane-bound ATPase functioning to convert the energy of the high-energy bond of ATP into charge separation. In all of these there is coupling to the extrusion of protons by the chemiosmotic process. These processes lead to a preponderance of protons near the outer face and a paucity on the membrane facing the cytoplasm (Fig. 14.2). My analysis of this case (Koch 1986a) showed that the potential would also fall exponentially with the Debye constant; i.e., the potential varies with distance from the outer surface of the membrane according to the *Potential with distance* equation, and therefore the pH varies according to:

$$pH(x) = pH(\infty) + \psi(x)/60, \qquad \textit{pH with distance}$$

where $\psi(x)$ is measured in millivolts, $pH(x)$ is the pH at a distance x from the membrane, and 60 is the conversion factor to convert electromotive force in millivolts into pH units (Fig. 14.3).

Detailed calculations demonstrated that the pH increments could be several units at the surface of an energy-transducing membrane relative to the bulk phase. During a steady state where the protons are being pumped out, but the resistance of the return path is high, the balance leaves a local excess of protons very near the outside of the membrane. For the case when the voltage is 1,200 mv (appropriate for the electrochemical combustion of H_2 and O_2) with a mechanism such that the electron pair from a hydrogen molecule during its transfer to oxygen causes the ejection of an average of three pairs of protons, there will 1,200/3 or 400 mv of protonmotive force developed across the membrane. Consequently, an internally energized cell membrane imbedded between two pH 7 environments would in the very near vicinity of the outer surface of the membrane have a pH that could be several units lower and an equally high pH shift on the inside of the membrane, causing a total span of 400/60 = 6.7 pH units across the membrane from bulk phase to bulk phase.

When the protons cannot easily return because of the relative impermeability of the phospholipid membrane to protons and when the ionic strength is the same on both sides, then 1/2 of the 400 mv or 200 mv is the contribution to $\psi(0)$ from chemiosmosis on each face. In addition, the local pH on both surfaces is lowered because of the bound negative charges on the phospholipids (see review in Koch 1986a) due to the Gouy-Chapman-Stern effect. This will not affect the arguments below because the lowering occurs on both surfaces. Fig. 14.3 shows the pH profile for a typical ionic environment on the assumption that another 160 mv are produced by the fixed charges, so that $\psi(0)$ is 360 mv. This corresponds to

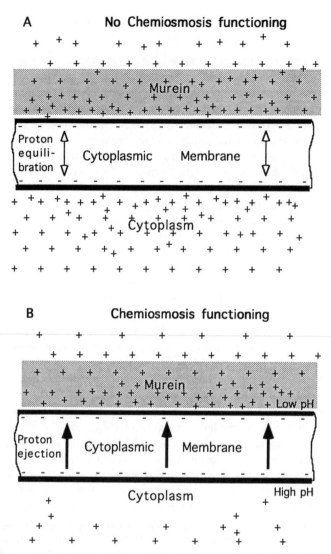

Figure 14.2. Distribution of charges around an energized and unenergized membrane. A phospholipid membrane with fixed negatively charged head groups is depicted by the minus signs inside the CM. The pluses indicate the location of countercharges and the contribution of protons from the ionization of water. The corresponding OH^- charges are not indicated. If the membrane were unenergized (as in A), the cloud of countercharges would hover near the surface according to the Gouy-Chapman-Stern model. However, when a chemiosmotic expulsion of protons is linked to an exergonic process of electron transport, photosynthesis, or ATPase action, then the cloud of plus charges is diminished on the cytoplasmic face and augmented on the periplasmic surface. This means that the pH in the murein decreases enough to cause the neutralization of carboxyl groups, largely on the DAP residues. This in turn decreases the charge attraction to amino groups and causes the murein layer to expand.

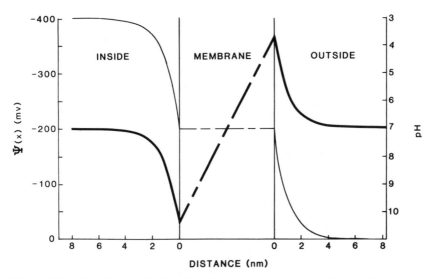

Figure 14.3. The theoretical pH profile of an energized membrane. The combined action of a dead-ended, "stalled," chemiosmotic process together with the effects of fixed negative head groups of the phospholipids is estimated to be 360 mv. This corresponds to a 6-pH unit decrease in pH. Assuming the bulk pH of the growth medium to be 7.0, the surface pH could be as low as 1. The approximate pH, assuming an environment of 0.01 M NaCl (solid line) and 0.1 M NaCl (dashed line), is shown with respect to the distance from the cytoplasmic membrane. In the absence of the chemiosmotic process where only the Gouy-Chapman-Stern mechanism acts, the pH change would be only half as much and the carboxyl groups of the murein would retain their negative charge.

a drop in pH of 360/60 = 6 pH units on the outer face. This is an amazing thought; these pH shifts could cause very dramatic chemical effects—although only on a nanometer scale. Admittedly, this estimated low pH of 1 when the bulk phase was pH = 7 is only approximate, because it is low enough to protonate even phosphate groups of the phospholipids.

On the inner face of the cytoplasmic membrane, there will be an increase in pH above the cytoplasmic pH due to the chemiosmotic drain of protons ejected to the exterior. This pH shift is partially compensated, however, because of the net negative fixed charges on the membrane, which tend to lower the pH in the neighborhood of the inner surface. Of course, these local pH changes could not be measured with ordinary instrumentation. Logically, if atom-sized measuring equipment were available, these pH values could be measured but only if the pH electrode as well as the reference electrode could be placed in a nanometer range from the energized membrane.

With less substantial pH drops, it has been shown that the purified sacculi of *Escherichia coli* would expand 4-fold when a net charge of either sign was developed (contingent on the internal hydrogen bonds being broken (see Figs.

10.8, 10.9, and Fig. 14.4). At the isoelectric point the sacculi have the area expected from cytological measurements for turgor-free bacteria. Two features of the pH-driven expansion process are of note: its cooperativity and its reversibility. The former feature shows that hydrogen bonds and local electrostatic attractions are weak, but important under turgor-free conditions; the latter shows that neither are covalent bonds ruptured nor is there a complex secondary structure. A typical surface area versus pH is shown in Fig. 10.9.

pH and the Ionization of the Peptidoglycan

Because peptidoglycan is a polyelectrolyte, the intact sacculus or fragments of wall have titration curves that in principle approach those of the component ionizable groups only at high ionic strength. An experimental titration curve is shown in Figs. 14.5 and 14.6, and a theoretical titration curve for peptidoglycan in the absence of electrostatic interaction is shown in Fig. 14.7. The latter is based on the composition of *E. coli* peptidoglycan together with literature values for the degree of crosslinking and on our measurements of the degree of acetylation of the hexosamines (Koch and Woeste 1992). The calculated curve also required pK values for the various groups; these values, taken from the literature, are given on page 278. It can be seen that even though the sacculus is a very large polyelectrolyte, this composite curve would apply quite well to the intact Gram-negative sacculus because most of its charge interactions are weak. It should be noted that *Myxococcus xanthus* peptidoglycan has a chemical composition (White *et al.* 1968) similar to other Gram-negative organisms. The difference with *E. coli*, however, is the presence of nontitratable glucose moieties (Johnson and White 1972). Additionally, there are patches of trypsin-sensitive material holding together regions formed of peptidoglycan.

Apposition of the Cytoplasmic Membrane and the Sacculus

There are many topics that recur often in this book and one of these is the spacing of the CM, M, and OM layers. In this section I will discuss this topic in depth even though it has been treated in Chapter 13. Because of the osmotic pressure difference between the cytoplasm of bacteria and their usual milieu, living cells develop a turgor pressure as the cytoplasmic membrane presses against the only structure in the cell capable of resisting stress, the covalently linked fabric of the sacculus. In this book I have tacitly assumed close apposition of membrane and sacculus and have supported this claim by the argument that at the sites of wall enlargement the cytoplasmic membrane must be very close to the sacculus. Such close proximity is, in part, due to several factors. The substrates for the wall polymerization pass through the cytoplasmic membrane and are available on the outer surface, and the enzymes that crosslink the peptide chains (the

A Expanded

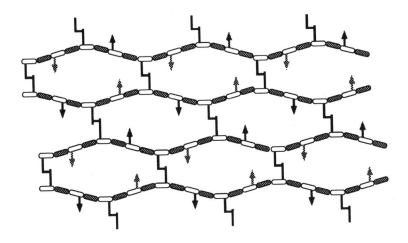

B Contracted

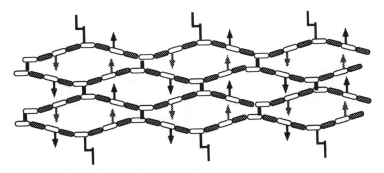

Figure 14.4. The elastic characteristic of the sacculus material. Change in the area is outlined. When the charge repulsion no longer functions the octapeptide compacts into a very short structure. A. Expanded; B. Contracted.

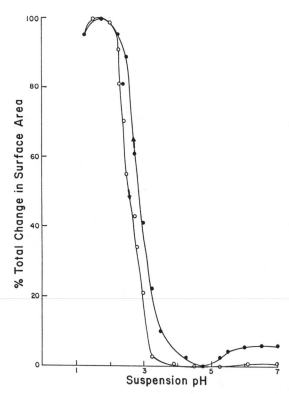

Figure 14.5. Expansion of the Gram-negative sacculus by acid titration. Figure modified from Woeste (Dissertation Indiana University, 1988). Purified sacculi from *E. coli* were examined by low-angle light scattering and the mean sacculus area computed (see Chapter 10) throughout the acid branch of the titration. Notice the similarity of the forward and reverse titrations. This means that the expansion or contraction show little hysteresis, i.e., wall expansion is reversible and not cooperative.

transpeptidases) are intrinsic proteins bound in the membrane and cannot be expected to act over a distance larger than a nanometer from the surface of the CM. Although little is known specifically about the wall metabolism of gliding bacteria, I will assume that knowledge obtained with *E. coli* will apply as these are all Gram-negative bacteria.

This attractive idea of close apposition has been brought into question. Because of the appearance of the envelope in certain types of electron microscope fixation procedures, it has been held that the peptidoglycan of the Gram-negative bacterium is a gel 14 nm thick spanning the space between the two membranes (Hobot *et al*. 1984). The dimensions are correct (see Chapter 13). This claim of the gel-like nature of the murein, however, can be countered because that would not be the expected property of a cytoplasmic membrane complex that serves to contain

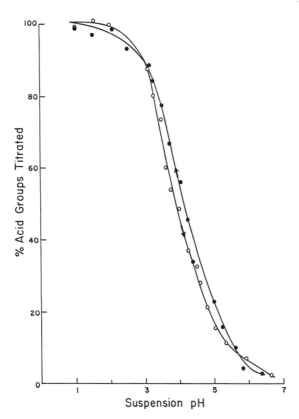

Figure 14.6. Experimental pH titration curve of the sacculus of *E. coli*. Figure taken from Woeste (Dissertation Indiana University, 1988) for the titration of intact purified sacculi initially washed in distilled water. Notice again the reversibility and lack of hysteresis.

the fluid cytoplasm under high pressure. It should be pulled tight and the stress-bearing peptidoglycan should be about 1 nm thick per layer. As addressed in Chapter 4, experiments and theory lead to the conclusion that the murein layer is largely only one layer thick.

There is a semiplausible counterargument. It has been claimed that the periplasmic space has an osmotic pressure similar to that of the cytoplasm, partly due to proteins and partly (under some conditions), due to membrane-derived oligosaccharides (MDOs) and their countercharges (Stock *et al.* 1977; Kennedy 1982). This implies that the stress due to osmotic differences with the environment is actually expressed on the outer membrane, which then transfers the stress to the peptidoglycan. A theoretical rebuttal to this idea is that cohesive forces in the outer membrane structure are not strong enough to resist temporary or permanent rupture over the area unsupported by links to the peptidoglycan (see Chapter

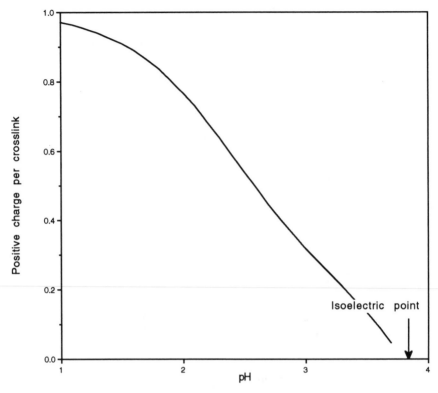

Figure 14.7. Theoretical acid titration curve for the sacculus of *E. coli*. Calculated on the assumption of the composition and p*K* values given on p. 278 in Chapter 10.

13). I submit this rebuttal merely as a conjecture because in the literature there are no adequate measurements of the mechanical properties of materials similar to those of the outer membrane, nor do we know the area of unsupported outer membrane surfaces present in living bacteria. Knowledge of the distribution of areas of outer membrane that are not attached would enable us to determine the pressure differential that could be supported without rupture or the formation of large enough gaps to allow salts, proteins, and MDOs to permeate outwardly. One can be sure that the more crosslinks of outer membrane to the peptidoglycan via lipoprotein, then the smaller the individual unsupported segments of surface area, and the greater the pressure needed for rupture. If the 100,000 lipoprotein molecules binding the outer membrane to the individual sacculi of *E. coli* were regularly arrayed and the surface area of the cell were 11.9 μm^2, then each lipoprotein would be associated with 119 nm^2 of outer membrane. If the area of a phospholipid molecule in a leaflet of the bilayer is 5 nm^2, then the average area of an unsupported domain is 238 phospholipid molecules held together by apolar bonds. Studies by Evans and Needham (1987) show that phospholipid membranes can resist very little tensile stress. The finding (Schwarz and Koch 1995) (see Chapter 13) that the osmotic pressure of the periplasm is little more than that in the medium, deduced from the observation that plasmolysis can be prevented by a relatively low concentration of periosmotic molecules that cannot penetrate the outer membrane is another reason to doubt that the periplasmic space actually has an appreciable osmotic pressure above that of the environment.

Review of Experiments on Gliding Motility

Excellent reviews of gliding motility are available (Burchard 1981; Castenholz 1982; Pate 1988). Reexamination of a variety of organisms as studied with a variety approaches to see whether the observations are consistent with the contraction/expansion model proposed here is the goal of this section. I begin by asking questions and rhetorically answering them.

Is chemiosmosis the only source of energy for gliding motility? Apparently yes. It has been generally found that uncouplers of the proton conduction type block motility in cyanobacteria (Castenholz 1973; Halfen and Castenholz 1971) and *Cytophaga* (Duxbury *et al.* 1980). The key references to supplement the current reviews by Pate (1988) and Khan (1988) in this area are Glagoleva *et al.* (1980), Ridgway (1977), and Larsen *et al.* (1974). The intermediate role of ATP hydrolysis seems to be ruled out.

Is a solid support surface really required? Yes, but there may be extenuating circumstances. Drews and Nultsch (1962) and Halfen and Castenholz (1971) have shown that a surface is not absolutely essential and that a viscous medium or a water-air interface will do. The former finding demonstrates that it is not the asymmetry of the air surface layer that is essential and the latter shows that the surface "skin" of a fluid may provide the necessary environment. The answer to this question does not help exclude the model under consideration nor a number of the others, but is fundamental to an understanding of the process.

What is unique about the substratum? Possibly only electronegative atoms. The speed and ability to glide depend on the physical nature of the surface. For example, *Cytophaga johnsonae* glides 10 times faster on glass than on an agar surface; actually, of course, it glides on the slime layer that may be upon either the glass or agar. The lack of "give" of the glass may be important. As mentioned, an ideal surface, glass covered with slime, should be firm, yet capable of being deformed, i.e., "being wedged into," but then that part of the cell must loosen when a cell part shrinks so that part can slip past the surface during the second part of the cycle. Thus, the substratum must be what is called a Stefan's adhesive. The key point is that hydrogen bonds are favored when the cell is pressed against the hydrophilic substratum.

What is the role of slime? It is a poor adhesive, but maybe it is a special kind of adhesive. Although the slime can be an indicator of movement, as when rings of slime were observed translating along a trichome of *Anabaena* (Walsby 1968), it plays an essential role in gliding motility. The slime is a complex material akin to mucopolysaccharides, which give it physical properties that would be desirable for organisms moving by the model presented here. Humphrey *et al.* (1979) present evidence that it is mainly a linear colloid. Again, for the present model, the substratum should allow (not too strong) attachment, and it should not form too permanent a bonding.

How fast must the cycling take place? Quite fast. Attempts have failed to observe discontinuous movements of *Oscillatoria* (Schultz 1955) and *Archangium* (Reichenbach 1965). This, of course, only proves that any jerkiness in movement in these cases is too fast to be observed or occurs with small amplitude in different sections of the cell or trichome at the same time.

Does the morphology or presence of external structures influence gliding motility? Very little. Organisms of quite a variety of shapes glide. They can exist as groups of cells and may be enveloped or sheathed. Cells in trichomes can move transversely (Pangborn *et al.* 1977), and cells of spiral-shaped organisms can move in a helical way (Reichenbach 1980). For the model presented here to be applicable to those cases, a more elaborate distribution of shunt sites and their control must be postulated.

Are there external or internal organelles of movement? The answer is not clear. There is no evidence for, and even some against, functional organelles. Burchard (1974) showed that a 50% lethal dose of ultrasound did not prevent motility. On the other hand, Lapidus and Berg (1982) and Pate and Chang (1979) have observed movements of polylatex beads and have speculated that this was due to their association with treads or spinning rotors, respectively. Recently, Lünsdorf and Reichenbach (1989) have found ultrastructures in the walls of *Myxococcus fulvus* that could be interpreted as part of a complex forming belts that encircle the cell in helical paths. It is too soon to understand how they would function.

Flagellar Motility

Flagellar motility is quite different from gliding motility because of the rotary movement of the flagella. It is clear that prokaryotes are able to convert chemical energy into mechanical work without the aid of mechano-proteins, as discussed in Chapter 2. Perhaps flagellar rotation can be explained by expansion/contraction of the peptidoglycan layer. In fact, the model given here can be modified to supersede the model of Berg and Khan (1983) for the mechanism of flagellar rotation. If a series of shunt sites was arranged on the circumference of a circle in the cytoplasmic membrane and sequentially activated, then the peptidoglycan would shrink immediately above each site and subsequently reexpand when the shunt closed. This process in turn could lead to rotation of the "rotor" part of the motor mechanism. In this structure, the rotor would be what has been designated the S-ring (MacNab 1987). Consequently, a mechanism of expansion/contraction of sequential parts of the sacculus could lead to rotation. Each site on the rotor would be like the specific mechanism detailed above, for the case of linear propagation of a single rod-shaped cell. For such a complex movement, each site must contract and expand in a precise order and timing.

The mechanism proposed here is sufficiently powerful so that with different shunting arrangements and temporal controls a large array of mechanical processes in prokaryotes can be explained. It is proposed that peptidoglycan is the contractile substance for prokaryotes. Not all prokaryotes have peptidoglycan; perhaps pseudo murein in motile archaea or other substances serve the role of contractile elements that drive motility in motile archaebacteria and mycoplasma. But I am not aware of *gliding* motility in any organism without murein.

15

Prokaryotic Perspective

KEY IDEAS

Biological evolution can adapt to new and different conditions with new strategies.

The three phases of a biotic evolutionary step are saltation, radiation, and refinement.

The saltation is the Yang phase, or macro-mutation, or "great leap forward."

The radiation is the diversification phase in which exploration of all new possibilities occurs.

The refinement is the Yin phase or adaptation phase in which a line of descent becomes modified to be in tune to the habitat and niche.

The time from the First Cell to the Last Universal Ancestor consisted of very many Yin phases, acting in both parallel and series, in which the development of cell biology occurred.

Two independent Yangs led to almost simultaneous, but different, solutions for the osmotic problem and led to the exoskeleton of bacteria and the endoskeleton of eukaryotes.

The Yang that created eubacteria required a number of parallel independent advances that together led to the bacterial exoskeleton strategy.

All cells, including bacteria, must not only grow and divide, but must also find new suitable habitats to continue the line.

Enterococcus hirae, Bacillus subtilis, Escherichia coli, Myxococcus xanthus, *and* Microcyclus aquaticus *have different and special strategies to cope with their own habitats and niches.*

Kinds of Evolution

"Evolution" is a word that came into biology from the discipline of geology and the word suggests that mountain ranges will erupt, will be worn down, only to erupt again. Abiotic change, even if called "evolution," is quite distinct from biotic evolution. If a mountain "figured out a better way to live," there would be no way for it to leave a legacy for later descendent mountains; it has no genetic system. Some aspects of abiotic evolution may be quick on the geological scale, like the development of random polymers after the precursors arise, but abiotic evolution can be slow and of long duration. For example, after the big bang, stars had to form and die and be reborn to cycle through several generations until one, like our own sun, could have a solar system with planets containing suitable elements and with a physical environment so that life forms could develop based on organic molecules. Thus, the progression of star types is an evolution in which a descendent star has properties generated by the nucleosynthesis of its predecessor, but again this is not "adaptive" as in the way Darwin used the term "evolution."

On the other hand, Darwin might have been content to use the term in the context of "social evolution." Such institutions as hunting groups of cave dwellers, armies, commercial banks, etc., arise and develop just as life forms do. These institutions are social organizations that arise and are propagated in both unchanged and changed form. They possess cultural "inheritance" that allows adaptive change. This, of course, is the key idea of biological evolution; entities arise from a predecessor, are copied (replicated), become dispersed, are capable of being mutated, and can pass the changes to their descendants. In all these cases the "offspring" usually inherit the previous organization intact, but may inherit it in a mutated form that may be better, but often is worse, in the particular external environment in which it happens to exist. Particularly important, some variants may have extended their abilities not only quantitatively, but also qualitatively beyond those of their "parents." Thus, this type of evolution has a one-to-one correspondence with the generally accepted process for the evolution of species. The distinction between the abiotic type of evolution is that the equivalent of organisms cannot adapt to conditions, whereas in biotic evolution organisms have ways to change and adapt to react to the ambient circumstances and to the exploitation of their environment in qualitatively different ways. This possibility in turn allows the organisms to change and expand into different environments and different growth strategies.

Biotic evolution has a special type of kinetics different than those detailed in Chapter 1. It has two component processes that I will label the Yang and the

Yin phases. The Yang is the rare but drastic change, macromutation, or saltation; and the Yin refers to the subsequent numerous small stages of refinement. In Chinese philosophy the Yang is the masculine principle that is aggressive, imaginative, and outspoken, whereas the Yin principle is the female one that embodies empathy, compassion, and gentleness. Yang and Yin mean much more than the English translation of saltation and refinement. But biotic evolution has a third component when viewed globally: this is diversification or radiation (see Fig. 15.1). After a Yang transition opens new opportunities, Yin refinement of the organisms may allow simultaneously many different paths of evolution. One can examine the fossil record and see many instances. Just one biological example: the first kind of animal that became terrestrial arose by a saltation or Yang event. (This event was the product of many small independent Yin changes in many different aspects of biology that together created the dramatic change; this idea

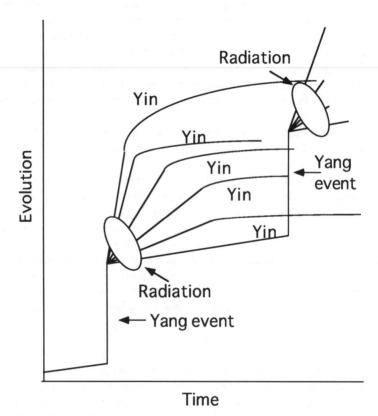

Figure 15.1. Yang, Yin, and radiation in "Darwinian" type evolution. A Yang event is a change that opens up new niches and habitats. The Yin events refine different strategies. Some lines may die out, but others may serve as the line in which a new Yang event occurs.

will be explored below.) After this saltation, the first land-based animal was able to radiate by generating a range of species adapted to different aspects of the new habitat or available niches. Diversification in many directions and invasion into a variety of previously unoccupied lifestyles required a different series of Yin events in the different lines of descent. Within each line, the descendants slowly adapted to their chosen niches and were perfected by small modifications, each of which allowed the sublines to compete with other sublines in a way that led some to succeed and some to fail. The key point is that the microadaptations by Yin tuned the diversified products to fit the range of new niches and habitats that were made accessible by the revolutionary Yang event. (A special point is that Yin refinements were necessary not only to optimize, but to set the stage for the next Yang.)

Other Yang events in history were the origin of cellular life, all eight of the categories listed in Fig. 1.1, the utilization of fire, the invention of the wheel, the development of the bow and arrow, the domestication of plants and animals, the manufacture of bronze, the 1492 discovery of the new world, the perfection of valved wind instruments, Watt's refinement of steam engine governor, the formulation of the theory of evolution, the production of personal computers by Apple and IBM, the combination of the use of restriction enzymes, bacteria, plasmids, and viruses to do gene cloning, and the construction of the polymerase chain reaction (PCR) technique from the available molecular genetic techniques.

Referring to the last point, I and many thousands of molecular biologists ought to have been capable of inventing PCR. We had all the tools developed by the Yin studies of biochemistry and microbial genetics in the late 1970s, but the Yang event was Gary Mulis' synthesis of these studies in a different way for a new purpose. This is the pattern of all the Yang events—they stitched together, in a new way, disparate entities that have existed before.

The relevance of evolution to this book should be clear. Only the original, abiotic kind of evolution took place as organic systems evolved to yield complex but "meaningless" molecules. Even the generation of catalysts and selection for quasi-species of RNA, although important, was not the Yang event that initiated life: these events were Yin events that tuned the system to the environment, but did not open a new form of exploitation. The Yang required the coming together of several spontaneous processes and situations that first gave rise to an entity that could adaptively improve itself not only to exploit the initial habitat and niche but to be capable of evolving to occupy other habitats and niches (see Fig. 15.2). Creation of life opened "new worlds to conquer." The ensuing Yin phase, starting from the First Cell, was long and difficult. No doubt it contained many microYang phases, but there were no Yangs or megaYangs and no diversification for a long time. The final product of the development of cell biology at the time of the Last Universal Ancestor was spectacular in terms of its competence to do many things well. Evolution had not produced a spectrum of different biotypes because the system was constrained so that each newly improved variety elimi-

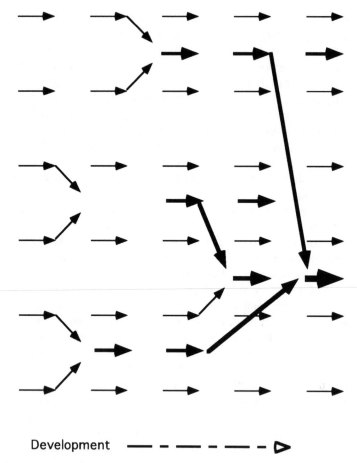

Development ▷

Figure 15.2. Systematic refinement (Yin) and new combinations with extraordinary new pathfinding (Yang) development. Most often Yang events depend on previous development adapted for a variety of special purposes. The Yang event in such cases is the combination of such diverse developments that together create a new function and provide opportunities that did not exist before.

nated competing lines instead of permitting diversity through alternative non-competing strategies.

The basic constraint preventing diversification was that there was only one limiting resource. I suggest that this was the quantity of utilizable energy that continuously could be generated by abiological processes (geological, meteorological, and astronomical). One can imagine that diverse habitats, such as high- and low-temperature environments, could have allowed stable diversity to develop, and tribes of individuals better adapted to one or the other temperature would exist. It may be that such diversification did develop, but the fact from molecular evolutionary biological studies is that, before the Last Universal Ancestor, multiple forms did not persist. In fact, there is some physiological evidence that high-temperature forms were important in the generation of archaea, but this aspect does not appear to be the environmental force that permitted the development of stable diversity.

The suggestion, championed here, is that the development of endo- and exoskeletons was the important factor that led to the split of the world ecosystem into the eubacterial and eukaryotic lines. These two independent, and different, solutions to the emerging problem of resisting the self-inflicted osmotic forces tending to rupture the cell were the two Yang events that broke the monophyletic degeneracy. This "osmotic" problem was gradually created by the predecessors of the Last Universal Ancestor as a result of their own success in developing and improving a powerful intermediary metabolism and in perfecting all the other parts of cellular biology. With the accumulation of high concentrations of intracellular substances, for all organisms since the Last Universal Ancestor, combating the osmotic or turgor pressure became an important aspect of life. It is so immense that aspects of it permeate most of the chapters of this book as well as all general and cell physiology texts. It is suggested that the "degeneracy-breaking" development was the independent achievement of alternative abilities to resist turgor pressure, either with an endoskeleton or with an exoskeleton (see Fig. 15.3). I assume from the existence of the branch in the phylogenetic tree that the eukaryotic and prokaryotic solutions to turgor stress were developed close together in time, so that these two approaches did compete with the prototype, but did not compete with each other. If at first they did not compete with each other, then after the Yin phases one line could not displace the other and the issue became moot because both approaches were sufficiently developed and had started to create a spectrum of different diverse strategies that now could maintain their independent existence. Subsequent developments could then take place creating very diverse eukaryotes and bacteria and a progressively more varied world biosphere. (The archaea may have developed a different strategy that made them the majority for a while, but they became generally less successful as the world environment became more oxidizing, although they are able to persist in special habitats and niches).

The Creation of the Bacterial Exoskeleton Strategy

At this point we leave the evolutionary trail of other organisms and focus on certain bacteria. As with other Yang events, the coming together of previously unrelated elements was the key to the new invention. One of the essential components for the new Yang event was the development of secretory mechanisms to allow large molecules (mostly proteins) to selectively be exported through the cytoplasmic membrane. Another was the combination of techniques acting to drive the reactions that link the oligopeptidoglycans together in a strange

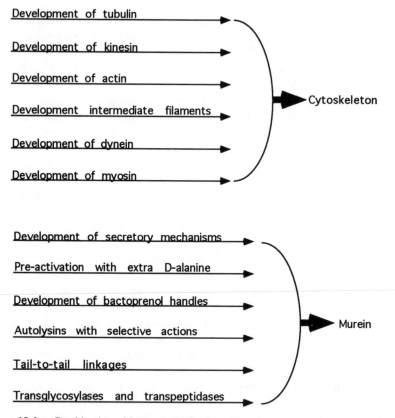

Figure 15.3. Combination of features yielding two new ways to combat turgor pressure. Two different strategies evolved to cope with a high internal osmotic pressure: The endoskeleton and the exoskeleton. (There are other techniques, such as forming a water vacuole and extruding it.) It can be seen that for the eukaryote way of life (above) and the eubacterial way (below), the development of a range of techniques was needed to build either an internal or an external skeleton to serve the needed function.

place; *viz.*, outside the cell proper. This required the development of pre-payment strategies for the energy costs by the inclusion of the extra D-alanine in peptide linkage and the export of the oligopeptide disaccharide with an attached pyrophosphate connecting it to the membrane-bound bactoprenol. A third element was the generation of a special class of secreted enzymes, the autolysins, that could be "issued general instructions so as to be pre-controlling" to actuate cleavage of the existing wall fabric to allow safe growth. Of course, to the student of microbiology, the wonderful chemical and structural nature of the wall fabric with its peptide "woof" and glycan "warp" has been the centerpiece for appreciating and extolling the bacterial way of life.

An effective exoskeleton that mitigates against turgor pressure is the hallmark of most eubacteria. The exceptions require additional studies. As challenges to the next generation of microbiologists, find out what the osmotic pressures of these exceptions are, what the osmotic pressure of their environment is, and how strong their walls are. As a specific challenge, because it is reasonable to assume that mycoplasma were once Gram-positive organisms that lost their wall, how did they solve the osmotic problem? In this connection the role of surface protein layers in the mycoplasma or S-layers in other organisms needs further study.

Lifestyle of Pertinent Microorganisms

In this book we have delved into the properties of a number of prokaryotes. In this last chapter, they will be re-examined with regard to their growth and survival, with emphasis on the Yin and Yang components of their strategies for continuing existence, not over evolutionary time but in their day-to-day existence. The issue is that in order to persist, an organism must be capable of completing its "culture cycle." By the "culture cycle" I mean the series of events from the time that an organism is introduced into a new environment when it can multiply to the time that eventually there is a "lucky" descent organism that finds a new environment. This requirement for colonization of new habitats is a necessary part of the cycle because all bacterial habitats are ephemeral. This colonization, like the evolution of a new strategy, can be classified as a kind of Yang event. But the Yin of bacterial growth is the myriad aspect of growing, colonizing, surviving, and exploiting habitat and nutrients when and where available. It involves the exquisite regulation of the physiological properties of bacteria.

In the rest of this chapter, the culture cycles of organisms discussed in earlier chapters will be presented. These organisms are not distributed across the list of bacterial species (Olsen *et al.* 1994), but they do provide a glimpse of the diversity of actual strategies used by modern prokaryotes.

A Fermentative, Fastidious, Gram-Positive Organism Forming Lactic Acid

Starting with *Enterococcus hirae* (which we will take to be the prototype of other lactic acid bacteria), imagine that one cell is dropped into a bowl of milk. The iron deficiency of the milk is no obstacle because the organism grows fermentatively and does not have or need respiratory enzymes. The initial presence of oxygen is also no problem because these organisms can deal with the oxidative stress by making fewer free radicals than do aerobes and by being able to detoxify some of them with peroxidases. Moreover, the oxygen is quickly consumed by obligate aerobes that then cannot grow further. The competition with many other organisms that might attempt to grow in the milk is minimal because the fermentation of the lactose of milk lowers the pH and blocks the growth of many species. Eventually, the pH falls enough to block the growth even of the streptococci, enterococci, and lactococci; each strain has its individual lower limit of pH for continued growth.

Milk and bodily tissues and exudates are suitable habitats for these organisms. Thus, such organisms as a class are adapted to grow in rich environments such as the epithelium of the respiratory tract, mouth or intestine, vagina, and in spoiling plants and dairy products, but they are limited because they have "elected" to forgo an iron-requiring respiration. They have also forgone many metabolic pathways, so they are dependent on the availability of many purines, pyrimidines, amino acids, and vitamins in their diet. Thus, like mammals, they have lost many metabolic processes possessed by the Last Universal Ancestor. It can be argued circularly for both cases, that the lack of independence from need for specialized nutrients resulting from the economies of deleting metabolic pathways is not important to their life strategy because their choice of lifestyle limits them to an environment that generally provides the needed finished organic molecules.

The Yang event under consideration is the unique event of a particular organism in finding a bowl of milk in a timely fashion. This is followed by explosive growth, which may be considered either as part of the Yang phase or part of the Yin execution of a preprogrammed consumption of resources. Although the Yang proper is the lucky finding of a hitherto sterile bowl of milk, subsequent Yin events in most culture cycles is the finding of many modest homes where small colonies can be maintained for long periods of time. For pathogenic streptococci, these colonies could be in the upper respiratory tract, in the gastrointestinal tract, or on plant surfaces. For the overall life style, the Yin goal is to spread to many limited sites so that some will be poised for (and be ready for) the next Yang event. Although there is movement from one habitat of limited size to spread to yet other limited sites, the occasional "jackpot" Yang opportunity leads to a many-fold expansion and an increased chance to start new colonies. The bowl of milk or uncontrolled blood infections (septicemias) are metaphors for any rich, unused habitat that can support extensive growth. Of course, these windfalls of resources are not capable of supporting indefinite growth. The organisms are programmed to find, exploit, and destroy suitable habitats, and then "hope" (more anthropomorphism and teleology) that some of them can find a new environment in which to do the same. These bacteria, like other bacteria and like man, are not altruistic, nor do they take a long-term view of protecting their environment.

Many of the Group D streptococci, renamed *Enterococci,* are able to survive at higher temperatures and in higher salt concentrations than are characteristic of the insides of an animal body. Thus they, like the *Staphylococci,* survive on a sweaty skin or a dried-out, milk-laden nipple (which is one key to their ability to reach rare Yang-type opportunities to generate a massive supply of propagules). Salt tolerance seems to be associated with organisms whose hosts can sweat and not with furry animals that largely dissipitate heat in some other fashion, say panting. Another adaptation is that of exploiting a host if it becomes compromised for some other reason. Pathogenesis, from the point of view of the microorganism, is simply another opportunity to generate many propagules.

A Gram-Positive, Rod-Shaped, Spore-Forming, Aerobic Bacterium

The strategy of *Bacillus subtilis* is quite the opposite of that of *E. hirae*. It is a soil organism dependent upon the recycling pathway of leaves and other plant materials. A leaf develops, carries out photosynthesis, and then dies at a programmed season. Even before the leaf dies, certain fungi have already infected it and are consuming the most edible parts. The tree also mobilizes and sends down from the leaves magnesium, other nutrilites, and sugars to the roots for storage until the following spring. The leaf in the fall is induced to fall off by the action of the plant hormone, abscisic acid. The leaf falls and becomes infected with a series of other fungi, then other organisms act, and eventually the leaf becomes totally recycled (unless it becomes lodged in an anaerobic environment, in which case it may eventually become converted into peat or coal). The most resistant part, humic acid, derived from the lignins, is finally "done in" by the "white rots." The lignins are polyaromatic molecules serving as the mastic phase in the cell wall structure of the plant. The white rots are a group of taxonomically unrelated fungi that aerobically generate free radicals to break down the polymerized and crosslinked aromatic rings that make up lignins. In this complex scheme, *B. subtilis*, in a hyena-like scavenger fashion, utilizes the thus-generated breakdown products of lignin and the primary degradation products of cellulose. These simple carbon compounds (such as succinic acid) and aromatic ring-containing compounds) are metabolized by an oxygen-based respiration. So *B. subtilis* survives utilizing a very specialized and limited diet of leftovers after other organisms have taken the choice parts.

This apparently poor strategy is actually a good strategy when combined with a second trick—that of being able to sporulate. This allows the organism, when deprived of fresh carbon-containing substrates, to start and finish the formation of the resistant endospore. The spores lie dormant and survive until conditions again become satisfactory, then they quickly germinate and rapidly grow.

Actually, some strains have a second trick, *viz.*, that of becoming transformed by any compatible DNA in their vicinity. However, they do this only when things get tough (and sporulation is about to commence). This allows a recombination of gene characters that may permit further growth. In this case, the strain can become fine tuned to the local environment or acquire a way to utilize a novel substrate, but more likely the cells become worse off and the resulting lines die off. Thus, transformation is not usually helpful and may destroy organisms in nature at a rate of a million or billion to one survivor, but they were slated to die anyhow. However, if a successful transformant can grow, it very quickly can recoup the million- or billion-fold lost.

An Enteric Facultative Anaerobe

The world and ways of *E. coli* are very specialized (Koch and Schaechter 1985). Under natural conditions large numbers are seldom found. If *E. coli* cells were to find the bowl of milk before the *E. hirae*, the milk will soon spoil and taste bad because the *coli* can grow aerobically, and, after the oxygen is consumed, grow anaerobically utilizing fermentation. Its proteolytic enzymes will cleave some bonds in the proteins of milk and then ferment the products. The pH stays closer to neutrality and the organisms can grow for a longer time and more plentifully before the acid content stops growth because *E. coli* possesses a mixed-acid fermentation pathway in which a mole of glucose produces one mole of carboxylic acid (lactic, acetic, and some formic acid) and liberates nearly a mole of CO_2, which escapes from the organism's milieu. In contrast, the homofermentative lactic acid bacteria *E. hirae* forms two moles of lactic acid.

Organisms that live in a flowing system need special mechanisms to prevent being washed out. Fish naturally and continuously swim upstream, but no microorganism can swim as fast as the flow of contents through the gastrointestinal tract (Koch 1987b). The problem of washout in the large intestine is less because of extensive mixing due to peristalsis, so there the free-living strict anaerobes persist. But *E. coli* is in the minority because it is not as adept in growing fermentatively and as rapidly as must be done in the bulk contents of the colon. Moreover, the anaerobes produce substances from fermentations, like propionic acid, that inhibit the growth of the enteric organisms. The importance of these fermentation products was demonstrated, with a close relative of *E. coli*, *Salmonella typhimurium*, many of which are normally required to kill a mouse. By eliminating the strict anaerobes with an antibiotic, it is found that one antibiotic resistant *S. typhimurium* organism is sufficient to kill the mouse, after the antibiotic is removed because that cell can grow sufficiently to produce a septicemia before the anaerobes regrow and their waste products then force the numbers of coliforms down.

So the key question is: How do the *E. coli* cells colonize and persist in the human gut? Most of the time the answer is that they don't. Viewing the gut as a chemostat, they should be washed out because the turnover is just too fast. The experimental fact is that strains come and go; particular strains do not have long-term persistence in humans or in pigs living a hygienic existence. In fact, in the developed world, even in the absence of antibiotic treatment, middle- and upper-class individuals at certain times have no detectable coliforms in their feces. Of course, poorer humans, babies, and animals are reinfected at much higher rates, and coliforms can be isolated easily and regularly from them, but the strains still are frequently replaced.

So in summary, the coliforms are simply transients in the colon. But there is another aspect that can be deduced from the work of Linton (Linton *et al.* 1978; Hartley *et al.* 1978; see Koch 1987b for earlier references). Under sanitary conditions a strain of *E. coli* appearing in the feces persists for a few days or a week and then disappears. There may be a period with few total coliforms, but then another strain, as adjudged by its antigenic determinants and pattern of antibiotic resistance, appears and goes through the same rise and fall. This process of new strains appearing and being lost continues indefinitely. Occasionally, a strain seen earlier will re-emerge. This pattern only has an explanation if one postulates that there are sacrosanct microhabitats in the ileum or appendix wall where *E. coli* sequesters itself out of the mainstream of the small intestine flow and there ekes out a nearly zero growth existence. Occasionally, the flowing phase of the small intestine becomes infected from these source spots. In the contents of the small intestine the cells can grow very rapidly, because the pH, oxygenation, and nutrients there are optimal. This causes an infection of the fluids entering the large intestine, and thus there is a time when the particular strain of *E. coli* can prosper and grow rapidly to significant numbers (or large numbers for some pathogenic strains) in the large intestine. This colonization of the colon is only temporary and because of washout, the descendants of these occasional releases do not persist. Then another accidental release may occur in which cells of a different strain from another sacrosanct region of the small intestine peel off and give rise to yet another temporary burst of organisms of another serotype and spectrum of genetic markers in the feces.

No doubt, the adventures of an emergent fecal organism after it enters the outside world would make Ulysses' journey seem tame. But some do find a new host and thus maintain the species. Near the end of their journey they must survive an acid trip through the stomach. But the Odyssey is not over until they find a sacrosanct home in the lower small intestine where they can remain for years. A real Yang event is the rare times when *E. coli* becomes pathogenic and creates a large biomass, and because there are billions and billions of cells there is a better chance that at least some will make it to a new animal host.

Support for the above interpretation of the life history given here is that the physiology of *E. coli* is tuned to the environment and nutrients likely to be found in the lower small intestine, which are quite different from those in the colon. Inducible enzymes like those of the *lac* operon or alkaline phosphatase would only serve a useful role in the small intestine and not in the large (see Koch 1987).

A Myxobacterium

The life story of *Myxococcus xanthus* occupies half of all seminars given on any aspect of the organism's biology; it has a complex and fascinating lifecycle. For any exposition from the natural history point of view, we should start with the myxospores. They are quite stable, but when conditions become favorable (a wet environment with sufficient nutrients), the spores germinate into rod-shaped vegetative cells that look like large *E. coli*. The vegetative cells move by gliding motility. After one has formed a slime trail as it moves, other cells may cross and then follow in the existing trail. This tendency leads to a form of aggregation that serves a key role in their life strategy. Each cell secretes hydrolytic enzymes (proteases, etc.). Consequently, the production and secretion of enough of a mixture of enzymes to lyse and digest other kinds of bacteria that may be present can occur only if enough cells follow along the same trail and become bunched. A major source of nutrition of the cells are small peptides resulting from the hydrolysis of protein. There is certainly a lot more that should be learned about how these proteases attack and destroy the food bacteria, but the point is that the *M. xanthus* cells are predatory and operates as a wolf pack. When there are no more food bacteria, a different mode of growth ensues. Then they aggregate much more strongly and pile on top of each other and differentiate into a fruiting body. This is interesting because it resembles apoptosis (programmed cell death) of higher cells in that certain cells die (commit suicide). In this case the dying cells contribute their cellular contents towards the development of the fruit, but do not contribute genetically. The remaining cells differentiate to form myxospores contained in the macroscopic fruiting bodies.

A Freshwater Heterotroph with Gas Vacuoles

Microcyclus (Ancylobacter) aquaticus, like all the organisms discussed here, is a product of the diversity of biotypes present in the world today. It is dependent on other life forms because it, like the others, is a modern heterotroph. Its special strategy is to float in freshwater lakes and grow on the low level of nutrients available there. It does not need to grow rapidly and cannot grow fast; it is therefore outcompeted by other faster-growing organisms in eutrophic environments and is incapable of growth in the absence of oxygen. So it occupies a special niche. Its special trick depends on its gas vacuoles that give it buoyancy. It may not be apparent why *Microcyclus aquaticus* has the capability to make gas vacuoles, but one can formulate an hypothesis for function of the vacuoles.

Microcyclus aquaticus is the *B. subtilis* of the freshwater lake. Nutrients enter the lake in the runoff from the surrounding land. To some extent the succinic acid and other organic degradation products of terrestrial plants have escaped consumption by the *B. subtilis* and other similar consumers in the soils and find their way into lakes. In addition, there is a food chain in the lake in which the degradation of aquatic higher plant materials may make simple organic compounds.

There is in addition to these two food chains sources of reduced carbon, however, an even more abundant source of oxidizable carbon in the lake environments. This arises from photosynthetic organisms in the water column. The problem any photosynthetic organism, be it a higher plant, an alga, or a cyanobacterium (blue-green alga), must face is what to do when there is plenty of light, but not enough fixed nitrogen, phosphorus, or essential minerals to support growth. Dry land plants face the same problem and, for this reason, have a number of mechanisms to dissipate the energy of the incoming light without destruction of their photosynthetic apparatus. These mechanisms include forming carotenes to dissipate free radicals and some pigment molecules to absorb light and convert the energy into heat. Terrestrial plants possess glyoxosomes and other plastids to oxidize photosynthetic products, with the net effect of recycling the carbon dioxide. They also carry out photo-oxidation with molecular oxygen. All inclusively, these processes dissipiate the light energy in small enough units (that can be thought of as "quanta") so that the energy does not bleach or break down the pigments needed for photosynthesis. Aquatic photosynthetic organisms have these protective mechanisms too, but in addition have the possibility of forming and then wasting the photosynthetic products by secreting them into the aqueous environment. These compounds include glycerol and a range of other simple organic compounds. Although the organisms have wasted energy and carbon dioxide, the result is that they can safely stay in the light by carrying out nondestructive futile photosynthesis as long as carbon dioxide is available. Consequently, near the surface of the lake during the daylight hours, there is a continuing supply of organic carbon spilling out of the primary producers available for consumption by heterotrophs. Therefore, it behooves an aquatic bacterial heterotroph to inhabit the surface layers where both organic carbon and oxygen are being generated, because one or both may be absent from deeper layers.

Walsby (1994) has made a related argument with respect to the physiology of the algae and cyanobacteria. His argument is that in sufficiently bright light photosynthesis will increase the osmotic pressure because osmolytes are created and are retained (or only slowly leaked to the environment). This has the simple consequence of spontaneously collapsing a proportion of the vesicles depending on the success of the photosynthesis. When the vesicles collapse there is an increase in the density of the cell, which causes it to sink out of the phototic zone and away from predators. Subsequently, the vesicles will reform to allow the cell again to resurface. He argues that this is the reason that the vesicles of both photosynthetic and nonphotosynthetic organisms exhibit a broad distribution of collapse pressures in order to allow them to fine tune their adjustment of depth in the water column.

Concluding Statement

The puzzles of how life formed, how it developed, how and why life remained monophyletic for so long, why it then became so diversified, how all life forms became so sophisticated, and how all the extant life forms work together (at least up until the present) are puzzles that may be near a fundamental solution in 1995 or at least before the end of the millennium.

A central purpose of this book has been to try to show how far prokaryote life forms could evolve without the magic of mechano-enzymes and mechano-proteins. The easy part was to show that bacteria can do without such sophisticated proteins. The harder and more important part has been to develop the logic that the First Cell, not only had no mechanoproteins, but could do without them. This is even though the development of mechanoproteins was an extremely important advance later. All the rest follows.

A second purpose of this book, and of my earlier papers, is to show that life did not need to start with functional fermentation, respiration, or photosynthesis, but that energy could be trapped by chemiosmosis alone.

As a good nonvitalist, I assume that these and all such problems did solve themselves. There are still many fundamental and hard-to-understand aspects of life. They would be easier to understand if we could appeal to divine intervention, but modern biologists with a conviction that is truly a "religious" fervor are not able to do that. Although all the questions are important, the prime question is: How did the First Cell start? This is the same as asking what Yang event permitted the establishment of a system that had an energy metabolism and obeyed the Darwinian set of three rules: accurate replication of a molecule that served a function, rare mutation with an altered action, and selection to tune the system to the ambient conditions and beyond. I wonder whether if life were to start on many planets just like earth was 4 billion years ago in a solar system just as our's was, what kinds of life would there be 4 billion years later. Would the planets all have an abundant diverse ecosystem? Would they have a small biomass and little diversity? Would they have had life which then floundered and disappeared, because it had not developed enough diversity?

Bibliography

Adler, H. I., W. D. Fisher, A. Cohen, and A. A. Hardegree. 1967. Miniature *Escherichia coli* deficient in DNA. *Proc. Natl. Acad. Sci. USA.* **57**:421–426.

Alberts, B., D. Bray, J. Lewis, M. Raff, K. Roberts, and J. D. Watson. 1994. *Molecular Biology of the Cell.* Third edition. Garland Publishing, Inc. New York.

Alemohammad, M. M., and C. J. Knowles. 1974. Osmotically induced volume and turbidity changes of *Escherichia coli* due to salt, sucrose, glycerol, with particular reference to the rapid permeation of glycerol into the cell. *J. Gen. Microbiol.* **82**:125–142.

Alton, T. H., and A. L. Koch. 1974. Unused protein synthetic capacity of *Escherichia coli* grown in phosphate-limited chemostats. *J. Mol. Biol..* **86**:1–9.

Anba, J., A. Bernadac, J.-M. Pages, and C. Lazdunski. 1984. The periseptal annulus in *Escherichia coli*. *Biol. Cell.* **50**:273–278.

Anderson, A. J., R. S. Green, A. J. Sturman, and A. R. Archibald. 1978. Cell wall assembly in *Bacillus subtillis*: localization of cell wall material during pulsed release of phosphate limitation, its accessibility to bacteriophage and concanavalin A, and its susceptibility to turnover. *J. Bacteriol.* **135**:886–899.

Archibald, A. R. 1976. Cell wall assembly in *Bacillus subtillis*: development of bacteriophage binding properties as a result of the pulsed incorporation of teichoic acid. *J. Bacteriol.* **127**:956–960.

Archibald, A. R., and H. E. Coapes. 1976. Bacteriophage SP50 as a marker for cell wall growth in *Bacillus subtilus*. *J. Bacteriol.* **125**:1195–1206.

Baldwin, W. W., and A. L. Koch. 1995. Unpublished work.

Bartnicki-Garcia, S. 1990. Role of vesicles in apical growth and a new mathematical model of hyphal morphogenesis, pp. 211–232. In I. B. Heath (ed.) *Tip Growth.* Academic Press, Inc., London.

Bartnicki-Garcia, S., F. Hergert, and G. Gierz. 1989. Computer simulation of fungal morphogenesis and the mathematical basis for hyphal (tip) growth. *Protoplasma* **153**:46–57.

————. 1990. A novel computer model for generation cell shape: Application to fungal morphogenesis, pp. 43–60. In P. J. Kuhn, A. P. J. Trinci, M. J. Jung, and L. G. Copping (eds.), *Biochemistry of Cell Walls and Membranes in Fungi*. Springer-Verlag, Berlin.

Bartnicki-Garcia, S., and G. Gierz. 1991. Predicting the molecular basis of mycelial-yeast dimorphism with a new mathematical model of fungal morphogenesis. pp. 28–48. In J. W. Bennett and L. L. Lasure (eds.), *More Gene Manipulations in Fungi*. Academic Press, Inc., London.

Bartnicki-Garcia, S., and E. Lippman. 1973. The bursting of the hyphal tip of fungi: Presumptive evidence for a delicate balance between wall synthesis and lysis in apical growth. *J. Gen. Microbiol.* **73**:487–500.

Bayer, M. E. 1991. Zones of membrane adhesion in the cryofixed envelope of *Escherichia coli*. *J. Struct. Biol.* **197**:268–280.

Bayne-Jones, S., and E. F. Adolf. 1933. Growth in size of microorganisms measured from motion pictures. III. *Bacterium coli*. *J. Cell Comp. Physiol.* **2**:329–348.

Begg, K. J., and W. D. Donachie. 1991. Experiments on chromosome separation and positioning in *Escherichia coli*. *The New Biologist* **3**:475–485.

Berg, H. C. 1983. *Random Walks in Biology*. Princeton University Press, Princeton, NJ.

Berg, H. C., and S. Khan. 1983. A model for the flagellar rotary motor, pp. 456–497. In H. Surd and C. Vieger (eds.), *Motility and Recognition in Cell Biology*. Walter de Gruyter. Berlin.

Best, J. B. 1955. The inference of intracellular properties from observed kinetic data. *J. Cell and Comp. Physiol.* **46**:1–27.

Beveridge, T. J. 1981. Ultrastructure, chemistry, and function of the bacterial wall. *Int. Rev. Cytol.* **72**:299–317.

Beveridge, T. J., and L. L. Graham. 1991. Surface layers of bacteria. *Microbiol. Rev.* **55**:634–705.

Bi, E. P., and J. Lutkenhaus. 1991. FtsZ ring structure associated with division in *Escherichia coli*. *Nature* **354**:161–164.

Bloom, M., E. Evan, and O. G. Mouritsen. 1991. Physical properties of the fluid lipid-bilayer component of the cell membranes: A perspective. *Quart. Rev. Biophys.* **24**:293–397.

Boye, E., and A. Løbner-Olesen. 1991. Bacterial growth studied by flow-cytometry. *Res. Microbiol.* **142**:131–135.

Boye, E., M. Marinus, and A. Løbner-Olesen. 1991. Quantitative DAM methyltransferase in *E. coli*. *J. Bacteriol.* **174**:1682–1685.

Boys, C. V. 1890. *Soap Bubbles and the Forces Which Mould Them*. Society of Christian Knowledge. Reprinted by Dover, New York (1959).

Brakenhof, G. J. 1979. Imaging modes in confocal scanning light microscopy (CSLM). *J. Microscopy* **117**:233–242.

Braun, V. 1975. Covalent lipoprotein from the outer membrane of *Escherichia coli*. *Biochim. Biophys. Acta* **415**:335–377.

Braun, V., H. Gnirke, U. Henning, and K. Rehn. 1973. Model for the structure of the shape-maintaining layer of the *Escherichia coli* cell envelope. *J. Bacteriol.* **114**:1264–1270.

Braun, V., and H. C. Wu. 1994. Lipoproteins, structure, function, biosynthesis and model for protein export. pp. 319–341 In J.-M. Ghuysen, and R. Hakenbeck. (eds.), *Bacterial Cell Walls.* Elsevier, Amsterdam.

Briehl, M. M., and N. H. Mendelson. 1987. Helix hand fidelity in *Bacillus subtilis* macrofibers after spheroplast regeneration. *J. Bacteriol.* **169**:5838–5840.

Brooks, R. F., D. C. Bennett, and J. C. Smith. 1980. Mammalian cell cycles need two random transitions. *Cell* **19**:493–504.

Buchanan, C. 1979. Altered membrane proteins in a minicell-producing mutant of *Bacillus subtilus.* *J. Bacteriol.* **139**:305–307.

Burchard, R. P. 1974. Studies on gliding motility in *Myxococcus xanthus. Arch. Microbiol.* **99**:271–280.

―――. 1981. Gliding motility of prokartyotes: Ultrastructure, physiology, and genetics. *Ann. Rev. Microbiol.* **35**:497–529.

―――. 1982. Evidence for contractile flexing of the gliding bacterium flexibacter FS-1. *Nature,* London **298**:663–665.

―――. 1984. Gliding motility and taxes. pp. 139–161. In E. Rosenberg (ed.), *Myxobacteria Development and Cell Interactions.* Springer-Verlag, New York.

Burdett, I. D. J., and M. L. Higgins. 1978. Studies of the pole assembly in *Bacillus subtillis* as seen in central, longitudinal, thin sections of cells. *J. Bacteriol.* **133**:959–971.

Burdett, I. D. J., and T. B. Kirkwood. 1983. How does a bacterium grow during its growth cycle? *J. Theor. Biol.* **103**:11–20.

Burdett, I. D. J., T. B. Kirkwood, and J. B. Whalley. 1986. Growth kinetics of individual *Bacillus subtilus* cells and correlation with nucleoid extension. *J. Bacteriol.* **167**:219–230.

Burdett, I. D. J., and R. G. E. Murray. 1974. Spetum formation in *Escherichia coli*: Characterization of septal structures and the effects of antibiotics on cell division. *J. Bacteriol.* **119**:303–324.

Burman. L. G., J. Raichler, and J. T. Park. 1983. Evidence for diffuse growth of the cylindrical portion of the *Escherichia coli* murein sacculus. *J. Bacteriol.* **155**:983–988.

Cairns-Smith, A. G. 1986. *Clay Minerals and the Origin of Life.* Cambridge University Press, Cambridge, England.

Campbell, A. 1957. Synchronization of cell division. *Bacteriol. Rev.* **21**:263–272.

Casjens, S., and J. King. 1975. Virus Assembly. *Annu Rev. Biochem.* **44**:555–611.

Castenholz, R. W. 1973. Movement, pp. 329–339. In N. G. Carr and A. B. Whitton (eds.), *The Biology of the Blue-Green Algae.* University of California Press, Berkeley and Los Angeles, CA.

―――. 1982. pp. 413–439. In N. G. Carr and A. B. Whitton (eds.), *The Biology of Cyanobacteria.* Blackwell Scientific Publishers, Oxford, England.

Chang, S., R. Mack, S. L. Miller, and S. L. Strathearn. 1983. Prebiotic organic syntheses and the origin of life, pp. 53–92. In J. W. Schopf et al. (eds.), *Earth's Earliest Biosphere: Its Origin and Evolution*. Princeton University Press, Princeton, NJ.

Chatterjee, A. P., A. Dasgupta, and A. N. Chatterjee. 1988. Spacial dependence of stress-distribution for rod-shaped bacteria. *J. Theor. Biol.* **135**:309–321.

Clark, D. J. 1968. The regulation of DNA replication and division cycle in *E. coli. Cold Sprg. Harb. Symp. Quant. Biol.* **33**:823–838.

Clarke-Sturman, A. J., A. R. Archibald, I. C. Hancock, C. R. Harwood, T. Merad, and J. A. Hobot. 1989. Cell wall assembly in *Bacillus subtilus*: Partial conservation of polar wall material and the effect of growth conditions on the pattern of incorporation of new material at the polar caps. *J. Gen. Microbiol.* **135**:657–665.

Cohen, G. N., and J. Monod. 1957. Bacterial permeases. *Bacteriol. Rev.* **21**:164–194.

Collins, J., and M. Richmond. 1962. Growth rate of *Bacillus cereus* between divisions. *J. Gen. Microbiol.* **28**:15–33.

Cook, W. R., F. Kepes, D. Joseleau-Petit, T. J. MacAlister, and L. I. Rothfield. 1987. A proposed mechanism for the generation and localization of new division sites during the division cycle of *Escherichia coli. Proc. Natl. Acad. Sci. USA.* **84**:7144–7148.

Cook, W. R., P. A. J. De Boer, and L. I. Rothfield. 1989. Differentiation of the bacterial cell division site. *Int. Rev. Cytol.* **118**:1–31.

Cooper, S. 1982. The contiuum model: Implications for G1-arrest and G(0), pp. 315–336. In C. Nicolini (ed.), *Cell Growth*. Plenum, New York.

———. 1988a. Leucine uptake and protein synthesis are exponential during the division cycle of *Salmonella typhimurium. J. Bacteriol.* **170**:436–438.

———. 1988b. What is the bacterial growth law during the division cycle? *J. Bacteriol.* **170**:5001–5005.

———. 1988c. Rate and topography of cell wall synthesis during the division cycle of *Salmonella typhimurium. J. Bacteriol.* **170**:422–430.

———. 1989. The constrained hoop: An explanation of the overshoot in length during a shift-up of *Escherichia coli. J. Bacteriol.* **171**:5239–5243.

———. 1991. *Bacterial Growth and Division*. Academic Press, Inc., San Diego, CA.

Cooper, S., D. Gally, Y. Suneoka, M. Penwell, K. Caldwell, and K. Bray. 1993. Petidoglycan synthesis in *Salmonella typhimurium*. pp. 161–168. In M. A . de Pedro, J.-V. Höltje, and W. Löffëlhardt (eds.), *Bacterial Growth and Lysis: Metabolism and Structure of the Bacterial Sacculus*. Plenum Press, New York.

Cooper, S., and M.-L. Hsieh. 1988. The rate and topography of cell wall synthesis during the division cycle of *Escherichia coli. J. Gen. Microbiol.* **134**:1717–1721.

Cooper, S., and C. E. Helmstetter. 1968. Chromosome replication and the cell division cycle of *Escherichia coli* B/$_r$. *J. Mol. Biol.* **31**:519–540.

Costerton, J. W. 1979. The role of electron microscopy in the elucidation of bacterial structure and function. *Ann. Rev. Microbiol.* **33**:459–479.

Cota-Robles, E. H. 1963. Electron microscopy of plasmolysis in *Escherichia coli. J. Bacteriol.* **85**:499–503.

Csonka, L. N. 1989. Physiological and genetic responses of bacteria to osmotic stress. *Microbiol. Rev.* **53**:121–147.

Csonka, L. N., and A. D. Hanson. 1991. Prokaryote osmoregulation: Genetics and physiology. *Ann. Rev. Microbiol.* **45**:569–606.

Cullis, P. R., and B. De Kruijff. 1979. Lipid polymorphism and the functional roles of lipids in biological membranes *Biochim. Biophys Acta.* **559**:399–420.

Cutler, R. G., and J. E. Evans. 1966. Synchronization of bacteria by a stationary-phase method. *J. Bacteriol.* **91**:469–476.

Darnell, J., H. Lodish, and D. Baltimore. 1990. *Molecular Cell Biology.* Scientific American Books, New York.

Dayrell-Hart, B., and R. P. Burchard. 1979. Association of flexing and gliding in *Flexibacter. J. Bacteriol.* **137**:1417–1420.

Deamer, D. W., and G. L. Barchfeld. 1982. Encapsulation of macromolecules by lipid vesicles under simulated prebiotic conditions. *J. Mol. Evol.* **18**:203–206.

de Boer, P. A. J., R. E. Crossley, and L. I. Rothfield. 1989. A division inhibitor and a topological specificity factor coded for by the minicell locus that determines proper placement of the division septum in *E. coli. Cell* **56**:641–649.

Delbrück, M. 1949. A physicist looks at biology. *Internat. Sym. Centre Natl. Researche Scientifique.* **8**:91–104.

Dennis, P. P. 1972. Stable ribonucleic acid synthesis during the cell division cycle in slowly growing *Escherichia coli. J. Biol. Chem.* **247**:204–208.

Dickson, M. R., S. Kouprach, B. A. Humphrey, and K. C. Marhsall. 1980. Does gliding motility depend on undulating membranes? *Micron* **11**:381–382.

Donachie, W. D. 1968. Relationship between cell size and time of initiation of DNA replication, *Nature*, London **219**:1077–1079.

Donachie, W. D., K. J. Begg, and N. F. Sullivan. 1984. Morphogenes of *Escherichia coli*, pp. 27–88. In R. Losick and L. Shapiro (eds.), *Microbial Development.* Cold Spr. Harb. Lab., Cold Spr. Harb., NY.

Doyle, R. J., J. Chaloupka, and V. Vinter. 1988. Turnover of cell walls in microorganisms. *Microbiol. Rev.* **52**:554–567.

Doyle, R. J., and A. L. Koch. 1987. The functions of autolysins in the growth and division of *Bacillus subtilus. Crit. Rev. Microbiol.* **15**:169–222.

Drews, G., and W. Nultsch. 1962. Spezielle Bewegungsmechanismen von Einzellern (Bacterien, Algen), pp. 876–919. In W. Ruhland (ed.), *Handbuch der Pflanzenphysiologie.* **XVII-2**: Springer-Verlag, Berlin.

Driehaus, F. 1989. Peptidoglycan structure and metabolism in *Escherichia coli*. Dissertation, University of Amsterdam, The Netherlands.

Duxbury, T., B. A. Humphrey, and K. C. Marshall. 1980. Continuous observation of bacterial gliding motility in a dialysis microchamber: The effect of inhibitors. *Arch. Microbiol.* **124**:169–175.

Dworkin, M. 1983. Tactic behavior of *Myxococcus xanthus. J. Bacteriol.* **154**:452–459.

Dworkin, M., K. H. Keller, and D. Weisberg. 1983. Experimental observation consistent with a surface tension model of gliding motility of *Myxococcus xanthus*. *J. Bacteriol.* **15**:1367–1371.

Ecker, R. E., and G. Kokaisl. 1969. Synthesis of protein, ribonucleic acid and ribosomes by individual bacterial cells in balanced growth. *J. Bacteriol.* **98**:1219–1226.

Ecker, R. E., and M. Schaechter. 1963. Ribosome content and the rate of growth of *Salmonella typhimurium*. *Biochim. Biophys. Acta.* **76**:275–279.

Eigen, M., and P. Schuster. 1979. *The Hypercycle*. Springer-Verlag, Berlin.

Errington, F. P., E. O. Powell, and N. Thompson. 1965. Growth characteristics of some gram-negative bacteria. *J. Gen. Microbiol.* **39**:109–123.

Evans, E., and D. Needham. 1987. Physical properties of surfactant bilayers membranes: Thermal transitions, elasticity, rigidity, cohesion and colloidal interaction. *J. Phys. Chem.* **91**:4219–4228.

Fan, D. P., B. E. Beckman, and H. L. Gardner-Eckstrom. 1975. Mode of wall synthesis in Gram-positive rods. *J. Bacteriol.* **123**:1157–1162.

Fein, J. E., and H. J. Rogers. 1976. Autolytic enzyme-deficient mutants of *Bacillus subtilus*. *J. Bacteriol.* **127**:1427–1442.

Ferguson, S. J. 1990. Periplasm underestimated. *Trends Biochem. Sci.* **15**:327–329.

———. 1992. The periplasm, pp. 311–330. In S. Mohan, C. Dow, and J. A. Cole (eds.), *Prokaryotic Structure and Function: A New Perspective*. Soc. Gen. Micro. Sym., Cambridge University Press, Cambridge, England.

Flügge, W. 1973. *Stresses in Shells*. 2d edition. Springer-Verlag, New York.

Frazer, A. C., and R. Curtis, III. 1974. Production, properties and utility of bacterial minicells. *Curr. Top. Microbiol. Immunol.* **69**:1–84.

Frehel, C., B. Ferrandes, and A. Ryter. 1971. Réactions d'oxydo-réduction au niveau des membranes of cytoplasmique et mésosomique de *Bacillus subtilus*. *Biochim. Biophys. Acta* **234**:226–241.

Gibson, C. W., L. Daneo-Moore, and M. L. Higgins. 1983. Initiation of wall assembly sites in *Streptococcus faecium*. *J. Bacteriol.* **154**:573–579.

Giesbrecht, P., J. Wecke, and B. Reinicke. 1976. On the morphogenesis of the cell wall of staphylococci. *Int Rev. Cytol.* **44**:225–318.

Giesbrecht, P., H. Labischinski, and J. Wecke. 1985. A special morphogenenetic wall defect and the subsequent activity of "murasomes" as the very reason for penicillin-induced bacteriolysis in staphylococci. *Arch. Microbiol.* **141**:315–324.

Glagoleva, T. N., A. N. Glagoleva, M. V. Gusev, and K. A. Nikitina. 1980. Proton motive force supports gliding in cyanobacteria. *FEBS Letts.* **117**:49–53.

Glauner, B., J. V. Höltje, and U. Schwarz. 1988. The composition of the murein of *Escherichia coli*. *J. Biol. Chem.* **263**: 10088–10095.

Glauner, B., J. V. Höltje. 1990. Growth pattern of the murein sacculus of *Escherichia coli*. *J. Biol. Chem.* **263**:18988–18996.

Goodell, E. W. 1985. Recyling of murein by *Escherichia coli*. *J. Bacteriol.* **163**:305–310.

Goodell, E. W., and U. Schwarz. 1983. Cleavage and resynthesis of peptide crossbridges in *Escherichia coli* murein. *J. Bacteriol.* **156**:136–140.

———. 1985. Release of cell wall peptides into culture medium by exponentially growing *Escherichia coli*. *J. Bacteriol.* **162**:391–397.

Graham, L. I., T. J. Beveridge, and N. Nanninga. 1991. Periplasmic space and the concept of the periplasm. *Trends Biochem. Sci.* **16**:328–329.

Gray, D. I., G. W. Gooday, and J. I. Prosser. 1990. Apical extension in *Streptomyces coelicolor* A3. *J. Gen. Microbiol.* **136**:1077–1084.

Griffith, A. A. 1920. The phenomena of rupture and flow in solids. *Phil. Trans. Royal Soc. (London)* **A221**:63–98.

Grover, N. B., C. L. Woldringh, and L. J. H. Koppes. 1987. Elongation and surface extension of individual cells of *Escherichia coli* B/$_r$: Comparison of theoretical and experimental size distributions. *J. Theor. Biol.* **129**:337–348.

Gruner, S. M. 1985. Intrinsic curvature hypothesis for biomembrane lipid composition: A role for nonbilayer lipids. *Proc. Natl. Acad. Sci. USA.* **82**:3665–3669.

Haldane, J. B. S. 1929. The origin of life. *The Rationalist Annual.* Reprinted in *The Origins of Life: The Central Concepts* D. W. Deamer and G. R. Fleischaker (eds.) pp. 73–81. Jones and Bartlett Publishers (1994).

Halfen, L. N. 1979. pp. 250–267. In W. Haupt and M. E. Feinleib (eds.), *Encyclopedia of Plant Physiology.* **7**:250. Springer-Verlag, Berlin and Heidelberg.

Halfen, L. N., and R. W. Castenholz. 1971. Energy expenditure for gliding motility in a blue-green algae. *J. Phycol.* **7**:258–260.

Hargreaves, W. R., and D. W. Deamer. 1978. Lysosomes from ionic single chain amphiphiles. *Biochem. J.* **17**:3759–3768.

Harold, F. M. 1990. To shape a cell: An inquiry into the causes of morphogenesis of micoorganism. *Microbiol. Rev.* **54**:381–431.

———. 1992. Personal communication.

Hartley, C. L., H. M. Clements, and K. B. Linton. 1977. *Escherichia coli* in the fecal flora of man. *J. Appl. Bacteriol.* **43**:261–269.

Harvey, R. J., A. G. Marr, and P. R. Painter. 1967. Kinetics of growth of individual cells of *Escherichia coli* and *Azotobacter agilis*. *J. Bacteriol.* **93**:605–617.

Hayes, J. M. 1983. Geochemical evidence bearing on the origin of aerobiosis, a speculative hypothesis. pp. 290–31. In The earth's earliest biosphere: its origins and evolution. Schopf, J. W. (ed.). Princeton University Press, Princeton NJ.

Helmstetter, C. E., and S. Cooper. 1968. DNA synthesis during the division cycle of rapidly growing *E. coli*. B/$_r$. *J. Mol. Biol.* **31**:507–518.

Henis, Y. I. 1993. Lateral and rotational diffusion in biological membranes, p. 279–239. In M. Shinitzky (ed.), *Biomembranes: Physical aspects*. VCH, Weinheim.

Hiemenz, P. C. 1986. Principles of colloidal and surface chemistry. 2d edition. Marcel Dekker, Inc, New York.

Hiemstra, H. 1987. Topology of lipoprotein assembly in the cell wall of *Escherichia coli*. Thesis, The University of Groningen, The Netherlands.

Hiemstra, H., N. Nanninga, C. L. Woldringh, M. Inouye, and B. Witholt. 1987. Distribution of newly synthesized lipoprotein over the outer membrane and the peptidoglycan sacculus on an *Escherichia coli lac-lpp* strain. *J. Bacteriol.* **169**:5434–5444.

Higgins, M. L., A. L. Koch, D. T. Dicker, and L. Daneo-Moore. 1986a. Autoradiographic studies of the synthesis of protein and RNA as a function of cell volume in *Streptococcus faecium*. *J. Bacteriol.* **167**:960–967.

Higgins, M. L., and G. D. Shockman. 1976. Study of a cycle of cell wall assembly in *Streptococcus faecalis* by three dimensional reconstruction of thin sections in cell, *J. Bacteriol.* **137**:1346–1358.

Hinshelwood, C. 1946. *The Chemical Kinetics of the Bacterial Cell*. Clarendon Press, Oxford, England.

Hiraga, S. 1992. Chromosome and plasmid partition in *Escherichia coli*. Annu Rev. Biochem. **61**:283–306.

Hobot, J. A., E. Carlemalm, W. Villiger, and E. Kellenberger. 1984. Periplasmic gel: A new concept resulting from the reinvestigation of the bacterial envelope ultrastructure by new methods. *J. Bacteriol.* **162**:143–152.

Hoffman, H., and M. E. Franks. 1965. Synchrony of division in clonal microcolonies of *Escherichia coli*. *J. Bacteriol.* **89**:513–517.

Holland, I. B., S. Cararegola, and V. Norris. 1990. Cytoskeletal elements and calcium: Do they play a role in the *Escherichia coli* cell cycle? *Research Microbiol.* **41**:131–136.

Höltje, J.-V. 1993. Three for one—A simple mechanism that guarantees a precise copy of the thin, rod-shaped sacculus of *Escherichia coli*, pp. 419–426. In M. A. de Pedro, J.-V. Höltje, and W. Löffelhardt (eds.), *Bacterial Growth and Lysis: Metabolism and Structure of the Bacterial Sacculus*. Plenum Press, New York.

Höltje, J.-V., and B. Glauner. 1990. Structure and metabolism of the murein sacculus. *Res. Microbiol.* **141**:75–89.

Höltje, J.-V., and W. Keck. 1988. Organization of the major autolysin in the envelope of *Escherichia coli*, pp. 181–188. In P. Actor et al. (eds.), *Antibiotic Inhibition of Bacterial Cell Surface Assembly and Function*. American Soc. Microbiol., Washington, D.C.

Höltje, J.-V., and U. Schwarz. 1985. Biosynthesis and growth of the murein sacculus, pp. 77–119. In N. Nanninga (ed.), *Molecular Cytology of Escherichia coli*. Academic Press, London.

Hooper, A. B., and A. A. Dispirito. 1985. In bacteria which grow on simple reductants, generation of a proton gradient involves extracytoplasmic oxidation of substrate. *Microbiol. Rev.* **49**:140–157.

Horowitz, N. H. 1945. On the evolution of biochemical syntheses. *Proc. Natl. Acad. Sci. USA* **31**:153–157.

Humphrey, B. A., M. R. Dickson, and K. C. Marshall. 1979. Physiochemical and *in situ* observations on the adhesion of gliding bacteria to surfaces. *Arch. Microbiol.* **120**:231–238.

Jacob, F., and J. Monod. 1961. Genetic regulatory mechanisms in the synthesis of proteins. *J. Mol. Biol.* **3**:283–356.

Jacob, F., S. Brenner, and F. Cuzins. 1963. On the regulation of DNA replication in bacteria. *Cold Sprg. Harb. Symp. Quant. Biol.* **28**:239–347.

Jensen, K. F., and Pedersen, S. (1990). Metabolical growth rate control in *Escherichia coli* may be a consequence of subsaturation of the macromolecular biosynthetic apparatus with substrate and catalytic components. *Microbiol. Rev.* **54**:89–100.

Johnson, R. Y., and D. White. 1972. Myxospore formation in *Myxococcus xanthus*: Chemical changes in the cell wall during cellular morphogenesis. *J. Bacteriol.* **112**:849–855.

Jolliffe, L. K., R. J. Doyle, and U. N. Streips. 1981. Energized membrane and cellular autolysis in *Bacillus subtilus*. *Cell* **25**:753–763.

Kaiser, D. 1979. Social gliding is correlated with the presence of pili in *Myxococcus xanthus*. *Proc. Natl. Acad. Sci. USA.* **76**:5952–5956.

Kell, A., and R. W. Glaser. 1993. On the mechanical and dynamic properties of plant cell membranes: Their role in growth, direct gene transfer and protoplast fusion. *J. Theor. Biol.* **160**:41–62.

Keller, K. H., Grady, M., and Dworkin, M. 1983. Surface tension gradients: A feasible model for gliding motility of *Myxococcus xanthus*. *J. Bacteriol.* **155**:1358–1366.

Kelly, C. D, and O. Rahn. 1932. The growth rate of individual bacterial cells. *J. Bacteriol.* **23**:147–153.

Kendall, D. G. 1948. On the role of variable generation time in the development of a stochastic birth process. *Biometrika.* **35**:316–330.

———. 1952. On the choice of mathematical model to represent normal bacterial growth. *J. R. Statist. Soc. B* **14**:41–44.

Kennedy, E. P. 1982. Osmotic regulation and the biosynthesis of membrane-derived-oligosaccharides in *Escherichia coli*. *Proc. Natl. Acad. Sci. USA* **78**:1092–1095.

———. 1987. Membrane-derived oligosaccharides, pp. 672–680. In F. C. Neidhardt, J. L. Ingraham, K. B. Low, B. Magasanik, M. Schaechter, and H. E. Umbarger (eds.), *Escherichia coli* and *Salmonella typhimurium Cellular and Molecular Biology.* vol. 1. American Soc. Microbiol., Washington, D.C.

Kepes, F., and A. Kepes. 1980. Synchronization automatique de la croissance de *Escherichia coli*. *Ann. Inst. Pasteur Microbiol. (Paris)* **131A**:3–16.

Khan, S. 1988. Analysis of bacterial flagellar rotation. *Cell Mot. Cytoskel.* **10**:38–46.

Khan, S., and H. C. Berg. 1983. Isotope and thermal effects in chemiosmotic coupling to the flagellar motor of Streptococcus. *Cell* **32**:913–919.

Kirkwood, T. B. L., and I. D. J. Burdett. 1988. Estimating the growth pattern of microorganisms in distinct stages of the cell cycle. *J. Theor. Biol.* **130**:255–273.

Knowles, C. J. 1971. Salt induced changes of turbidity and volume of *E. coli*. *Nature New Biology* **229**:154–155.

Koch, A. L. 1961. Some calculations on the turbidity of mitochondria and bacteria. *Biochim. Biophys. Acta* **51**:429–441.

———. 1962. The evaluation of the rates of biological processes from tracer kinetic data. I. The influence of labile metabolic pools. *J. Theor. Biol.* **3**:283–303.

———. 1966. On evidence supporting a deterministic process of bacterial growth. *J. Gen. Microbiol.* **43**:1–5.

———. 1967. Kinetics of permease catalyzed transport. *J. Theor. Biol.* **14**:103–130.

———. 1970. Overall controls on the biosynthesis of ribosomes in growing bacteria. *J. Theor. Biol.* **28**:203–231.

———. 1971. The adaptive responses of *Escherichia coli* to a feast and famine existence. *Adv. Microbial Physiol.* **6**:147–217.

———. 1972. Enzyme evolution: The importance of untranslatable intermediates. *Genetics* **72**:297–316.

———. 1973. Incorporation of permease into the membrane throughout the growth cycle. Abstr. 73rd Ann. Meeting Amer. Soc. Microbiol. *Abstr. G* **257**:69.

———. 1974. The pertinence of the periodic selection phenomenon to procaryote evolution. *Genetics* **77**:127–142.

———. 1976. How bacteria face depression, recession, and derepression. *Perspectives Biol. Med.* **20**:44–63.

———. 1977. Does the initiation of chromosome replication regulate cell division? *Adv. Microbial Physiol.* **16**:49–98.

———. 1979. Microbial growth in low concentrations of nutrients, pp. 261–279. In M. Shilo (ed.), *Strategies in Microbial Life in Extreme Environments*. Dahlem Konferenzen-1978, Berlin.

———. 1980a. Selection and recombination in populations containing tandem multiplet genes. *J. Mol. Evol.* **14**:273–285.

———. 1980b. Inefficiency of ribosomes functioning in *Escherichia coli* growing at moderate rates. *J. Gen. Microbiol.* **116**:165–171.

———. 1981. Growth, pp. 179–207. In P. Gerhardt, R. G. E. Murray, R. N. Costilow, E. W. Nester, W. A. Wood, N. R. Krieg, and G. R. Phillips (eds.), *Manual of Methods for General Bacteriology*. American Soc. Microbiol., Washington, D.C.

———. 1982a. The shape of the hyphal tips of fungi. *J. Gen. Microbiol.* **128**:947–951.

———. 1982b. Multistep kinetics: Choice of models for growth of bacteria. *J. Theor. Biol.* **98**:401–417.

———. 1982c. On the growth and form of *Escherichia coli*. *J. Gen. Microbiol.* **128**:2527–2540.

———. 1982d. Spacial resolution of autoradiograms of rod-shaped organisms. *J. Gen. Microbiol.* **128**:2541–2546.

———. 1983. The surface stress theory of microbial morphogenesis. *Adv. Microbial Physiol.* **24**:301–366.

———. 1984a. Evolution *vs.* the number of gene copies per primitive cell. *J. Mol. Evol.* **20**:71–76.

————. 1984b. Shrinkage of growing *Escherichia coli* cells through osmotic challenge. *J. Bacteriol.* **2**:914–924.

————. 1984c. Turbidity measurements in microbiology. *ASM News* **50**:473–477.

————. 1984d. How bacteria get their shapes: The surface stress theory. *Com. Mol. Cell Biophys.* **2**:179–196.

————. 1985a. Primeval cells: Possible energy-generating and cell-division mechanisms. *J. Mol. Evol.* **21**:270–277.

————. 1985b. How bacteria grow and divide in spite of internal hydrostatic pressure. *Can. J. Microbiol.* **31**:1071–1084.

————. 1985c. The macroeconomics of bacterial growth, pp. 1–42. In M. M. Fletcher and G. D. Floodgate (eds.), *Bacteria in Their Natural Environment*. Soc. Gen. Microbiol, London.

————. 1986a. The pH in the neighborhood of membranes generating a protonmotive force. *J. Theor. Biol.* **120**:73–84.

————. 1986b. The basis of synchronization by repetitive dilution of a growing culture. *J. Theor. Biol.* **123**:333–346.

————. 1987a. Evolution from the point of view of *Escherichia coli.*, pp. 85–103. In P. Calow (ed.), *Evolutionary Physiological Ecology*. Cambridge University Press, Cambridge, England.

————. 1987b. Why *Escherichia coli* should be renamed *Escherichia ilei*, pp. 300–305. In A. Torriani-Gorini, F. Rothman, S. Silver, A. Wright, and E. Yagil (eds.), *Phosphate Metabolism and Cellular Regulation in Microorganisms*. American Soc. Microbiol., Washington, D.C.

————. 1988a. The sacculus, a nonwoven, carded, stress-bearing fabric. pp. 43–59. In P. Actor, L. Daneo-Moore, M. L. Higgins, M. R. J. Salton, and G. D. Shockman (eds.), *Antibiotic Inhibition of Bacterial Cell Surface Assembly and Function*. American Soc. Microbiol., Washington, D.C.

————. 1988b. Biophysics of bacterial wall viewed as a stress-bearing fabric. *Microbiol. Rev.* **52**:337–353.

————. 1988c. Speculations on the growth strategy of prosthecate bacteria. *Can. J. Microbiol.* **34**:390–394.

————. 1988d. Why can't a cell grow infinitely fast? *Can. J. Microbiol.* **34**:421–426.

————. 1988e. Partition of autolysins between the medium, the internal part of the wall, and the surface of the wall of Gram-positive rods. *J. Theor. Biol.* **134**:463–472.

————. 1989a. Calculation of surface area of sacculi from low-angle light scattering measurements. *J. Microbiol. Methods* **9**:139–150.

————. 1989b. The origin of the rotation of one end of a cell relative to the other end during the growth of Gram-positive rods. *J. Theor. Biol.* **141**:391–402.

————. 1990a. Recent extensions of the surface stress theory. pp. 39–63, In R. K. Poole and M. J. Bazin (eds.), *Microbial Growth Dynamics*. Oxford University Press, Oxford.

————. 1990b. Diffusion: The crucial process in many stages of the biology of bacteria. *Adv. Microbiol. Ecology* **11**:37–70.

———. 1990c. Additional arguments for the key role of "smart" autolysins in the enlargement of the wall of gram-negative bacteria. *Res. Microbiol.* **141**:529–541.

———. 1990d. Growth and form of the bacterial cell wall. *Amer. Sci.* **78**:327–341.

———. 1990e. The relative rotation of the ends of *Bacillus subtilus. Arch. Microbiol.* **153**:569–573.

———. 1990f. The sacculus contraction-expansion model for gliding motility. *J. Theor. Biol.* **142**:95–112.

———. 1990g. The surface stress theory for the case of *E. coli*: The paradoxes of gram-negative growth. *Res. Microbiol.* **141**:119–130.

———. 1990h. Positioning of the cell division site. *Res. Microbiol.* **141**:136–139.

———. 1991a. The wall of bacteria serves the role that mechano-proteins do in eukaryotes. *FEMS Microbiol. Rev.* **88**:15–26.

———. 1991b. Evolution of ideas about bacterial growth and their pertinence to higher cells, pp. 561–575. In O. Arino, D. E. Axelrod, and M. Kimmel (eds.), *Mathematical Population Dynamics.* Marcel Dekker, Inc., New York.

———. 1992. Differences in the formation of poles of *Enterococcus* and *Bacillus. J. Theor. Biol.* **154**:205–217.

———. 1993a. Microbial genetic responses to extreme challenges. *J. Theor. Biol.* **160**:1–21.

———. 1993b. Biomass growth rate during the cell cycle. *CRC Critical Rev.* **19**:17–42.

———. 1993c. The growth law of *Bacillus subtilus. Antonie van Leeuwenhoek* **63**:45–53.

———. 1993d. Stresses on the surface stress theory. In M. A. de Pedro, J.-V. Höltje, and W. Löffëlhardt (eds.), pp. 427–443, *Bacterial Growth and Lysis: Metabolism and Structure of the Bacterial Sacculus.* Plenum Press, New York.

———. 1994a. Growth measurements, pp. 248–277. In Phillip Gerhardt, editor-in-Chief, *Methods for General and Molecular Bacteriology.* American Soc. Microbiol., Washington, D.C.

———. 1994b. Development and diversification of the Last Universal Ancestor. *J. Theor. Biol.* **168**:269–280.

———. 1995a. The problem of apical growth. *J. Theor. Biol.* **171**:137–150.

———. 1995b. Did intracellular pathogens begin before the time of the Last Universal Ancestor? *Quarterly Rev. Biol.* (submitted).

———. 1995c. 102. The similarities and difference of individual bacteria within a clone. In F. C. Neidhardt, J. L. Ingraham, E. C. C. Lin, K. B. Low, B Magasanik, M. Schaechter, and H. E. Umbarger (eds.) *Escherichia coli* and *Salmonella typhimurium Cellular and Molecular Biology.* 2d ed., vol. 2. American Soc. Microbiol., Washington, D. C. (In press).

———. 1995d. The Monod model and its alternatives. In J. A. Robinson and G. A. Milliken (eds.), *Mathematical Models In Microbial Ecology.* Chapman & Hall, New York (In press).

————. 1995e. The geometry of plasmolysis spaces formation in bacteria and the role of endocytosis, tubular structures, and Scheie structures in their formation. *J. Theor. Biol.* (in press).

————. 1995f. The differences and similarities of individual bacteria within a clone. In Neidhardt, F. C., J. L. Ingraham, E. E. Lin, K. B. Low, B Magasanik, M. Schaechter, and H. E. Umbarger (eds.). 1995. *Escherichia coli* and *Salmonella typhimurium* cellular and molecular biology. Second edition. Vols. 1 and 2, *Amer. Soc. Microbiol.*, Washington, D.C. (in press).

Koch, A. L., and I. D. J. Burdett. 1984. The Variable-T model for Gram-negative morphology. *J. Gen. Microbiol.* **130**:2325–2338.

————. 1986a. Normal pole formation during total inhibition of wall synthesis. *J. Gen. Microbiol.* **132**:3441–3449.

————. 1986b. Biophysics of pole formation of Gram-positive rods. *J. Gen. Microbiol.* **132**:3451–3457.

Koch, A. L., and C. S. Deppe. 1971. *In vivo* assay of protein synthesizing capacity of *Escherichia coli* from slowly growing chemostat cultures. *J. Mol. Biol.* **55**:549–562.

Koch, A. L., and R. J. Doyle. 1985. Mechanism of inside-to-outside growth and turnover of the wall of Gram-positive rod. *J. Theor. Biol.* **117**:137–157.

————. 1986. The growth strategy of the Gram-positive rod. *FEMS Microbiol. Rev.* **32**:247–254.

Koch, A. L., and E. Ehrenfeld. 1968. The size and shape of bacteria by light scattering measurements. *Biochim. Biophys. Acta* **165**:262–273.

Koch, A. L., and M. L. Higtins. 1982. Cell cycle dynamics inferred from the static properties of cells in balanced growth. *J. Gen. Microbiol.* **128**:2877–2892.

————. 1984. Control of wall band splitting in *Streptococcus faecalis* ATCC 9790. *J. Gen. Microbiol.* **130**:735–745.

————. 1991. Simulation of the streptococcal population dynamics, pp. 577–591. In O. Arino, D. E. Axelrod, and M. Kimmel (eds.), *Mathematical Population Dynamics*. Marcel Dekker, New York.

Koch, A. L., M. L. Higgins, and R. J. Doyle. 1981a. Surface tension-like forces determine bacterial shapes: *Streptococcus faecium. J. Gen. Microbiol.* **123**:151–161.

————. 1982b. The role of surface stress in the morphology of microbes. *J. Gen. Microbiol.* **128**:927–945.

Koch, A. L., and J.-V. Höltje 1995. A physical model for the precise location of the division site during growth of rod-shaped bacterial cells. *Microbiology* (submitted).

Koch, A. L., G. Kirchner, R. J. Doyle, and I. D. J. Burdett. 1985. How does a *Bacillus* split its septum right down the middle? *Ann. Inst. Pasteur Microbiol.* **136A**:91–98.

Koch, A. L., S. L. Lane, J. Miller, and D. Nickens. 1987. Contraction of filaments of *Escherichia coli* after disruption of the cell membrane by detergent. *J. Bacteriol.* **166**:1979–1984.

Koch, A. L., H. L. T. Mobley, R. J. Doyle, and U. N. Streips. 1981b. The coupling of wall growth and chromosome replication in Gram-positive rods. *FEMS Microbiology Letts*. **12**:201–208.

Koch, A. L., and M. F. S. Pinette. 1987. Nephelometric determination of osmotic pressure in growing gram-negative bacteria. *J. Bacteriol*. **169**:3654–3663.

Koch, A. L., and M. Schaechter. 1962. A model for statistics of the cell division process. *J. Gen. Microbiol*. **29**:435–454.

―――. 1985. The world and ways of *E. coli*. pp. 1–125. In A. L. Demain and N. A. Solomon (eds.), Biology of Industrial Microorganisms. Vol. 1, Addison-Wesley, Reading, Mass.

Koch, A. L., and T. Schmidt. 1991. The first cellular bioenergetic process: Primitive generation of a protonmotive force. *J. Mol. Evol*. **33**:297–304.

Koch, A. L., and H. Schwarz. 1995. Phase microscopic observations of the types of plasmolysis and the structural role of the layers of the envelope of Gram-negative bacteria. *Microbiology*. (Submitted).

Koch, A. L., and C. H. Wang. 1982. How close to the theoretical diffusion limit do bacterial uptake systems function? *Arch. Microbiol*. **131**:36–42.

Koch, A. L., and S. W. Woeste. 1992. The elasticity of the sacculus of *Escherichia coli J. Bacteriol*. **174**:4811–4819.

Koch, A. L., and C. L. Woldringh. 1994. The inertness of the poles of a Gram-negative rod. *J. Theor. Biol*. **171**:415–425.

Kolter, R., D. A. Siegele, and A. Tormo. (1993). The stationary phase of the bacterial life cycle. *Annu. Rev. Microbiol*. **47**:855–874.

Kooijman, S. A. L. M. 1993. *Dynamic energy budgets in biological systems*. Cambridge University Press, Cambridge, England.

Koppes, L. J. H., C. L. Woldringh, and N. B. Grover. 1987. Predicted steady-state cell size distributions for various growth models. *J. Theor. Biol*. **129**:325–335.

Kubitschek, H. E. 1962. Normal distribution of cell generation rate. *Exp. Cell Res*. **26**:439–450.

―――. 1968. Linear cell growth in *Escherichia coli*. *Biophys. J*. **8**:792–804.

―――. 1970. Introduction to research with continuous cultures. Prentice-Hall, Englewood Cliffs, NJ.

―――. 1971. Control of cell growth in bacteria: Experiments with thymine starvation. *J. Bacteriol*. **105**:472–476.

―――. 1986. Increase in cell mass during the division cycle of *Escherichia coli*. B/$_r$A. *J. Bacteriol*. **168**:613–618.

―――. 1987. Buoyant density variation during the cell cycle in microorganisms. *CRC Crit. Rev. Microbiol*. **14**:73–97.

―――. 1990. Cell growth and abrupt doubling of membrane proteins in *Escherichia coli* during the division cycle. *J. Gen. Microbiol*. **190**:599–606.

Kuznetsov, E. N. 1991. *Underconstrained Structural Systems*. Springer-Verlag, Berlin.

Labischinski, H., G. Barnickel, H. Bradaczek, and P. Giesbrecht. 1979. On the secondary and tertiary structure of murein. *Eur. J. Biochem.* **95**:147–155.

Labischinski, H., G. Barnickel, D. Naumann, and P. Keller. 1983. Conformational and topical aspects of the three-dimensional architecture of bacterial peptidoglycan. *Ann. Inst. Pasteur Microbiol.* **136A**:45–50.

Labischinski, H., E. W. Goodell, A. Goodell, and M. L. Hochberg. 1991. Direct proof of a "more-than-single-layered" peptidoglycan architecture of *Escherichia coli* W7: A neutron small-angle scattering study. *J. Bacteriol.* **173**:751–756.

Lapidus, I. R., and H. C. Berg. 1982. Gliding motility of *Cytophaga* sp. strain U67. *J. Bacteriol.* **151**:1358–1366.

Larsen, S. H., J. Adler, J. J. Gargus, and R. W. Hogg. 1974. Chemo-mechanical coupling without ATP: The source of energy for motility and chemotaxis in bacteria. *Proc. Natl. Acad. Sci. USA.* **71**:1239–1243.

Lederberg, J., and E. M. Lederberg. 1952. Replica plating and indirect selection of bacterial mutants. *J. Bacteriol.* **63**:399–406.

Leduc, M., C. Frehel, E. Sigal, and J. V. van Heijenoort. 1989. Multilayer distribution of peptidoglycan in the periplasmic space of *Escherichia coli*. *J. Gen. Microbiol.* **135**:1243–1254.

Linton, A. H., B. Handley, and A. D. Osborn. 1978. Fluctuations in *Escherichia coli* O-serotypes in pigs throughout life in the presence and absence of antibiotic treatment. *J. Appl. Bacteriol.* **44**:285–298.

Lündsdorf, H., and H. Reichenbach. 1989. Ultrastructural details of the apparatus for gliding motility of *Myxococcus fulvus* (Myxobacterales). *J. Gen. Microbiol.* **135**:1633–1641.

Luria, S., and M. Delbrück. 1943. Mutations of bacteria from virus sensitivity to virus resistance. *Genetics.* **28**:491–511.

Lutkenhaus, J. 1993. FtsZ ring in bacterial cytokinesis. *Molecular Microbiol.* **9**:403–409.

Maaløe, O., and N. O. Kjeldgaard. 1966. *Control of Macromolecular Synthesis.* Benjamin, New York.

MacAlister, I. J., W. R. Cook, R. Weigand, and L. I. Rothfield. 1987. Membrane-murein attachment at the leading edge of the division septum: A second membrane-murein structure associated with the morphogenesis of the gram-negative division septum. *J. Bacteriol.* **169**:3945–3951.

MacAlister, T. J., B. MacDonald, and L. I. Rothfield. 1983. The periseptal annulus: An organelle associated with cell division in Gram-negative bacteria. *Proc. Natl. Acad. Sci. USA.* **80**:1372–1376.

MacLean, F. I., and R. J. Munson. 1961. Some environmental factors affecting the length of *Escherichia coli* organisms in continuous culture. *J. Gen. Microbiol.* **25**:17–27.

MacRae, T. H., and H. D. McCurdy. 1976. Evidence for motility-related fimbriae in the gliding microorganism *Myxococcus xanthus*. *Can. J. Microbiol.* **22**:1589–1593.

Manor. H., and R. Haselkorn. 1967. Size fractionation of exponentially growing *Escherichia coli*. *Nature (London)* **214**:983–986.

Mark, H. 1943. Elasticity and strength, pp. 990–1052. In E. Ott (ed.), *Cellulose and its Derivatives*. Interscience, New York.

Marr, A. G., P. R. Painter, and E. H. Nilson. 1969. Growth and division of individual bacteria. *Symp. Soc. Gen. Microbiol.* **19**:237–261. Cambridge University Press, Cambridge, England.

Mason, D. J., and D. M. Powelson. 1956. Nuclear division as observed in live bacteria. *J. Bacteriol.* **71**:474–479.

Matin, A., E. A. Auger, P. H. Blum, and J. E. Schultz. 1989. Genetic basis of starvation survival in nondifferentiating bacteria. *Ann. Rev. Microbiol.* **43**:293–316.

Matts, T. C., and C. J. Knowles. 1971. Stopped-flow studies of salt-induced turbidity changes of *Escherichia coli*. *Biochim. Biophys. Acta* **249**:583–587.

McNab, R. M. 1987. Motility and chemotaxis, pp. 732–759. In F. C. Neidhardt, J. L. Ingraham, K. B. Low, B. Magasanik, M. Schaechter, and H. E. Umbarger (eds.), *Escherichia coli* and *Salmonella typhimurium Cellular and Molecular Biology*. vol. 1, American Soc. Microbiol., Washington, D.C.

Meinhardt, H. 1982. *Models of Biological Pattern Formation*. Academic Press, London.

Mendelson, N. H. 1976. Helical growth of *Bacillus subtilus*: A new model for cell growth. *Proc. Natl. Acad. Sci. USA*. **73**:1740–1744.

———. 1978. Helical *Bacillus subtilis* macrofibers: Morphogenesis of a bacterial multicellular macroorganism. *Proc. Natl. Acad. Sci. USA* **75**:2478–2482.

———. 1978. The helix clock: A potential biomechanical cell cycle timer. *Microbiol. Rev.* **46**:341–375.

Mendelson, N. H., D. Favre, and J. J. Thwaites. 1984. Twisted states of *Bacillus subtilus* reflect structural states of the cell wall. *Proc. Natl. Acad. Sci. USA*. **81**:3562–3566.

Mendelson, N. H., and J. J. Thwaites. 1988. Studies of *Bacillus subtilus* macrofiber twist states and bacterial thread biomechanics: Assembly and material properties of cell walls, pp. 109–125. In P. Actor, L. Daneo-Moore, M. L. Higgins, M. R. J. Salton, and G. D. Shockman (eds.), *Antibiotic Inhibition of Bacterial Cell Surface Assembly and Function*. American Soc. Microbiol., Washington, D.C.

Meng, K. E., and R. M. Pfister. 1980. Intracellular structure of *Mycoplasma pneumoniae* revealed after membrane removal. *J. Bacteriol.* **144**:390–399.

Merad, T., A. R. Archibald, I. C. Hancock, C. R. Harwood, and J. A. Hobot. 1989. Cell wall assembly in *Bacillus subtilus*: Visualisation of old and new material by electron microscopic examination of samples selectively stained for teichoic acid and teichuronic acid. *J. Gen. Microbiol.* **135**:645–655.

Metzler, D. E. 1977. *Biochemistry: The Chemical Reactions of Living Cells*. Academic Press, New York.

Miguélez, E. M., C. Hardisson, and M. B. Manzanal. 1995. Incorporation and fate of N-acetyl-D-glucosamine during hyphal growth in *Streptomyces*. (unpublished).

Miguélez, E. M., M. C. Martin, M. B. Manzanal, & C. Hardisson. 1988. Hyphal growth in streptomyces. pp. 490–495. In: *Biology of Actinomycetes '88*. (Y. Okami, T. Beppa, and J. H. Ogawara) Japan Scientific Societies Press. Tokyo, Japan.

Milkman, R. 1973. Electrophoretic variation in *E. coli* from natural sources. *Science*. **182**:1024–1026.

Miller, S. L., and L. E. Orgel. 1973. *The Origins of Life on Earth*. Prentice-Hall, New York.

Mirelman, D. 1979. Biosynthesis and assembly of cell wall peptidoglycans, pp. 115–116. In M. Inouye (ed.), *Bacterial Outer Membranes: Biogenesis and Functions*. John Wiley & Sons, Inc. New York.

Mitchell, P. 1956. *Proc. R. Phys. Soc.* **25**:32.

Mitchell, P., and J. Moyle. 1956. Osmotic function and structure in bacteria. *Symp. Soc. Gen. Microbiol.* **6**:150–180.

Mitchison, J. M., and Vincent, W. S. 1965. Preparation of synchronous cell cultures by sedimentation. *Nature* **205**:987.

Mobley, H. L. T., A. L. Koch, R. J. Doyle, and U. N. Streips. 1984. Insertion and fate of the cell wall in *Bacillus subtilus*. *J. Bacteriol.* **158**:169–179.

Mulder, E., and C. L. Woldringh. 1989. Actively replicating nucleoids influence positioning of division sites in *Escherichia coli* filament-forming cells lacking DNA. *J. Bacteriol.* **171**:4303–4314.

———. 1993. Plasmolysis bays in *Escherichia coli*: Are they related to development and positioning of division sites? *J. Bacteriol.* **175**:2241–2247.

Nanninga, N. 1991. Cell division and peptidoglycan assembly in *Escherichia coli*. *Mol. Microbiol.* **5**:791–795.

Nanninga, N., C. L. Woldringh, and L. J. H. Koppes. 1982. Growth and division of *Escherichia coli*, pp. 225–270. In C. Nicolini (ed.), *Cell Growth*. Plenum Publishing, New York.

Neidhardt, F. C., J. L. Ingraham, E. E. Lin, K. B. Low, B. Magasanik, M. Schaechter, and H. E. Umbarger (eds.). 1995. *Escherichia coli and Salmonella typhimurium Cellular and Molecular Biology*. Second edition. vols. 1 and 2, American Soc. Microbiol., Washington, D.C.

Neidhardt, F. C., J. L. Ingraham, K. B. Low, B. Magasanik, M. Schaechter, and H. E. Umbarger (eds.). 1987. *Escherichia coli and Salmonella typhimurium Cellular and Molecular Biology*. vols. 1 and 2. American Soc. Microbiol., Washington, D.C.

Neidhardt, F. C., J. L. Ingraham, and M. Schaechter. 1990. *Physiology of the Bacterial Cell*. Sinauer Associates, Sunderland, MA.

Neidhardt, F. C., and B. Magasanik. 1960. Studies on the role of ribonucleic acid in the growth of bacteria. *Biochim. Biophys. Acta* **42**:99–116.

Neimark, H. C. 1977. Extraction of an actin-like protein from the prokaryote *Mycoplasma pneumiaae*. *Proc. Natl. Acad. Sci. USA*. **74**:4041–4045.

Nikaido, H. 1992. Porins and specific channels of bacterial outer membranes. *Molecular Microbiol.* **6**:435–442.

———. 1993. Transport across the bacterial outer membrane. *J. Bioenerg. Biomembrane.* **25**:581–589.

Niki, H., A. Jaffé, R. Imamura, T. Ogura, and S. Hiraga. 1991. The new gene *mukB* codes for a 177 kd protein with coiled-coil domains involving chromosome partioning of *E. coli. EMBO J.* **10**:183–193.

Norris, V. 1992. Phospholipid domains determine the spatial organization of the *Escherichia coli* cell cycle: The membrane tectonics model. *J. Theor. Biol.* **154**:91–107.

Ogden, G. B., M. J. Pratt, and M. Schaechter. 1988. The replicative origins of the *E. coli* chromosome binds to cell membranes only when hemimethylated. *Cell* **154**:127–135.

Olsen, G. J., C. R. Woese, and R. Overbeek, 1994. The winds of (evolutionary) change: Breathing new life into microbiology. *J. Bacteriol.* **176**:1–6.

Oparin, A. I. 1936. *The Origin of Life.* Dover, New York.

Ou, L.-T., and R. E. Marquis, 1970. Electromechanical interaction in cell walls in gram-positive cocci. *J. Bacteriol.* **101**:92–101.

Painter, P. R., and A. G. Marr. 1968. Mathematics of microbial populations. *Ann. Rev. Microbiol.* **22**:519–548.

Pangborn, J., D. A. Kuhn, and J. R. Woods. 1977. Dorsal-ventral differentiation in Simonella and other aspects of its morphology and ultra structure. *Arch. Microbiol.* **113**:197–204.

Park, J. T. 1949. Uridine 5′ pyrophosphate derivative: I. Isolation from *Staphylcoccus aureus. J. Biol. Chem.* **194**:877–884.

Pate, J. L. 1988. Gliding motility in prokaryotic cells. *Can. J. Microbiol.* **34**:459–465.

Pate, J. L., and L.-Y. E. Chang. 1979. Evidence that gliding motility in prokartyotic cells is driven by rotary assemblies in the cell envelopes. *Curr. Microbiol.* **42**:59–64.

Pinette, M. F. S., and A. L. Koch. 1987. Variability of the turgor pressure of individual cells of a gram-negative heterotroph *Ancylobacter aquaticus. J. Bacteriol.* **169**:4737–4742.

———. 1988. Biophysics of ampicillin action on a gas-vacuolated gram-negative rod, pp. 157–163. In P. Actor, L. Daneo-Moore, M. L. Higgins, M. R. J. Salton, and G. D. Shockman (eds.), *Antibiotic Inhibition of Bacterial Surface Assembly and Function.* American Soc. Microbiology, Washington, D.C.

Pooley, H. M. 1976. Turnover and spreading of old wall during surface growth of *Bacillus subtilus. J. Bacteriol.* **125**:1127–1138.

Postgate, J. R., and Hunter, J. R. 1964. Accelerated death of *Aeobacter aerogenes* starved in the presence of growth-limiting substate. *J. Gen. Micro.* **34**:459–473.

Powell, E. O. 1955. Some features of the generation time of individual bacterial *J. Gen. Microbiol.* **42**:16–44.

———. 1958. An outline of the pattern of bacterial generation times. *J. Gen. Microbiol.* **18**:382–417.

———. 1964. A note on Koch and Schaechter's hypothesis about growth and fission of bacteria. *J. Gen. Microbiol.* **37**:231–249.

Previc, E. D. 1970. Biochemical determination of bacterial morphology. *J. Theor. Biol.* **27**:471–497.

Pringsheim, E. G. 1949. The relationship between bacteria and myxophyceae. *Bacteriol. Rev.* **13**:47–98.

Pritchard, R. H. 1974. On the growth and form of a bacterial cell. *Phil. Trans. Roy. Soc. (London)* **B267**:303–336.

Prosser, J. I. 1990. Comparison of tip growth in prokaryotic and eukaryotic filamentous microorganisms, pp. 233–259. In I. B. Heath (ed.), *Tip Growth*. Academic Press, London.

Rahn, O. 1931–1932. A chemical explanation of the variability of the growth rate. *J. Gen. Physiol.* **15**:257–267.

Reichenbach, H. 1965. Untersuchungen an *Archangium violaceum*. *Arch. Kikrobiol.* **52**:376–403.

———. 1980. Encycl. Cinematogr. Film e2424. G. Wolf (ed.), pp. 3–21. Inst. Wiss. Film, Göttingen.

Reinhardt, M. O. 1892. Das Wachtum der Pilzhyphen. Ein Beitrag zur Kenntnis des Flächenwachstums vegetalischer Zellmembranen. *Jahrbücher Wissenshaft Botanik.* **23**:479–566.

Richenberg, H. W., G. N. Cohen, G. Buttin, and J. Monod. 1956. La galactosidase-perméase d'*Escherichia coli*. *Ann. Inst. Pasteur.* **91**:829–857.

Ricklefs, R. E. 1990. *Ecology*. 3d edition, Freeman, New York.

Ridgway, H. F. 1977. Source of energy for gliding motility in *Flexibacter Polymorphus*: Effects of metabolic and respiratory inhibitors on gliding moments. *J. Bacteriol.* **131**:544–556.

Ridgway, H. F., and R. A. Lewin. 1988. Characterization of gliding motiility in *Flexibacter polymorphus*. *Cell Mot. Cytoskel.* **11**:46–63.

Roark, R. J., and W. C. Young. 1975. *Formulas for Stress and Strain*. Fifth edition. McGraw-Hill, New York.

Rogers, H. J., H. R. Perkins, and J. B. Ward. 1980. *Microbial Cell Walls and Membranes*. Chapman & Hall, London.

Rosenthal, R., and J. M. Krueger. 1988. Promotion of sleep by gonococcal peptidoglycan fragments. Structural requirements for somnogenic activity, pp. 581–590. In J. T. Poolman *et al.* (eds.), *Gonococci and Meningococci*. Kluwer Academic Publishers, Dordrecht.

Ryter, A. 1967. Relationship between synthesis of the cytoplasmic membrane and nuclear segregation in *B. subtillis*. *Folia Microbiologica* **12**:283–290.

Ryter, A., Y. Hirota, and U. Schwarz. 1973. Process of cellular division in *Escherichia coli*. *J. Mol. Biol.* **78**:185–195.

Salton, M. R. J. 1964. *The Bacterial Wall*. Elsevier, Amsterdam.

Salpeter, M. M., L. Bachmann, and E. E. Salpeter. 1978. Resolution in electron microscope autoradiography: IV. Application to analysis of autoradiography. *J. Cell Biol.* **76**:127–145.

Sargent, M. G. 1974. Nuclear segregation in *Bacillus subtilus*. *Nature (London)*. **250**:252–254.

———. 1975. Control of cell length in *Bacillus subtilus*. *J. Bacteriol*. **123**:1218–1234.

Schaechter, M., O. Maaløe, and N. O. Kjeldgaard. 1958. Dependency on medium and temperature of cell size and chemical composition during balanced growth of *Salmonella typhimurium*. *J. Gen. Microbiol*. **19**:592–606.

Schaechter, M., J. P. Williamson, J. R. Hood, Jr., and A. L. Koch. 1962. Growth, cell, and nuclear divisions in some bacteria. *J. Gen. Microbiol*. **29**:421–434.

Schaeppi, J.-M., H. M. Poooley, and D. Karamata. 1982. Identification of cell wall subunits in *Bacillus subtilus* and analysis of their segragation during growth. *J. Bacteriol*. **149**:329–337.

Scheie, P. O. 1969. Plasmolysis of *Escherichia coli* B/$_r$ with sucrose. *J. Bacteriol*. **98**:335–340.

Scheie, P. O., and H. Dalen. 1968. Spatial anisotropy in *Escherichia coli*. *J. Bacteriol*. **96**:1413–1414.

Scheie, P. O., and R. Rehberg. 1972. Response of *Escherichia coli* to high concentrations of sucrose in a mineral medium. *J. Bacteriol*. **109**:229–235.

Schleifer, K. H., and O. Kandler. 1972. Peptidoglycan types of bacterial cell wall and their taxonomic implications. *Bacteriol. Rev*. **36**:407–477.

Schopf, J. W. (ed.) 1983. *Earth's Earliest Biosphere*. Princeton University Press, Princeton, N. J., pp. 53–92.

Schultz, G. 1955. Bewegunstudien sowie electronmikrokopische Membranuntersuchen an Cyanophyceen. *Arch. Mikrobiol*. **21**:335.

Schwarz, H., and A. L. Koch. 1995. Electron microscopic observations of the types of plasmolysis and the structural role of the layers of the envelope of Gram-negative bacteria *Microbiology*. (Submitted).

Schwarz, U., A. Ryter, A. Rambach, R. Hellio, and Y. Hirota. 1975. Process of cellular division in *Escherichia coli*: Differentiation of growth zones in the sacculus. *J. Mol. Biol*. **98**:749–759.

Schwarz, U., and B. Glauner. 1988. Murein structure data and their relevance for the understanding of murein metabolism in *Escherichia coli*, pp. 33–40. In P. Actor, L. Daneo-Moore, M. L. Higgins, M. R. J. Salton, and G. D. Shockman (eds.), *Antibiotic Inhibition of Bacterial Cell Surface Assembly and Function*. American Soc. Microbiol., Washington, D.C.

Skarstad, K., H. B. Steen, and E. Boye. 1985. *Escherichia coli* DNA distributions measured by flow cytometry and compared with theoretical computer simulations. *J. Bacteriol*. **163**:661–668.

Smith, J. A., and L. Martin. 1973. Do cells cycle? *Proc. Natl. Acad. Sci. USA*. **70**:1263–1267.

Sonnenfeld, E. M., T. J. Beveridge, A. L. Koch, and R. J. Doyle. 1985. Asymmetric distribution of charges on the cell wall of *Bacillus subtilus*. *J. Bacteriol*. **163**:1167–1171.

Spratt, B. G. 1975. Distinct penicillin binding proteins involved in division, elongation, and shape of *Escherichia coli* K12. *Proc. Natl. Acad. Sci. USA.* **72**:2999–3003.

Spratt, B. G., L. D. Bowler, A. Edelman, and J. K. Broome-Smith. 1988. Membrane topology of penicillin-binding protein 1b and 3 of *Escherichia coli* and the production of water-soluble forms of high-molecular-weight pencillin-binding proteins, pp. 292–300. In P. Actor, L. Daneo-Moore, M. L. Higgins, M. R. J. Salton, and G. D. Shockman (eds.), *Antibiotic Inhibition of Bacterial Cell Surface Assembly and Function.* American Soc. Microbiol., Washington, D.C.

Stanier, R. Y., M. Doudoroff, and E. A. Adelberg. 1970. *The Microbial World.* 3d edition. Prentice-Hall, Englewood Cliffs, NJ.

Stanier, R. Y., and C. B. van Niel. 1962. The concept of a bacterium. *Arch. Mikrobiol.* **42**:17–35.

Stanier, R. Y., M. Doudoroff, and E. A. Adelberg. 1970. *The Microbial World,* 3d edition. Prentice-Hall, Englewood Cliffs, NJ.

Stock, J. B., B. Rauch, and S. Roseman. 1977. Periplasmic space in *Salmonella typhimurium* and *Escherichia coli. J. Biol. Chem.* **252**:7850–7861.

Tandeau de Marsac, N., D. Mazel, D. A. Bryant, and J. Houmard. 1985. Molecular cloning and nucleotide sequence of a developmentally regulated gene from the cyanobacterium *Calotrix* PCC 76011: A gas vesicle protein gene. *Nucleic Acids Res.* **13**:7223–7236.

Tanford, C. 1980. *The Hydrophobic Effect—Formation of Micelles and Biological Membranes.* 2d edition. Wiley, New York.

Templin, M. F., D. H. Edwards, and J.-V. Höltje. 1992. A mureinhydrolase is the specific target of bulgecin in *Escherichia coli. J. Biol. Chem.* **267**:2003.

Thompson, D'A. W. 1917. *On Growth and Form.* Cambridge University Press, Cambridge, England.

———. 1942. *On Growth and Form.* 2d edition. Cambridge University Press, Cambridge, England.

Tilbey, M. J. 1977. Helical shape and wall synthesis in a bacterium. *Nature* (London). **266**:450–452.

Timoshenko, S. P., and J. N. Goodier. 1970. *Theory of Elasticity.* 3d edition. McGraw-Hill, New York.

Tipper, D. J., and J. L. Strominger, 1965. Mechanism of action of penicillins: A proposal based on their structural similarity to acyl-D-alanyl-D-alanine. *Proc. Natl. Acad. Sci.* **54**:1133–141.

Tipper, D. J., and A. Wright. 1979. The structure and biosynthesis of bacterial cell walls, pp. 291–426. In J. R. Sokatch and L. N. Ornston (eds.), *The Bacteria.* vol. 7. Academic Press, London.

Trueba, F. J. 1981. A morphometric analysis of *Escherichia coli* and other rod-shaped bacteria. Dissertation at University of Amsterdam, The Netherlands.

Trueba, F. J., and C. L. Woldringh. 1980. Changes in cell diameter during the division cycle of *Escherichia coli. J. Bacteriol.* **142**:869–878.

Tyson, J. J. 1985. The coordination of cell growth and division—Intentional or incidental. *BioEssays.* **2**:72–76.

Tyson, J. J., and O. Diekmann. 1986. Sloppy size control of the cell division cycle. *J. Theor. Biol.* **118**:405–426.

van Wielink, J. E., and J. A. Duine. 1990. How big is the periplasmic space? *TIBA* **15**:136–137.

Verwer, R. W. H. 1979. Cytological study of the cell envelope and of the cell cycle of *Escherichia coli.* Dissertation, University of Amsterdam, The Netherlands.

Verwer, R. W. H., and N. Nanninga. 1980. Pattern of *meso*-DL-2,6,-diaminopimelic acid incorporation during the division cycle of *Escherichia coli. J. Bacteriol.* **144**:327–336.

Verwer, R. W. H., N. Nanninga, W. Keck, and U. Schwarz. 1978. Arrangements of glycan chains in the sacculus of *Escherichia coli. J. Bacteriol.* **136**:723–729.

Verwer, R. W. H., E. H. Beachey, W. Keck, A. M. Stoub and J. E. Poldermans. 1980. Orientation fragmentation of *Escherichia coli* sacculi by sonication. *J. Bacteriol.* **141**:327–333.

Vitkovic, L. 1987. Wall turnover of *Bacillus subtilus* Ni15 is due to a decrease in teichoic acid. *Can. J. Microbiol.* **33**:566–568.

von Meyenburg, K., and F. G. Hansen, 1987. Regulation of chromosome replication, pp. 1555–1577. In F. C. Neidhardt, J. L. Ingraham, K. B. Low, B. Magasanik, M. Schaechter, and H. E. Umbarger (eds.), *Escherichia coli and Salmonella typhimurium Cellular and Molecular Biology.* vol. 2. American Soc. Microbiol., Washington, D.C.

Wächterschäuser, G. 1988a. Pyrite formation, the first energy source for life: A hypothesis. *Syst. Appl. Microbiol.* **10**:207–210.

———. 1988b. Before enzymes and templates: Theory of surface metabolism *Microbiol. Rev.* **52**:452–484.

———. 1993. Ground work for evolutionary biochemistry: The iron-sulphur world. *Prog. Biophys. Mol. Biol.* **58**:85–201.

Walsby, A. E. 1968. Mucilage secretion and the movement of blue-green algae. *Protoplasma.* **65**:223–238.

———. 1971. The pressure relationships of gas vacuoles. *Proc. R. Soc. London B* **178**:301–326.

———. 1980. The water relations of gas-vacuolated prokaryotes. *Proc. R. Soc. London B* **208**:73–102.

———. 1994. Gas vesicles. *Microbiological Reviews.* **58**:94–144.

Ward, J. E., Jr., and J. Lutkenhaus. 1985. Overproduction of FtsZ induces minicell formation in *E. coli. Cell.* **42**:941–949.

Weibull, C. 1960. Movement, pp. 153–205. In I. C. Gunsalus and R. Y. Stanier (eds.), *The Bacteria.* vol 1. Academic Press, New York and London.

Weidel, W., and H. Pelzer. 1964. Bag-shaped macromolecules—A new outlook on bacterial cell walls. *Adv. Enzymol.* **26**:193–232.

Wessels, J. G. H. 1983. Cell wall synthesis in apical hyphal growth. *Int. Rev. Cytol.* **104**:37–79.

———. 1990a. Role of cell wall architecture in fungal tip growth generation, pp. 12–29. In I. B. Heath (ed.), *Tip Growth in Plants and Fungal Walls.* Academic Press, London.

———. 1990b. Fungal growth and development: A molecular perspective, pp. 17–48. In D. L. Hawksworth (ed.), *Frontiers in mycology.* CAB International Reghisberg, Germany.

Wessels, J. G. H., J. H. Sietsma, and A. S. M. Sonnenberg. 1983. Wall synthesis and assembly during hyphal morphogenesis of *Schizophyllum. J. Gen. Microbiol.* **129**:1607–1616.

White, D. 1995. *The Physiology and Biochemistry of Prokaryotes.* Oxford Press, Oxford.

White, D., M. Dworkin, and D. J. Tipper. 1968. Peptidoglycan of *Myxococcus xanthus*: structure and relationship to morphogenesis. *J. Bacteriol.* **95**:2186–2197.

Wielink, J. E., and Duine, J. A. 1990. How big is the periplasmic space? *Trends Biochem. Sci.* **15**:136–137.

Wientjes, F. B., and N. Nanninga. 1989. Rate and topography of peptidoglycan synthesis during the cell division in *Escherichia coli*: Concept of a leading edge. *J. Bacteriol.* **171**:3412–3419.

Wientjes, F. B., C. L. Woldringh, and N. Nanninga. 1991. Amount of peptidoglycan. *J. Bacteriol.* **173**:7684–7691.

Wiggins, P. M. 1990. Role of water in some biological process. *Microbiol. Rev.* **54**:432–449.

Wise, E. M., and J. T. Park. 1965. Penicillin: Its basic site of action as an inhibitor of a peptide cross-linking reaction in cell wall mucopeptide synthesis. *Proc. Natl. Acad. Sci. USA.* **54**:75–81.

Woese, C. 1979. A proposal concerning the origin of life on the planet earth. *J. Mol Evol.* **13**:95–101.

Woldringh, C. L. 1973. Effect of cation on the organization of the nucleoplasm in *Escherichia coli* prefixed with osmium tetyroxide or glutaraldehyde. *Cytobiol.* **8**:97–111.

———. 1976. Morphological analysis of nucleoid separation and cell division during the life cycle of *Escherichia coli. J. Bacteriol.* **125**:248–257.

Woldringh, C. L., N. B. Grover, R. F. Rosenberger, and A. Zaritsky. 1980. Dimensional rearrangement of rod-shaped bacteria following nutritional shift-up. II. Experiments with *Escherichia coli* B/r. *J. Theor. Biol.* **86**:441–454.

Woldringh, C. L., E. Mulder, P. G. Huls, and N. O. E. Vischer. 1991. Toporegulation of bacterial division according to the nucleoid occlusion model. *Res. Microbiol.* **142**:309–320.

Woldringh, C. L. and Nanninga, N. 1985. Structure of nucleoid and cytoplasm in the intact cell, pp. 161–197. In L. Nanninga (ed.), *Molecular Cytology of Escherichia coli.* Academic Press, New York.

Woldringh, C. L., P. Huls, E. Pas, G. J. Brakenhoff, and N. Nanninga. 1987. Topography of peptidoglycan synthesis during elongation and polar cap formation in a cell division mutant of *Escherichia coli* MC4100. *J. Gen. Microbiol.* **133**:575–586.

Woolley, P., and B. F. C. Clark 1989. Homologies in the structure of G-binding proteins: An analysis based on elongation factor EF-TU. *Biotechnology* **7**:913–920.

Yi, Q.-M., and J. Lutkenhaus. 1985. The nucleotide sequence of the essential cell-division gene *ftsZ* of *Escherichia coli*. *Gene* **36**:241–247.

Zaritsky, A., N. Grover, J. Naaman, C. L. Woldringh, and R. F. Rosenberger. 1982. Growth and form in bacteria. *Comments Mol. Cell Biophys.* **1**:237–260.

Zaritsky, A., C. L. Woldringh, and D. Mirelman. 1979. Constant peptidoglycan density at different rates. *FEBS Letts.* **98**:29–32.

Index

---

*Items listed in italics are equations given in text. In this index Cytoplasmic membrane is abbreviated CM; Murein, M; and Outer membrane OM.